THE PHYSIOGRAPHY OF SOUTHERN ONTARIO

NOTE

Please note that the large coloured maps referred to in the Prefaces are unfortunately no longer available, and therefore have not been supplied with this volume. In place of the large maps, a generalized smaller map has been inserted in the envelope at the back of the book.

SECOND EDITIO

L. J. Chapman D. F. Putnam

THE PHYSIOGRAPHY OF SOUTHERN ONTARIO

Published for the ONTARIO RESEARCH FOUNDATION
by UNIVERSITY OF TORONTO PRESS 1966

Reprinted 1967, 1969

Printed in Canada

SBN 8020 6071 4

FOREWORD TO THE FIRST EDITION

It is a very real privilege to write a brief introduction to this volume. In doing so I think with respect and gratitude of my two friends who are responsible for it. For them it represents many years of patient and thoughtful effort; summers spent in the field with note books, camera, and instruments; winters in discussion, analysis, and at the drawing-board. Secondly, I remember with gratitude a former colleague, Mr. T. D. Jarvis, and our first Chairman, Sir Joseph Flavelle, who gave this project their enthusiastic support. In more recent years we are indebted for financial support to the Research Council of Ontario and particularly to its Chairman, Dr. R. C. Wallace of Queen's University.

It is our hope that this book and maps will be of assistance to those who are teaching geography in our schools, to those engaged in teaching and in extension work in the field of agriculture, and to all who are interested in conservation. It is possible also that many others like myself will derive satisfaction from learning something about the geological history of this province. The beauty of the landscape will not be diminished but enriched by the background which this work affords.

H. B. SPEAKMAN, *Director*
Ontario Research Foundation

Toronto, 1951

PREFACE TO THE FIRST EDITION

This is a description of the surface of Southern Ontario. It deals with the major features controlled by the underlying rock structures and in particular describes the local land forms composed of unconsolidated materials. It is accompanied by four coloured maps upon which the classification and distribution of these land forms is shown. The map and report constitute the results of field work carried on since 1934, under the auspices of the Ontario Research Foundation.

Southern Ontario, for the purposes of this discussion, is the agriculturally developed portion of the province which, for the most part, lies to the south of the Canadian Shield, including the southwestern peninsular area bounded by the Great Lakes together with the lowland between the Ottawa and St. Lawrence rivers. It is underlain chiefly by stratified limestones and shales of Palaeozoic age, the original horizontal attitude of which has been but little disturbed. The limestone boundary is not strictly adhered to, however, because certain areas underlain by Precambrian rocks are included, notably some outcrops in Frontenac, Leeds, and Renfrew counties and adjacent areas floored by deep clay deposits. Included also is Manitoulin Island, because it is composed of flat plains of Palaeozoic rocks, even though it is usually thought of as part of northern Ontario.

Southern Ontario is an area of modest relief. The lowest land along the Ottawa River is barely 150 feet above sea level, while the highest points on the Blue Mountain south of Collingwood just fail to reach 1,800 feet above sea level. These occur on a rocky ridge near the village of Singhampton.

It is a glaciated region. In four successive cold periods of the Pleistocene, vast masses of ice moved across it, scouring the bedrock, breaking and pulverizing the dislodged pieces, overriding and moulding the resultant debris. The Palaeozoic limestones, sandstones, and shales were fairly easily eroded and, except in a few well-known marginal areas, the deep overburden stands in striking contrast to the scanty, discontinuous, sandy mantle on the harder rocks of the Canadian Shield. The depth of the overburden, the gentle slopes, and the high content of limestone and clay are responsible for the development of some highly productive and durable soils. It is no accident that these 30,000 square miles, or 19,000,000 acres, produce 95 per cent of the agricultural wealth of the province and one-quarter of that of the whole dominion.

Physiographic research has not received the attention in Canada that it deserves. Only one regional study has appeared heretofore, J. W. Goldthwait's masterly *The Physiography of Nova Scotia*, which was published in 1924. It has been a source of inspiration to the present writers. While most of the facts herein assembled are the result of our own field work, full use was made of the reports of previous investigators. Information about the bedrock is largely due to the long continued efforts of the Geological Survey of Canada, supplemented by the special studies of outside geologists. Among those to whom we are most indebted are A. W. G. Wilson, M. Y. Williams, C. L. Stauffer, M. E. Wilson, A. E. Wilson, J. F. Caley, and G. M. Kay. Investigation of the Pleistocene deposits has not received the same continued emphasis as the bedrock. However, there have been a number of outstanding individual contributors to the knowledge of surface geology in Ontario, among whom might be mentioned J. W. Spencer, G. K. Gilbert, A. L. Fairchild, F. Leverett, F. B. Taylor, A. P. Coleman, J. W. Goldthwait, W. A. Johnson, G. M. Stanley, R. E. Deane, C. P. Gravenor, A. Dreimanis, A. K. Watt, P. F. Karrow, N. R. Gadd, and C. I. Dell. It is notable that many of these are American geologists whose work in the adjoining states led them to see the need for having supplementary Canadian data.

The sorting of the highly variable surface deposits of a glaciated area is in itself a worthy task, but in this instance it was undertaken as part of a wider objective. Interested in climate and soils and their effects on crops, the writers studied the landforms and geological materials because this information throws some light on the character of the soils and their distribution. Since the acceptance of the profile as the basis of classification, efforts to establish soils studies as an independent science have led to an undue minimizing of the geological materials as a factor in soil formation. Especially in recently glaciated areas the characteristics of the parent materials are of primary importance. The distribution pattern of these soils is complex, and in order to have a clear mental picture of it the glacial history must be known.

It is perhaps in order to say a few words about the way in which our physiographic survey was carried on. The field mapping was done on a scale of one inch to a mile. Contoured sheets of the National Topographic series were used, and in areas where these were not available county road maps produced by the Ontario Department of Highways were used instead. In such areas, elevations were checked by the use of aneroid barometers. Later, much detail was obtained by reference to aerial photographs as they became available.

As can be seen by reference to our preliminary publications, Southern Ontario was regarded as being composed of three physiographic divisions, eastern, south-central, and southwestern Ontario and each was undertaken as a project, the central region first, then eastern Ontario, and eventually the much larger district west of the Niagara escarpment.

Traversing was done by automobile and in most sections all of the traversable east-west roads of the province were covered systematically as well as many north-south ones. Before this was done, however, several weeks were spent in preliminary reconnaissance in each section, and afterwards a great deal of checking in the field was necessary during the final drafting of the master maps. Since the roads of the province are, on the average, about one mile apart, it means that observations have been made along two sides of every square mile in the area. Occasionally when it was felt any particular block was not sufficiently visible from the road, foot traverses were made. In certain areas considerable walking was necessary because of poor roads. A few aeroplane flights, also, were made to assist in the interpretation of aerial photographs and to take photographs for illustrations.

When vertical aerial photographs are viewed under a stereoscope all the prominent landforms stand out and can be mapped. Escarpments in the bedrock, most of the moraines and the spillways associated can be identified in this way, as can drumlins, eskers, and the higher abandoned shorecliffs of the glacial lakes. The pattern exhibited by the photographs also serves as a guide. Even faint relief can produce a definite pattern, giving clues to the topography that would be missed on a ground survey. For instance, those till plains which are faintly drumlinized (fluted) show up in uniform colours broken by dark parallel streaks. Undrumlinized till plains exhibit a marbled pattern similar to that of till moraines. Aerial photography is undoubtedly the finest aid to geographic investigation which has been developed, and it is unfortunate that good vertical photographs were not available for all of Southern Ontario when the survey was being made.

The map (in four parts) accompanying this book is the first reasonably accurate presentation of the surface features of Southern Ontario. The base is derived from the old "Geographical Series," on the scale of 1:250,000, published by the Dominion of Canada Department of the Interior. It lacks many of the cartographical improvements of the National

Topographic Series, but complete base maps of the latter were not available.

We feel impelled to caution our readers that the map deals in landforms only. It is not a generalized soils map, although several factors basic to a classification of soils may be inferred from it. The report describes these landforms and their composition, and is not meant as a complete account of Pleistocene geology. The various substages of the Wisconsin glaciation as they apply in this region are not discussed, and neither are their ages, partly because these are indefinite and contentious at present. It leaves untouched the lower beds of the drift, the classification of which require a painstaking study of local stratigraphy.

Although we believe the map to be reasonably detailed for the scale used, we must point out that it is a broad scale generalization. The glacio-fluvial classification of land forms was made as simple as possible without obscuring all important distinctions. Terminal moraines, for instance, are differentiated according to whether they consist chiefly of glacio-fluvial sands and gravels or of till. Of course, there are many cases of association of the two types; small kames are common along all till-moraines, and kame-moraines have inclusions of sandy till and even of boulder clay. In association with the moraines a network of abandoned glacial drainage channels encircles the high ground west of the escarpment. Practically all of these have been mapped with the aid of aerial photographs; hence their positions are accurately charted. Each, however, is a complex system in itself involving steep-cut bluffs, terraces, meander scrolls, and long swamps which cannot be shown on the scale employed. On the other hand many minor details can be shown. The outlines of the drumlins are definite enough to show whether they are oval or of a more attenuated shape. Among the drumlins a few short eskers have been missed, but the mapping of the main eskers is nearly complete. Along the ancient shorelines a distinction is made between cliffs and beaches, and such features as wave-built gravel bars are shown wherever possible, regardless of whether they are referable to any of the established water levels. However, in some places parallel beaches marking successive water levels lie so close together that it is impossible to show them all individually on a map of this scale. The lacustrine plains are separated into two main categories: the sandy deltas and shore terraces, and the clay plains. There is also a separate category for those areas in which a shallow lacustrine clay serves to modify a till plain. The limestone plains with little or no drift enjoy a separate category, as does also the shallow drift over shale. The slopes of the Niagara escarpment, however, are mapped without distinguishing its component rock formations. Finally, the locations of the larger peat bogs are mapped since they, in themselves, are distinctive landscape types. Thus, altogether fifteen landform categories have been differentiated and find cartographic expression upon the master map.

The stream valleys do not appear on the map, except in so far as they are indicated by the streams themselves, but they are discussed. One full section deals with the relation of the rivers to the pre-existing land forms, and the nature of the dissection produced to date. In some cases a few facts are included about the amount and reliability of the stream flow and other attributes of the larger rivers.

We might well have contented ourselves with this analytical result since it provides the maximum detail available from the survey. However, analysis is only half of the geographic process. The other half is the preparation of a much simpler map dividing the area into convenient divisions each having distinctive characteristics. Moreover, the fact that such regions can be named appeals to most people. For instance we have noticed that the Peel plain has personality because it has a name by which it can be set apart from other clay plains in Southern Ontario. Some boundaries are not so distinctive as others and some geographer may feel obliged to shift them. Also, some of the names which we have applied may not stick and more apt ones will be found. In spite of these shortcomings we have identified, named, and described fifty-two minor natural regions which may be recognized in Southern Ontario.

PREFACE TO THE SECOND EDITION

This second edition of *The Physiography of Southern Ontario* is due to the steady and increasing demand for the book. It provides an opportunity to make some changes and additions in the light of studies made since 1950. The interpretations of glacial history at several stages have been altered. Some information on the composition of the glacial tills, clays, sands, and gravels is now available. Statistics have been brought up to date and regional descriptions improved. However, the text has not been changed greatly.

The large maps are the same and so regrettably include a sprinkling of cartographic errors. A generalized, smaller map is included with this edition. Being less cumbersome than the four large map sheets it should aid the reader and be convenient to carry when travelling. It was prepared for the Atlas of Canada and is used here through the courtesy of the Geographical Branch, Canada Department of Mines and Technical Surveys.

The writers wish to express their thanks to Mr. T. D. Jarvis and Dr. H. B. Speakman for their encouragement and support particularly from 1934 to 1950 during the course of the survey. Professor F. F. Morwick and Dr. O. M. McConkey of the Ontario Agricultural College gave much valuable help. We owe a great deal to Miss Maynard Grange of the Foundation for her editorial advice and to Miss Eleanor Harman and the editorial staff of the University Press. The fine co-operation of the Geographical Branch, Canada Department of Mines and Technical Surveys has made possible the inclusion of the small coloured physiographic map with this edition. Finally we are grateful for the continued support of the Ontario Research Foundation and the financial support of the Ontario Government.

We hope that this book will continue to serve as a technical reference and also to be enjoyed by all those interested in the landscape.

L. J. C.
D. F. P.

CONTENTS

4 PHYSIOGRAPHIC REGIONS 172

THE PHYSIOGRAPHY OF SOUTHERN ONTARIO

THE BEDROCK

1

The bedrocks of Southern Ontario are among the oldest beds to harbour the petrified remains of plants or animals. They rest upon the far more ancient Precambrian rocks similar to those which appear beyond the northern boundary. The fossils are those of salt-water organisms. This, along with the stratification of the sandstones, shales, and limestones, leaves no room for doubt as to the marine origin of the deposits. The great depth of the beds indicates a long period of inundation during which sand and clay or marl accumulated on the ocean floor. These deposits became cemented, under the pressure of overlying strata, forming solid rock. Finally, this part of the continent rose above sea level. The bedrocks consist of limestones, shales, sandstones, and dolomites overlapping each other so that they appear in concentric belts.

These strata, which were originally horizontal, have been faintly but noticeably warped in the Great Lakes region. Owing to pressure changes within the earth, broad domes were pushed up in the Adirondacks, Algonquin Park, and Wisconsin, while between them appeared basins in Michigan, New York, and the Ottawa valley. The broad arch running from the Dundalk area southwestward has been related by some geologists to the Algonquin Park dome (86). A small swell in the rocks, called the Cincinnati Arch, underlies Essex county. The main elements of the bedrock structure are illustrated by a diagram in Figure 2.

In the vicinity of the Bay of Quinte the rocks were further wrinkled so as to form three broad steps rising eastward (86). However, after this topography was cut by river valleys running southwestward and later planed by the glaciers, very little of this rise-and-tread remains.

Apart from the Ottawa valley there are no major faults or dislocations due to breaking of the bedrock in Southern Ontario. This lack of faults carries with it the comforting implication of a freedom from earthquakes. The mild earthquake of the summer of 1943, which centred at Cornwall, was a fresh reminder of the instability of the bedrock in the Ottawa basin. The locations of the main faults in that area are represented in Figure 3 as black lines, according to the mapping of Wilson, Stewart, and Caley (149), and Kay (86).

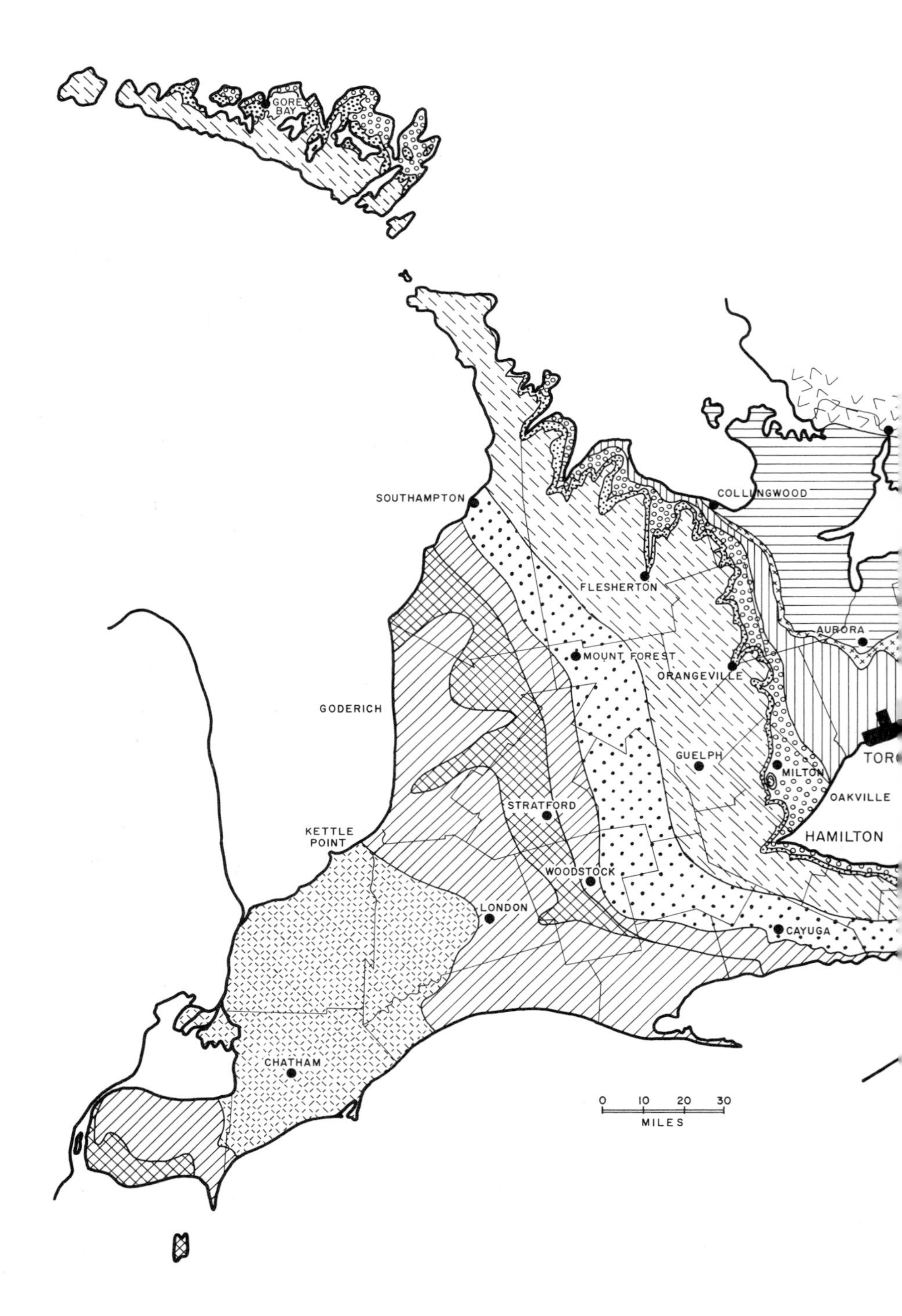
GORE BAY
SOUTHAMPTON
COLLINGWOOD
FLESHERTON
AURORA
MOUNT FOREST
ORANGEVILLE
GODERICH
GUELPH
MILTON
OAKVILLE
STRATFORD
KETTLE POINT
HAMILTON
WOODSTOCK
LONDON
CAYUGA
CHATHAM
0 10 20 30
MILES

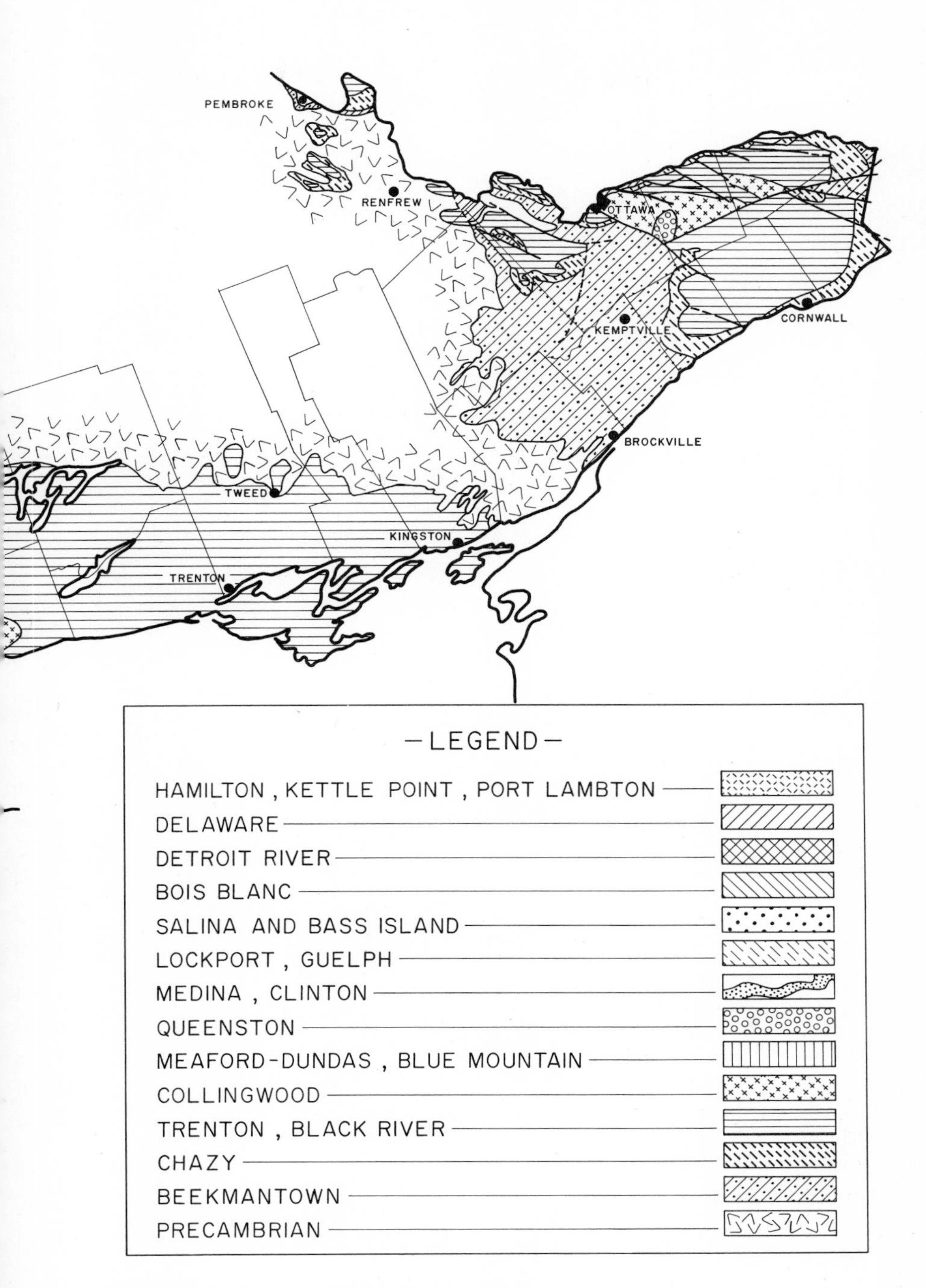

1. *Map of bedrock (after Canadian Geological Survey)*

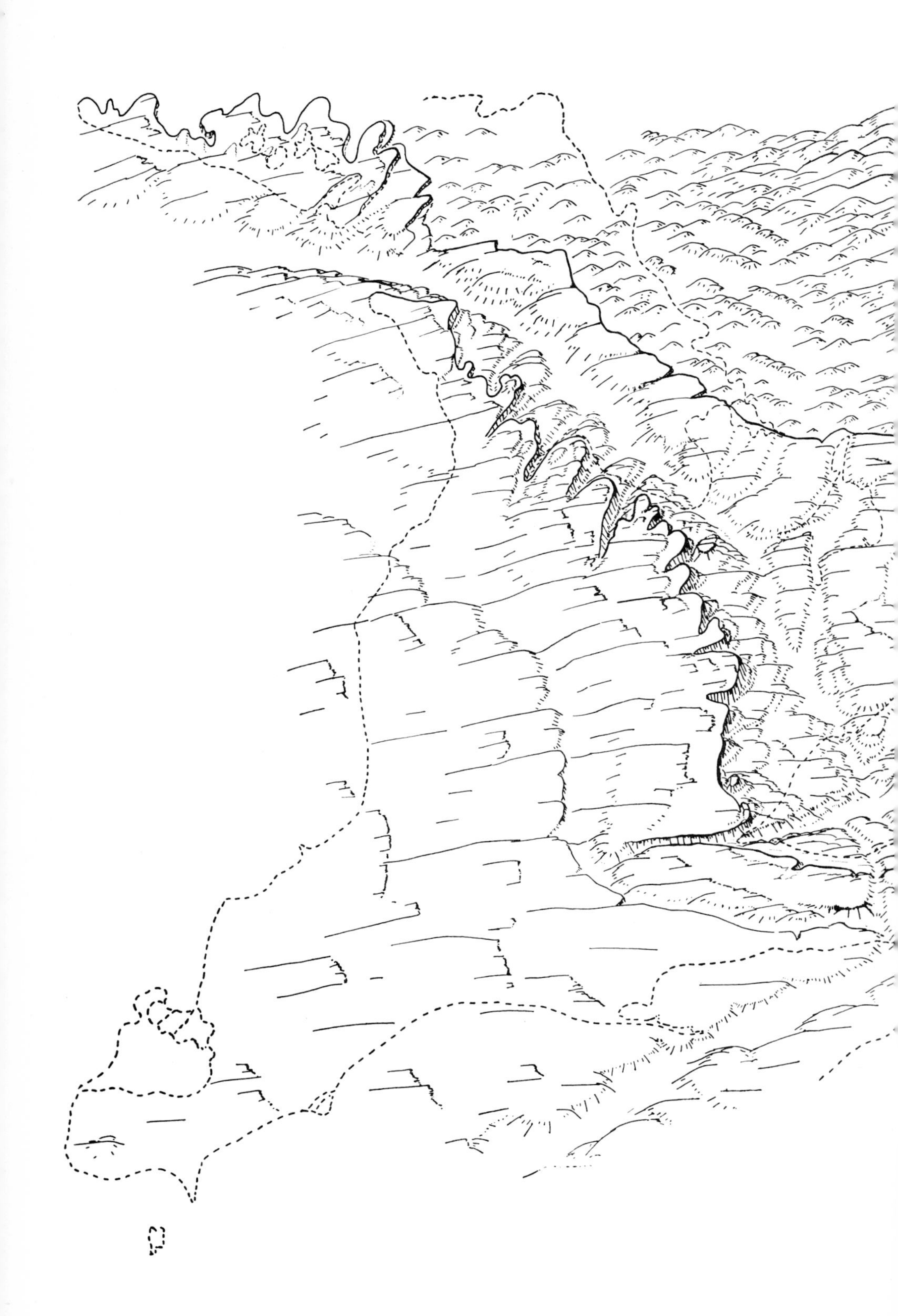

2. *Bedrock topography*

The Ottawa basin is bounded on the west by the Frontenac Axis, an arch in the rock between Algonquin Park and the Adirondacks. This is visibly expressed in the Thousand Islands, where the Precambrian rocks appear on the surface, the overlying limestones appearing to the east and west. Many of the Thousand Islands are granite knobs. The basin is sharply bounded on the north, just beyond the Ottawa River, by the edge of the Shield which rises abruptly to a height of about 1,000 feet. This is the result of faulting, the block to the south having dropped down so as to place the overlying sedimentary rocks in a protected position where they escaped the full forces of erosion. This down-dropped block

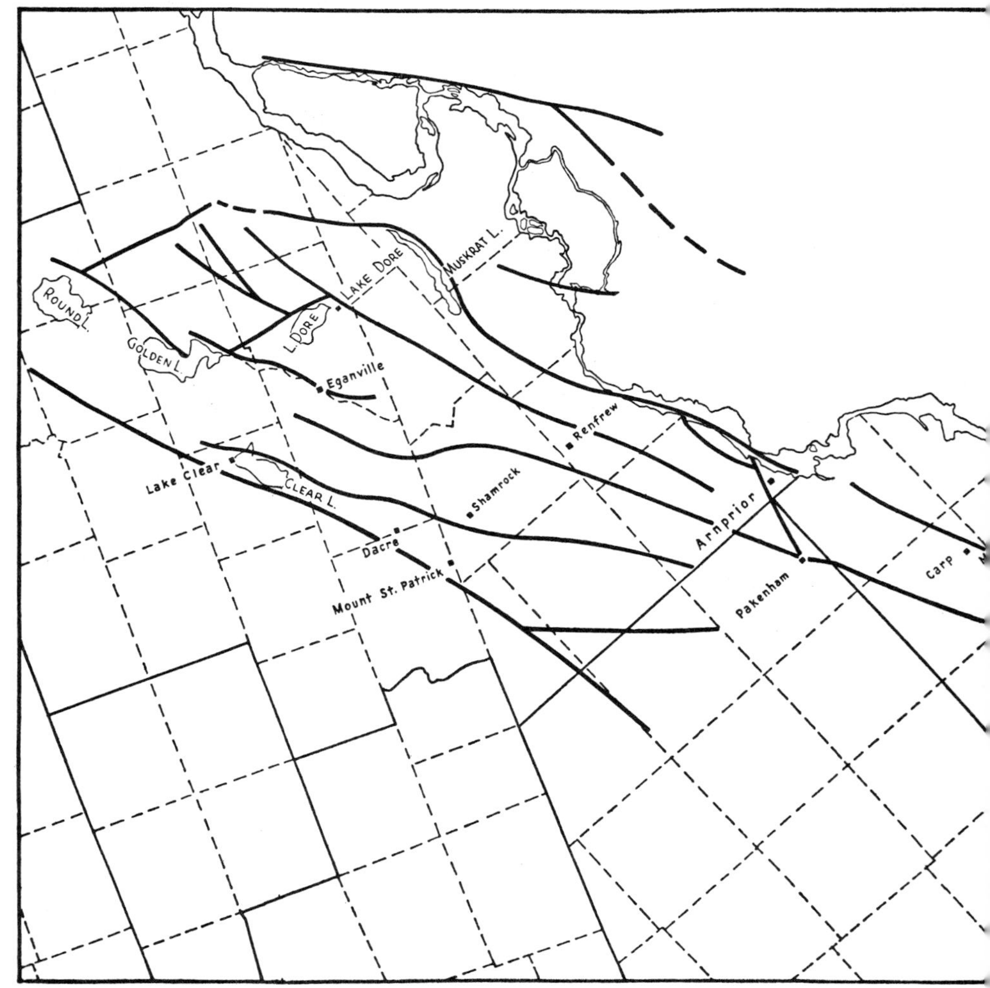

3. *Faults in the Ottawa lowland (after Wilson and Kay)*

is bounded on the west and south by faults, and is tilted towards the west. The western boundary is easily seen as a bluff facing the northeast which crosses Gloucester township from Ottawa to Russell village. On the south another fault is buried under clay and sand; it appears in northern Glengarry, but is not conspicuous.

West of Ottawa the main fault passes through the hamlet of Hazeldean and forms the northern flank of the Carp valley. Along this fault, Precambrian outliers overlook the Trenton limestones of the lowland; prominent also is the ridge of Nepean sandstone from which so much fine building stone has come to the city of Ottawa.

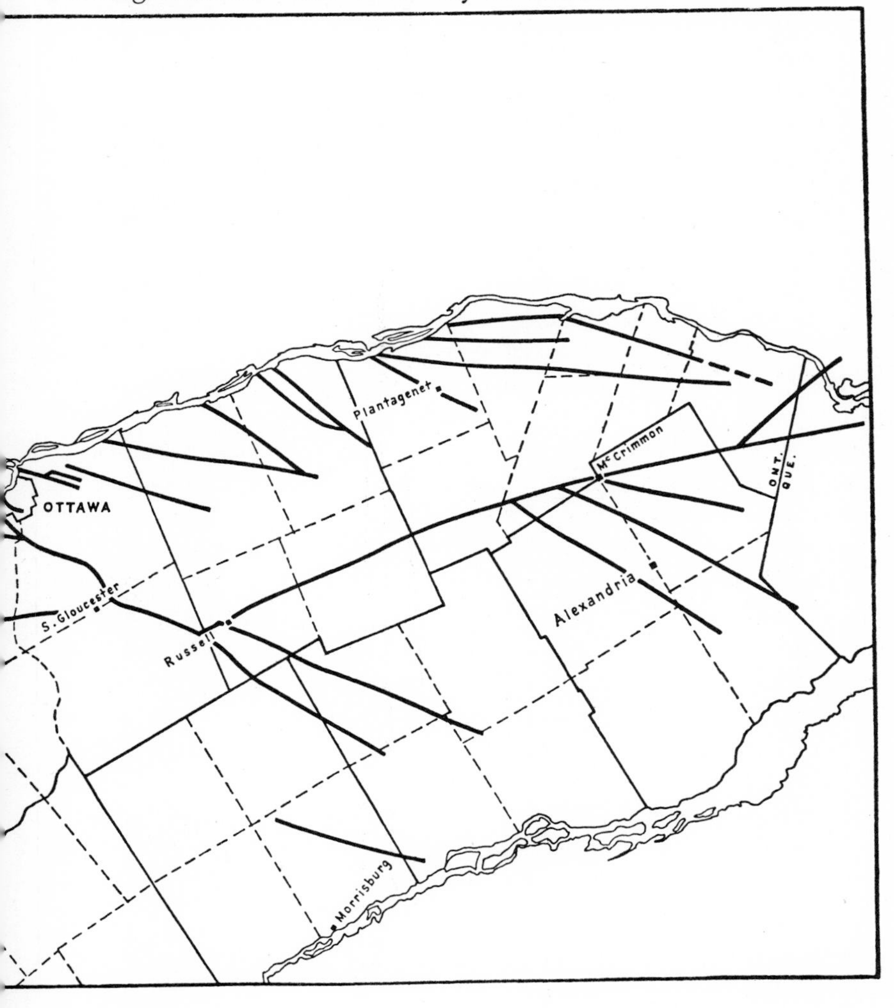

In Renfrew county there are prominent scarps on both sides of the Ottawa valley, the southwestern one lying south of Calabogie Lake, Mt. St. Patrick, and Clear Lake. Thus a block 35 miles in width is downdropped, forming what has been called the Ottawa-Bonnechère "graben" (86). Within it are several minor breaks. There are three south-facing escarpments bounding blocks that are tilted towards the north, and in each case Palaeozoic limestones are preserved on their northern flanks. These are the Muskrat scarp along the northeast side of Muskrat Lake, the Dore scarp extending from Lake Dore past Renfrew towards Arnprior, and the series which can be seen intermittently along the north side of the Bonnechère River upstream from Northcote Station, and northeast of Golden and Round Lakes.

Farther south in Renfrew the main area of farmland is confined roughly by the Pakenham fault whose north-facing scarp can be traced from near Eganville to Pakenham. A few miles southwest two parallel scarps face south and north respectively, while between them enough soil material accumulated to encourage the settlements of Mt. St. Patrick, Dacre, and Clear Lake.

The alternate strips of good and poor land in the Ottawa valley are due to the shallow sandy drift or rock outcrop on the brows of all these scarps and the deep drift in the depressions.

In addition to warping and faulting, there is a third method of fashioning the contour of the bedrock surface—by erosion, chiefly of running water. Before the advent of the glacial periods the Palaeozoic rocks of this area had been subjected for about 250,000,000 years to stream erosion which transported clay and sand to the sea. The surfaces of the more easily eroded rocks, such as the shales, were lowered, leaving the harder dolomite or limestone to form uplands. Particularly in the large basin centred in Michigan, the various rock formations are found in concentric belts with the strata dipping towards the centre; that is, towards the southwest in Ontario. This resulted in the development of curved lowlands bounded inwardly by escarpments. Known as cuesta and vale topography, it is common in England, northern France, and several other parts of the world.

Before glaciation, judging by the relief on similar rocks in Kentucky and Pennsylvania just outside the glaciated area, there were deep V-shaped valleys on the harder limestones and more rounded topography in areas of softer rocks.

There can be no doubt that the master rivers of preglacial times followed the lowlands, while tributaries led into them. Drainage from the upper Great Lakes area passed through the Georgian Bay depression and

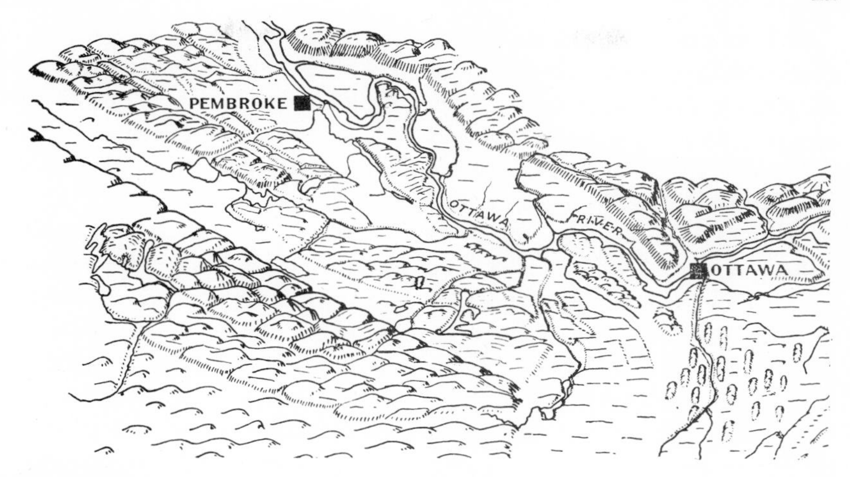

4. *Physiographic diagram of the Ottawa-Bonnechère graben*

continued directly towards Toronto, where it met a stream in the Lake Ontario basin. From there its course is not known definitely. It probably flowed eastward, as the concepts of Spencer (120) and Grabau (64) that the Dundas valley extends to the Erie basin have never been confirmed. There is no trace of it beyond Copetown.

By far the greatest topographic break produced by differential erosion of harder and softer rock in Ontario is the Niagara escarpment. It rises in New York near Rochester, skirts Lake Ontario to Hamilton, then passes north to Collingwood. The Blue Mountain south of Georgian Bay, the Bruce peninsula, and Manitoulin Island with its neighbours to the west constitute its extension in Ontario. Thence, it crosses the upper peninsula of Michigan and swings southward along the margin of that lake. It forms the Door peninsula in Wisconsin but disappears west of Chicago. Thus, it partly encircles the Michigan basin into which the beds dip at the rate of 20 feet or more to the mile.

The broad arch of the rocks from Windsor to Dundalk has already been mentioned, and the highest part of the Niagara escarpment is on the crest of it south of Collingwood. There, the cuesta stands about 1,000 feet above the lowland to the east, while it rises only 350 feet above Lake Ontario between Hamilton and Niagara Falls. In other words, the brow of the cuesta is over 1,700 feet above sea level near Singhampton but slopes down to 600 feet on the Niagara peninsula and to less than 800 feet on the Bruce peninsula. The "build" of southwestern Ontario is due to these two bedrock structures, the Dundalk (Algonquin) arch and the Niagara escarpment.

The only deep preglacial river valleys in the bedrock to be seen in the whole area are found along the Niagara escarpment. The Dundas valley, running west from the tip of Lake Ontario, is one of the longest, measuring ten miles to the point where it becomes filled with drift. The Credit Forks, the Hockley valley, and the Devil's Glen are well known for their scenic beauty, while the Mad, Pine, Noisy, and Pretty rivers all flow through notches in the escarpment south of Collingwood. The Beaver valley, extending thirty miles from Thornbury to Flesherton, is the best example of all. Kimberley village in the narrow upper reach of the valley sits 500 feet below the rim. Nearer Georgian Bay the valley suddenly broadens into Collingwood township to the east, and it is this section that has gained a reputation for fine farms and Northern Spy apples. The broad valley drained by the Bighead River which empties at Meaford resembles the lower part of the Beaver valley, but lacks the long extension into the heart of the upland. On the Bruce peninsula the two important valleys are below the level of Georgian Bay, affording Owen Sound and Wiarton excellent harbours. All of these valleys have been rounded to a U-shape by glaciers and have received coatings of boulder clay of variable depths.

Although the Niagara escarpment is outstanding, another escarpment, the Onondaga, crosses the Niagara peninsula from Fort Erie to Hagersville, beyond which it is buried by drift. Capped by limestone, it confines a lowland worn in the Salina formation consisting of limestone, shale, and sandstone interbedded. The Onondaga escarpment is not very high in Ontario, and several sections east of Hagersville lie buried under the clay of Haldimand and Welland counties. Another scarp appears at the edge of the Black River limestone overlooking the border of the Shield. Seldom more than 50 feet high, it occasionally grows to 75 feet. The Kawartha Lakes and a series of smaller lakes are found in the depression in front of this escarpment. Low scarps appear also in the Trenton formation, notably north of Lindsay and Peterborough. In the smaller Ottawa basin the cuesta-form relief is not appreciably developed.

2 GLACIAL GEOLOGY

The amount of erosion accomplished by the great Pleistocene glaciers is not subject to accurate measurement. Some measure of the depth of planing can be obtained by noting the depth of the drift which is the end-product of glacial erosion, and in Southern Ontario this averages between 75 and 100 feet. Most active erosion of the bedrock occurred on the brows of the escarpments and along the lowland routes taken by the principal streams of ice. The basins of Lakes Ontario, Huron, and Superior extend to below sea level; as such scooping-out could not have been the work of a river, it is attributed to the glaciers.

In those sections north of Lake Ontario where there is little or no overburden all the minor valleys are planed off the limestone, but a series of shallow valleys running southward still remain to mark the sites of bigger preglacial river valleys.

PRE-WISCONSIN BEDS

Pre-Wisconsin drift has been identified in Ohio and other states to the west but it is rarely seen in Ontario. In recent years the beds formerly considered to be Pre-Wisconsin in age at Scarborough Bluffs and other widely distributed lower beds in Ontario have come to be classified as Wisconsin. At Toronto, only the thin layer of till on the bedrock at the Don Valley brickyard (146) and lower fill seen in an excavation at the railroad marshalling yards south of Maple are Pre-Wisconsin, probably Illinoian. Also, some of the lower tills exposed in the high bluffs of Lake Erie are now dated as early Wisconsin (35). This means that Pre-Wisconsin deposits are very rare and may be disregarded when describing the surface features.

It was stated earlier that the build of Southern Ontario is mainly due to the bedrock. The notable exception to this rule is the massive ridge of glacial drift running east and west through the counties north of Lake Ontario. It forms the drainage divide south of Lake Scugog and Rice Lake and extends from the Niagara escarpment in Caledon township to the Trent valley. At its deepest in King township there is roughly 800 feet of overburden piled on the bedrock. The surface deposits are mostly sand and gravel, but dense, compact till sometimes appears in deep roadcuts and gullies within thirty or forty feet of the surface. Beyond this the structure of this moraine is unknown. From its great bulk it is surmised that the earlier glaciers may have laid down the basal core and that the earlier glaciers followed the same routes as the last one in this area.

It is well known that during the retreat of the last glacier extensive lowland areas were submerged in glacial lakes for a time. These are now clay or sand plains. The second to last glacier also may have been attended by glacial lakes during its retreat. The last glacier overrode the older beds of sand and clay. Apart from the shaley tills, the clayey tills are mostly due to the overriding of older clays of earlier Wisconsin or Pre-Wisconsin age. The brown clay till found in a broad belt east and south of Lake Huron is one example. The heavy till of the Niagara peninsula is another. The overridden clays are compacted and the strata are often contorted.

The sands and gravels when overridden resulted in sandy till. The till between Lake Simcoe and Georgian Bay is of this type. Even the sandy ridge in northern Oro township may be regarded as an overridden kame-moraine. The sandhills of Waterloo county might also be regarded similarly in spite of the included kames and outwash. In each of these cases the underlying sand and gravel is regarded as earlier Wisconsin rather than Pre-Wisconsin deposits.

5. *Striae on limestone, Lion's Head*

6. *Drumlin near Bradford*

WISCONSIN BEDS

While the Pre-Wisconsin glaciation is known only from fragmentary bits of evidence, the structural work done by the Wisconsin glacier stands intact on the surface where it may be seen. This monograph is the product of a systematic examination of the surface drift. The features are shown in some detail on the accompanying maps. The interpretation in terms of the last glacier's retreat is soon to be discussed and a few main points about the qualities of the materials will be given. But, before beginning that story, it may be in order to introduce the features from which it is pieced together and to give a bird's-eye view of their distribution.

The forward motion of the glacier produced two types of markings which point out the direction of movement. They are scratches (striae) on the bedrock and mouldings of the surface of the till sheet. The scratches are not retained on soft rock such as shale, nor were they always made on the hardest rock. In Southern Ontario they are best represented on massive limestones and dolomites. They are best preserved on rocks covered with some drift where scaling off of the surface by frost action has been prevented. These markings can only be seen where the rock is exposed and in most of Southern Ontario it is the moulding of the till sheet which shows the direction of glacial movement.

As the continental glaciers accumulated and advanced they had a planing action on the bedrocks. Soil and rock were carried forward, mixed together, and milled. The resulting rocky grist is a heterogeneous mass of stones and pebbles in a sand, silt, and clay matrix. It is not regularly stratified as are sediments deposited in running and standing water. Since it resembles tilled soil it is called glacial till. Layers of till in the overburden on bedrocks provide prime evidence of glaciation, especially when the till contains some rocks other than the local bedrock from "upstream."

In Southern Ontario most of the area has a covering of drift on the bedrock. Where the upper deposit is a sheet of till the surface is often moulded, especially where the till is medium in texture. Smooth oval hills appear, all aligned in the direction of movement of the last ice sheet. They are called *drumlins*. Drumlins may vary from impressive hills with steep sides to hardly noticeable undulations. The biggest ones are 150 feet high and over a mile in length. Some long low specimens reach a length of two miles. At the other extreme is a smoothed plain scored with shallow

7. *Flutings on a drumlin east of Peterborough*

grooves or "flutings." Restricted drainage in the flutings has resulted in the accumulation of humus in the surface soil and on aerial photographs these appear as dark parallel streaks.

The recession of the glacier was halted periodically by readvances. In these cooler, or moister, periods the ice lobes overrode the recently deposited drift, usually building a moraine at the terminus of each advance. Moraines are made either of till which has been pushed into place by the glacier or of coarsely stratified gravel and sand deposited at the ice front by drainage issuing from the melting ice. A common trait of all moraines is a knobby surface with undrained depressions of irregular shape between the knobs. Associated with the moraines are the abandoned stream channels formed by glacial drainage. Nearly all of these channels are occupied by streams at present, but it is apparent that these are much too small to have carved the main valley or deposited the great gravel

8. *Aerial view of a fluted till plain. The dark parallel streaks are caused by the darker soil in the shallow grooves or flutings.*

9. *Beach gravel. The rounded pebbles, evenly graded in the various strata, are typical*

10. *Beach ridges: above, ancient marine strands near Carleton Place; below, a gravel beach at Charwell Point on Lake Ontario*

terraces. Proof of this point is provided by the finding of channels here and there in which there are no streams.

The continental glacier completely disrupted the normal drainage of the Great Lakes region. The St. Lawrence valley was dammed early in the advance and reopened late in the recession of the glacier; thus the water in the Great Lakes' basins rose until it found an outlet to the Mississippi valley or eastward through the Mohawk valley in New York state. Minor lakes or pondings were often created where the glacier faced high land. The glacier produced a second effect in the submergence of the St. Lawrence and Ottawa valleys. Depression of the earth's crust caused by the great mass of ice resulted in submergence of the lowlands bordering the upper Great Lakes and caused postglacial submergence of the St. Lawrence–Ottawa lowlands by the sea. After the recession of the glacier the earth's crust rebounded, tilting the beaches of the glacial lakes towards the northeast, uncovering these borderlands, and raising the St. Lawrence–Ottawa lowland above the sea.

Beach ridges of gravel or sand and wave-cut bluffs with bouldery terraces at their bases are easy to identify. The longer-lived glacial lakes left distinct shore features and sandy deltas at the mouths of streams, but the shorter-lived lakes left only faint discontinuous beaches. The stratified silt and clay of the lake beds provide important supporting evidence of prolonged inundation and sometimes the only conclusive evidence.

Moraines. Figure 11 is a map of the moraines and larger glacial drainage channels. It distinguishes between the gravelly or sandy moraines left by melt-water issuing from the glacier and those made of till produced by a thrust of the glacier. The latter, but not the former, mark the terminal points of advances by the glacier. This small map is designed to give a general picture of the arrangement of the moraines and to attach the names given them either by F. B. Taylor (131) or the present writers.

The moraines of Southern Ontario were built either by a broad tongue of ice advancing through the lowlands of Lakes Ontario and Erie, or by the icesheet advancing from the north to meet this ice tongue and pushing forward as a lobe in the basin of Lake Huron. North of Lake Ontario the bulky Oak Ridges moraine stands at the meeting point of the two conflicting movements of ice, while west of the Niagara escarpment interlobate depositions are found in several places between Orangeville and London. These are sandy moraines with drainage channels running through them lengthwise. The upland of peninsular Ontario has an elaborate array of moraines and spillways around it, the centre-piece in this series being the horseshoe-shaped Port Huron morainic system. It will be noted on this

TARA STRANDS
BANKS
GIBRALTAR
SINGHAMPTON
CORN HILL
ORO SANDHILLS
WYOMING
SAUGEEN KAMES
ORANGEVILLE
SEAFORTH
MITCHELL
MILVERTON
ELMIRA
WATERLOO SANDHILLS
EASTHOPE
LAKESIDE
PARIS
GALT
TRAFALGAR
NIAGARA
FORT ERIE
LUCAN
INGERSOLL
WESTMINSTER
NORWICH
TILSONBURG
CHARING CROSS
BLENHEIM
LEAMINGTON

THE MORAINES
OF
SOUTHERN ONTARIO

0 10 20 30
MILES

11. *The moraines of Southern Ontario*

map that the Paris and Galt moraines end at Caledon instead of Singhampton and Gibraltar as in Taylor's initial work. Thus the moraines of the Ontario–Erie lobe are separated from those of the Georgian Bay and Lake Simcoe ice sheet. While its nose was under water, the Ontario lobe built a series of concentric moraines around the western end of Lake Ontario and later produced the Trafalgar moraine.

The scattered morainic knobs along the northern border of the limestone between Kingston and the Kawartha Lakes have been considered collectively as the Dummer moraine. East of Kingston there are no distinct continuous moraines to mark the position of the ice margin across this lowland at any stage. The stony till ridges and intervening plains in Glengarry are mapped as till plains, but later work tends to show that the ridges themselves are moraines somewhat smoothed by wave action. The sands in Grenville county and nearby parts of Dundas and Leeds appear to be ice-margin deposits partly reworked and spread out by the waves. Irregular, poorly sorted sand and gravel devoid of marine fossils appears in some of the low hills. Similar fragmentary, sandy morainic deposits with reworked strata on the surface containing marine fossils occur near Maxville, Kemptville, and just south of Ottawa (60).

12. *Moraine on Black River limestone near Erinsville*

13. *Stream gravel near Arkona, showing the typical subangular pebbles and stones and lack of even grading*

Spillways. The drainage channels associated with the moraines, are harder to recognize than the moraines themselves; the latter stand up above the surrounding terrain, whereas the spillways are entrenched. Moreover they are usually occupied by streams, which raises the question of whether they are normal stream valleys or glacial spillways. The typical spillway is a broad trough floored wholly or in part by gravel beds at one or more levels, often with a cedar swamp in the lowest part of the valley. It sometimes shows a peculiar disregard for existing grades, since it flowed along an ice front. The present stream has its own channel in the bottom of the larger valley. It is fairly common to find a spillway that now is unoccupied by any stream.

Drumlins. Because drumlins are so prevalent in Ontario the term will no doubt become widely used. In early days these glacial hills were called sowbacks, whalebacks or lenticular, mammillary, and elliptical hills by different writers. The name drumlin, Celtic for little hill, was first used by H. M. Close in 1866 when discussing a group in northern Ireland, and later it was generally adopted by geologists.

The process of drumlin formation has been the subject of much speculation. The concentric bedding revealed in section through drumlins is evidence that they were built by a plastering-on process (47). On the other hand, the gouges between them reveal that the building also entailed some scooping-out.

We have consistently noted, although this is not always supported in

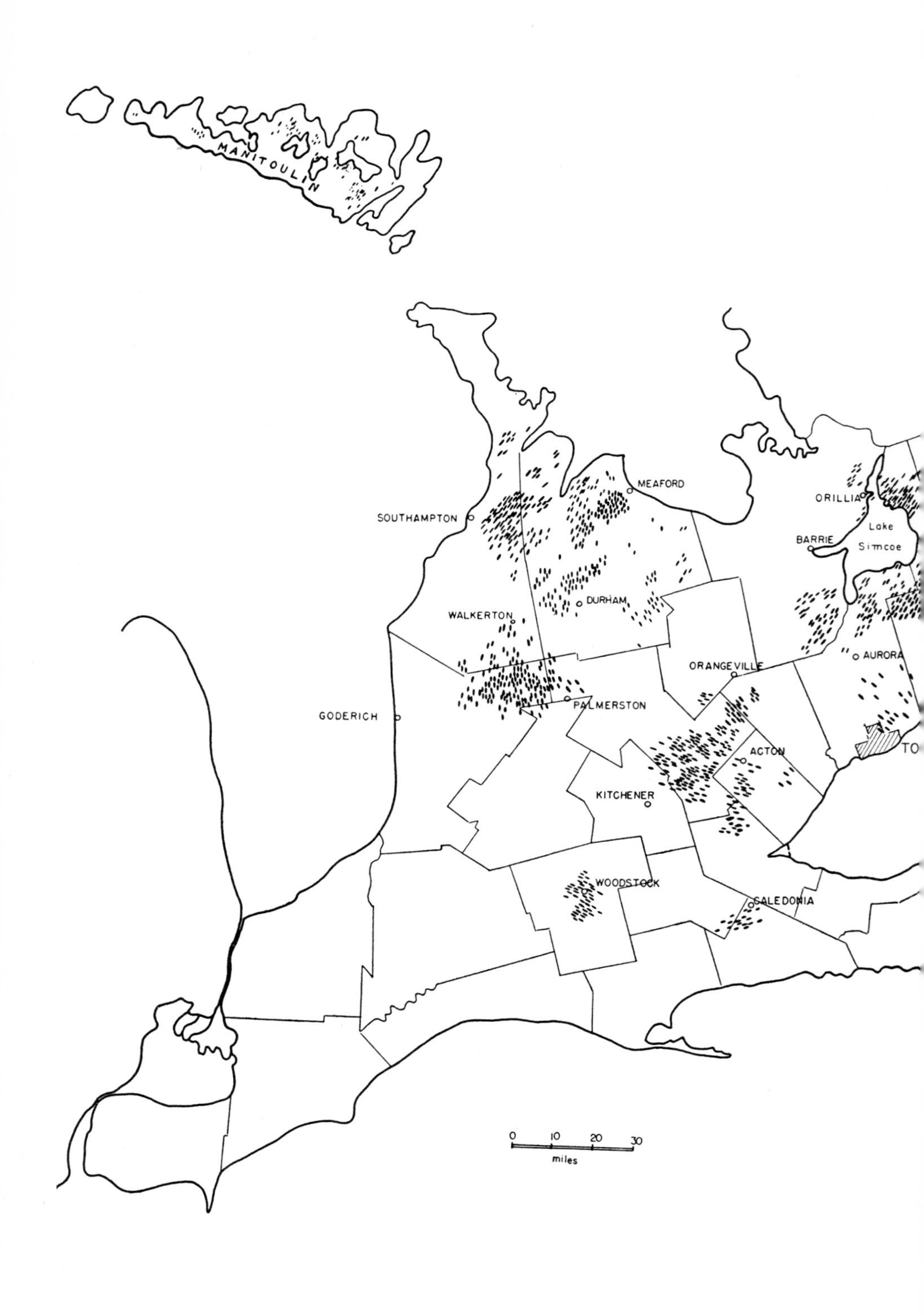
MANITOULIN
SOUTHAMPTON
MEAFORD
ORILLIA
Lake Simcoe
BARRIE
WALKERTON
DURHAM
ORANGEVILLE
AURORA
GODERICH
PALMERSTON
ACTON
KITCHENER
WOODSTOCK
CALEDONIA
0 10 20 30
miles

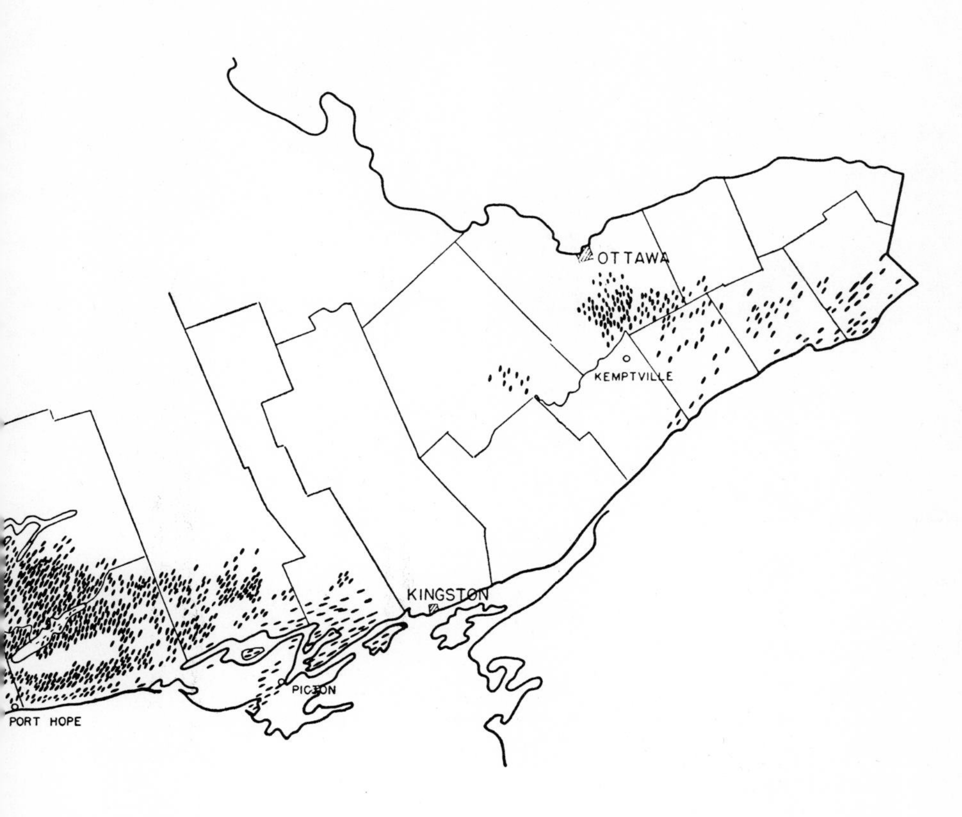

14. *Drumlin fields of Southern Ontario*

15. *A drumlin landscape*

geological reports, that drumlins are formed only of loamy till (Figure 16). Neither heavy boulder clay nor sandy tills are productive of drumlins in southern Ontario. Kames appear sometimes on the slopes of drumlins, especially on their tails, and sandy outwash or a spread of wind-blown silt and fine sand is often found on the surface.

Drumlins are of interest because they point in the direction of movement of the glacier. The small map (Figure 14) gives the alignment of the drumlins and shows where they occur, but the oval dots are merely symbols and do not represent drumlins accurately placed or drawn to scale. The various shapes of the drumlins in different areas are shown.

The first drumlins to appear during the retreat of the glacier were those between Orangeville and Kitchener. All except the few north of the Orangeville moraine are the products of the Ontario–Erie ice lobe. The large group north of the Guelph area, the scattered few northwest of Hamilton, and those near Caledonia show nicely the deployment of the ice lobe as it pushed westward from the Ontario basin. Near Orangeville the drumlins have a bearing of 35 degrees north of west, near Moffat they point directly west, while at Caledonia they bear 45 degrees to the south.

The first drumlins to be uncovered by the Georgian Bay lobe were those near Orangeville, which bear 60 degrees east of south. In the big field between Listowel and Owen Sound the drumlins point almost directly south, while the Arran group and those on the Bruce peninsula

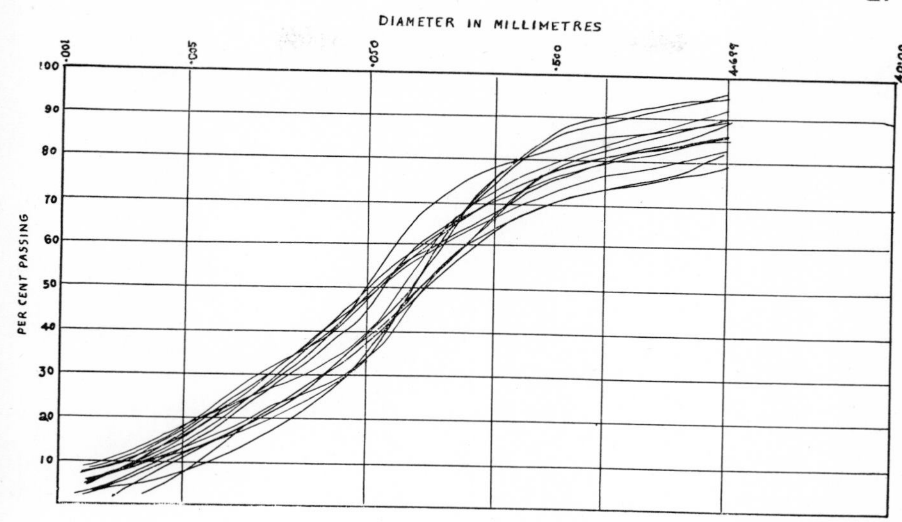

16. *Texture curves of tills from drumlines (Ontario Department of Highways data)*

bear to the west as much as 45 degrees. The latest thrust of the glacier over the Bruce peninsula was obviously governed by Colpoys Bay and Owen Sound. On Manitoulin Island most of the drumlins point 45 degrees west of south.

North of Lake Ontario 4,000 drumlins have been counted, the majority of them north of the interlobate moraine. This is a great field of drumlins, rivalling the one south of Lake Ontario in New York State which Fairchild regarded as the best display in the world. No doubt many more lie on the floor of Lake Ontario. Lake Scugog and Rice Lake are in the heart of this field and the several islands in Rice Lake are drumlins. Near Peterborough they are so closely clustered that it is difficult to separate them on a map. South of the Oak Ridges moraine most of the drumlins, formed by the Ontario ice lobe, are long, thin types pointing northward.

The relation of the drumlins to the interlobate moraine should be noted. In the Rice Lake area they have the same alignment north and south of the moraine, while beyond the eastern end of it the drumlin field is continuous nearly to the shore of Lake Ontario. Here are encountered the long, thin forms produced by a thrust directly into the Ontario basin from the St. Lawrence lowland.

The drumlins of eastern Ontario number about five hundred. They are most numerous in North Gower and Osgoode townships and more widely scattered in Dundas and Stormont counties. These point almost directly

OWEN SOUND
BARRIE
DURHAM
SHELBURNE
TEESWATER
GODERICH
LISTOWEL
BROCK'S SCHOOL
BRAMPTON
GUELPH
BRODHAGEN
STRATFORD
HAMILTON

17. *Eskers in Southern Ontario*

south. In southern Glengarry and Stormont there are hills of stony till which have the general form of drumlins but lack the usual smoothly moulded surfaces. It is suggested that they may represent drumlins built from the northeast by the Malone ice lobe, overridden later from the northwest by the Fort Covington advance. (98)

Eskers. Eskers are knobby, crooked ridges of gravel which are roughly in line with the direction of movement of the glacier. They are surficial deposits, resting on the till, and they consist of gravel, cobbles, or sand in irregular strata. The sides are at the angle of repose which increases with the coarseness of the materials, and their height is usually between 25 and 75 feet in southern Ontario. It is thought that the mother streams were confined to crevasses or tunnels at the base of the melting glacier. When associated with drumlins eskers may lie along the hillsides but they prefer to follow low ground. Some run only a few rods and are indistinguishable from kames, while the longest one in southern Ontario was traced for 55 miles.

Figure 17 is a map of eskers. Two sections of Southern Ontario were especially favourable for esker formation; the first is between Shelburne and Goderich and the second on the drumlinized plain north of the Oak Ridges moraine. It is usually possible to recognize eskers on contoured maps and they can be traced with ease and accuracy on aerial photographs. In eastern Ontario only a few have been recognized and these are partly levelled by wave action and their crests remoulded in the shape of beaches or bars. The total length of all the eskers in the area under study is approximately 400 miles. They give rise to poor soil but are valuable sources of gravel.

Glacial lakes. The succession of glacial lakes which were dammed to high levels in the basins of Lake Erie, Huron, and Simcoe have long been recognized and a great deal of information about them was summarized in 1915 by Leverett and Taylor (92). A convenient summary in tabular form is found on page 469 of that monograph. Again in 1958 Hough marshalled all the available information (74). In the Lake Ontario basin the shoreline of Lake Iroquois was mapped in good detail by A. P. Coleman (23). We have mapped the other beaches in similar detail. Soil and physiographic surveys have added greatly to our knowledge of the sediments in the abandoned lake floors, and this also has been helpful in

19. *Aerial photographs of the Atwood esker. The white patches near the roads are gravel pits. (R.C.A.F. photo)*

18. *Esker near Atwood, south of Listowel. This is a small esker, and the gravel is unusually fine.*

the interpretation of glacial history. The main shorelines are shown in Figure 20.

The highest glacial lake of the Erie and Huron basins emptied westward from Fort Wayne, Ohio, down the Wabash River to the Mississippi. It is called Lake Maumee. During the early life of this lake the Huron and Erie ice lobes contacted each other just west of London, but they separated during its existence. The little lake plain between London and Thamesford is either an isolated contemporary of Lake Maumee, or the inner reaches of an ice-confined bay. A later, lower stage of this lake probably extended into the valley of Dingman Creek and the depression north of Coldstream.

Sixty feet below the upper Maumee watermark the beach of Lake Whittlesey may be expected. A poorly developed beach in Ontario, it serves as a boundary between till plain and lake plain from Brantford to Exeter. At those two points it meets the arms of the Port Huron morainic system which mark the positions of the ice barriers at that stage. The outlet of Lake Whittlesey was at Ubly on the Thumb of Michigan, whence it emptied into a smaller lake in the Saginaw Bay depression and then drained across Michigan towards the Mississippi.

The Arkona beach, 30 to 40 feet below the Whittlesey, is also poorly defined in Ontario. According to Taylor, based on evidence seen in Michigan, Lake Arkona preceded Lake Whittlesey and the beach was therefore submerged and mutilated by the waters of the later lake. The Arkona beach and the upper beach of Lake Warren are separated by an elevation of about 15 feet and are seldom far apart. Although Lake Arkona takes its name from an Ontario village it did so through an error; the original beach at Arkona was found later to belong to Lake Whittlesey (93).

Lake Warren stood at the level of the water in the Saginaw Bay depression which drained by way of the Grand River valley into a lake in the Chicago area. Apparently it had two stages, the earlier being ten feet above the latter. Twin beaches are characteristic of this shoreline.

Between the Warren water level and those of Lakes Algonquin and Iroquois there are some fragmentary beaches at several levels. Leverett and Taylor recognized three lakes named Wayne, Grassmere, and Lundy (Elkton or Dana). An early stage of Lake St. Clair which left its mark on the topography and soil of Kent county should be mentioned at this time along with the delta and clay flats east of Dunnville. Both are confined roughly by 600-foot contours.

In the area north of Toronto the Ontario ice lobe dammed water against the Oak Ridges moraine and the Niagara escarpment. Similarly the glacier in the Lake Simcoe area held a lake on the northern slope of that moraine. These are the Peel and Schomberg "pondings."

Lake Algonquin occupied the Huron and Michigan Lakes basins and owes its existence to the depression of the earth's crust north of Grand Bend in Ontario, including the lowland southeast of Collingwood and around Lake Simcoe. The outlet at first was at Chicago by way of the Illinois River towards the Mississippi. Later, there were outlets at Kirkfield and Sarnia (Port Huron) and all three were correlated with the main Algonquin beach (92). This shoreline is much better developed than the Whittlesey and Warren beaches and no doubt it persisted longer. Indeed, there are several minor beaches at lower levels that are nearly as strong as the Warren beaches.

The Algonquin shoreline is mostly undercut and obliterated between Sarnia and Point Clark. From there to Port Elgin it is present as a high bluff within a mile or two of the Huron shore. At Port Elgin a large sand and gravel bar is built across the Saugeen valley. The shoreline runs inland east of Hepworth and consists of alternating stretches of bluffs and beaches. The Sauble River left deltaic sands around Hepworth, the deeper drier sands being subject to blowing. From Wiarton to Collingwood the Algonquin bluffs and beaches are nearly continuous and are close to Georgian Bay. In the Bighead and Beaver valleys there are good deltas.

GEORGIAN BAY
LION'S HEAD
NIPISSING
OWEN SOUND
PORT ELGIN
ALGONQUIN
PAISLEY
WARREN
ELMWOOD
WALKERTON
BEACH
COLLINGWOOD
NIPISSING
NIP.
ALGONQUIN
ANGUS
GODERICH
WARREN
EXETER
WHITTLESEY
TORONTO
CAMPBELLVILLE
WARREN
BRANTFORD
ARKONA
MAUMEE
LONDON
WYOMING
ARKONA
WARREN
WHITTLESEY
TILLSONBURG
WHITTLESEY
ALVINSTON
ST. THOMAS
WARREN
SIMCOE
EARLY ST. CLAIR
DRESDEN
RIDGETOWN
WARREN BEACHES
LAKE ST. CLAIR
ESSEX
LEAMINGTON

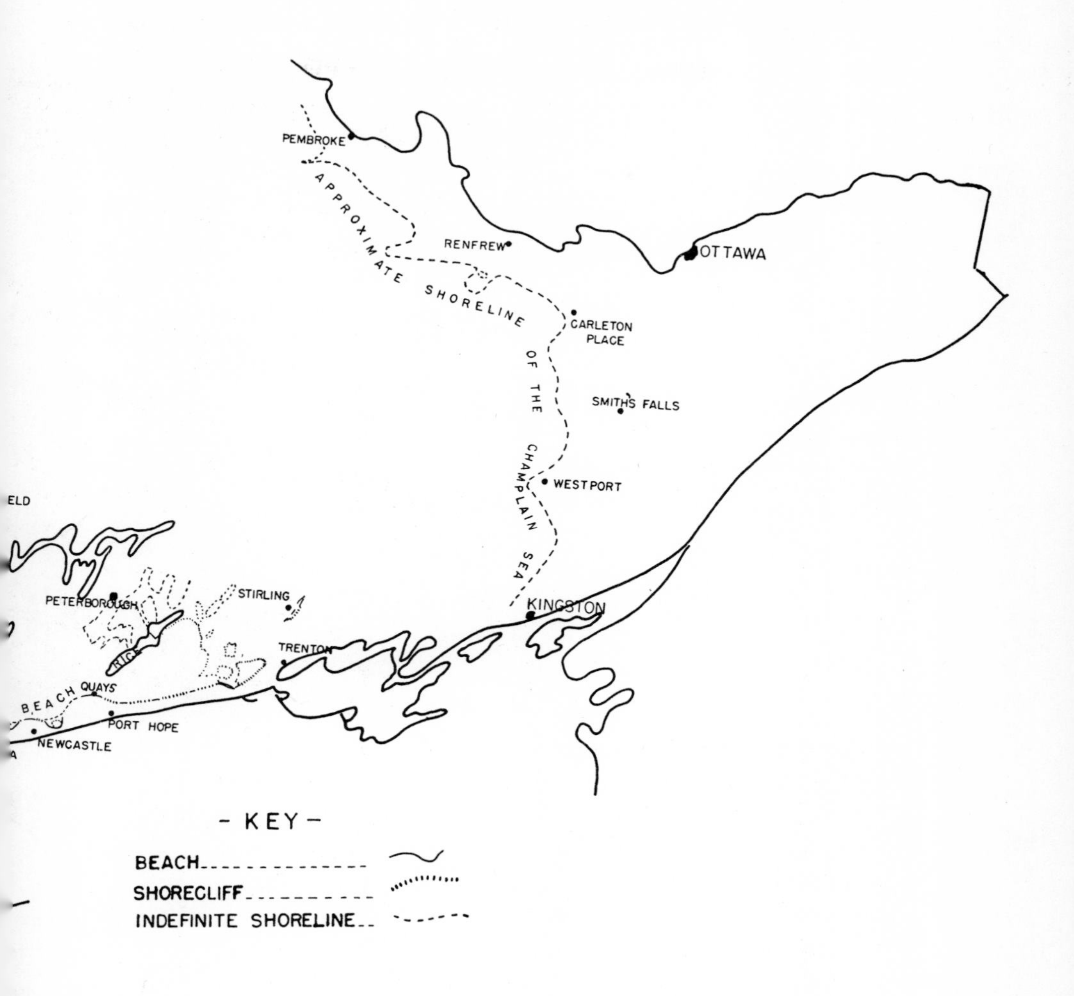

20. *Former and present shorelines in Southern Ontario*

21. *Varved clay, Holland Landing. These are thick varves, the lighter silty layers predominating over the darker clay layers.*

The Nottawasaga flats are sandy except around Stayner where there is clay. Other clay plains occur around Elmvale and in Orillia, Mara, and East Gwillimbury townships. Elsewhere the Algonquin lake bed in this area is sandy. The shoreline can be traced to Kirkfield where there is an outlet through Balsam Lake into the Trent valley.

The uplands between Lake Simcoe and Georgian Bay were islands in Lake Algonquin. The upper stage left bold bluffs and boulder-terraces around these islands, while several distinct but minor beaches form steps down the hillsides.

The Nipissing beach ranks with the main Algonquin beach in strength. Usually it is near the present shore and shows up mostly as a wave-cut bluff up to 90 feet high with boulder pavement on the terrace along the base. Gravel beaches were built in many sections; in fact sometimes a series of strands lie across the terrace between the Nipissing shorecliff and the present shore. South of Grand Bend, at Sauble Beach, and at Wasaga Beach lines of sand dune were built during and since Nipissing time. With this bouldery wave-swept border must be included the silty flats behind Lion's Head and the Minesing swamp with its adjacent flats. In the latter case there was an isolated lake south of the ridge at Edenvale which eventually cut down to Nipissing level.

Lake Iroquois was the forerunner of the present Lake Ontario and, because of the studies of Dr. A. P. Coleman, is the best known of the glacial lakes in Southern Ontario. Its shoreline is very prominent, equalling that of Lake Ontario in its development. Those beaches below the elevation of the Iroquois beach are not well developed (103).

The Champlain Sea, which occupied the St. Lawrence–Ottawa lowland following the recession of the glacier, invaded the Lake Ontario basin for a short period. The western shoreline of this body of water is indefinite. Like others before us, we searched in vain for a beach along the slope between Kingston and Renfrew, isolated gravel strands seen near Westport and White Lake being the only evidence we could find. The boundary shown on the map is therefore approximate. At lower levels there are many gravel bars known to be related to salt water because they contain marine shells in abundance. The marine beaches appear to have formed at any level where gravelly material was available, but separate continuous water planes cannot be traced.

THE RECESSION OF THE WISCONSIN GLACIER

At its maximum the Wisconsin glacier covered all of Ontario and extended to southern Ohio. It was not until it had melted back about 150 miles, uncovering nearly all of Ohio, that the first land was uncovered in Southern Ontario. Beginning at that point the next few paragraphs with the aid of thirteen stage maps will outline the retreat of the glacier from Southern Ontario. It deals with the position of the ice front, the drainage from the melting ice, and the glacial lakes at successive stages. The value of this fascinating story lies in the fact that it serves to explain the origin of the complex landscape of the region. Without this geological history it would be hardly possible to grasp and hold a mental picture of the various land forms and types of soil found throughout the province.

For the sake of clarity an outline of the successive stages in the retreat of the Wisconsin glacier as now interpreted will be given with a minimum of discussion. The reasons for making changes from previous concepts are given in separate papers (12) (13). Some points, however, are still vague and controversial. Several minor events have been omitted when they have had little or no effect on the upper drift.

The melting and retreat of the glacier was not simply an uncovering of the land from south to north. The wasting glacier consisted of several lobes and these first split apart near Orangeville as shown in Figure 22*a*. The surfaces of the lobes were highest in the centre and sloped towards

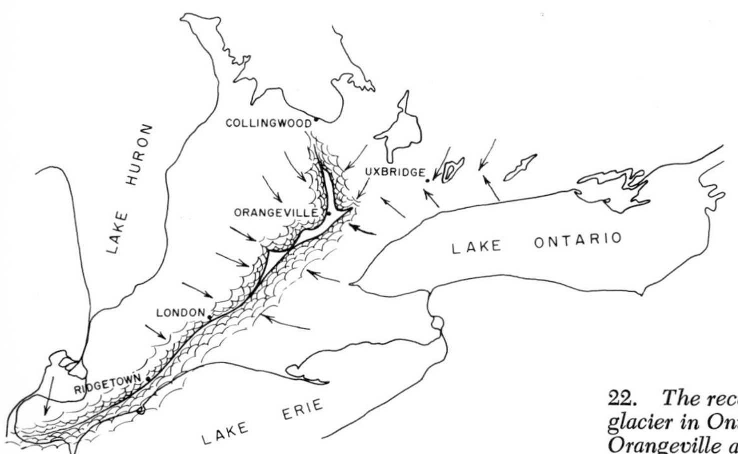

22. *The recession of the Wisconsin glacier in Ontario* (a) *Stage 1: at the Orangeville and Waterloo moraines*

the edges; which means that the relief was often opposite to that of the present land surface. Drainage flowing into the crease between the lobes brought in sand and gravel and built the Orangeville moraine. The sand contained a good deal of calcite while the gravel includes a sprinkling of siltstone both from east of the Niagara escarpment. The Waterloo moraine was built about the same time of similar materials, but the two are not connected as might have been expected. Moreover, the southward extension of the Waterloo sandhills implies a southward advance of the Huron ice lobe not shown in Figure 22*a*. The details of the glacier's movements in this area are still not clear.

Following the first split, the front of the southern ice lobe retreated and then readvanced over the Guelph and Woodstock drumlin fields and overrode most of the Waterloo moraine. After that it retreated to the Ingersoll moraine, while glacial drainage passed laterally across the Guelph drumlin field, cutting channels across the slope and leaving extensive beds of gravel in the hollows. In Oxford county three deep spillways were cut across the slope of the till plain, the first east of St. Mary's, the second past Embro, and the third past Woodstock and Ingersoll. Trout Creek, the Middle Thames, and the south branch of the Thames now occupy these valleys which all emptied into a small lake east of London.

In its retreat from the north side of the Orangeville moraine the Georgian Bay section of the glacier disclosed several drumlins and a short esker around Marsville and the smooth till plain to the north. Drainage from this ice front flowed down the path of the Grand River. It is apparent that this lobe of the glacier operated separately from the lobe in the basin of Lake Huron. First the Huron lobe thrust forward to build

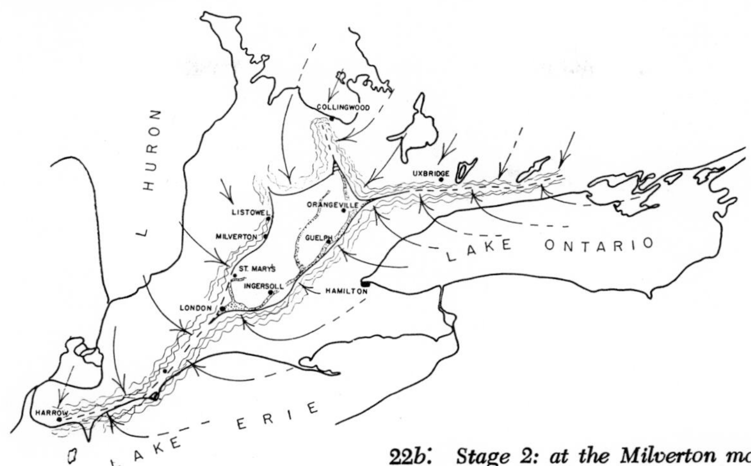

22b: *Stage 2: at the Milverton moraine*

the Milverton moraine as shown on the second stage map (Figure 22*b*). It then retreated and the Georgian Bay lobe readvanced overriding the Milverton moraine north of Listowel (Figure 22*c*). It left a sheet of gritty, stony till and smoothed the till surface while scoring it with shallow grooves or flutings (Figure 8). Only in the section from Grand Valley to Shelburne did it build a moraine, and there only a faint one, to mark the limit of this thrust. The Georgian Bay lobe then retreated and the Huron lobe advanced to the Mitchell moraine. A point that may not be evident from the fourth stage map (Figure 22*d*) is that the Milverton moraine is overridden south of Fullarton village. Evidently, as the Huron lobe pushed forward at this stage the frontal part underwent expansion. The course of the Mitchell moraine and the trend of the eskers north of Seaforth both indicate that a centre of pressure existed in the Huron ice lobe south of Goderich.

During the building of the Mitchell moraine, drainage collected in a shallow basin north of Brodhagen while the overflow went down the path of the North Thames River. Incidentally, two other shallow basins held

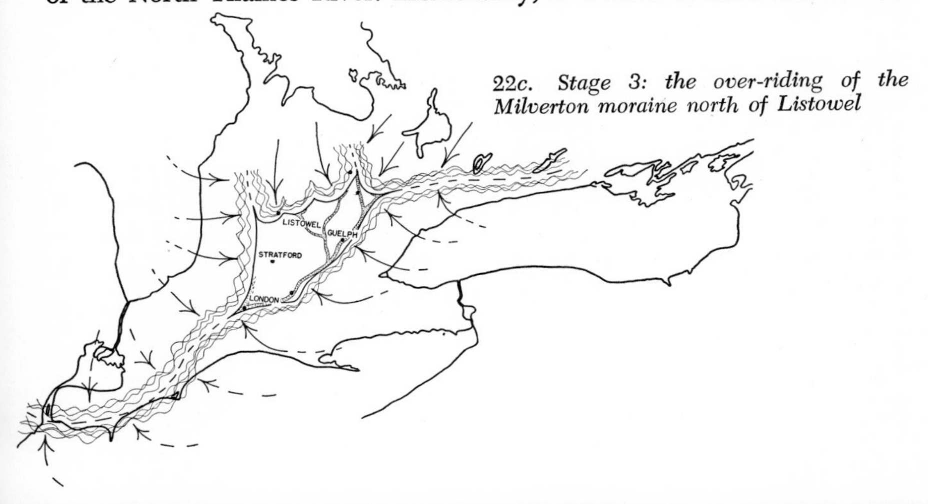

22c. *Stage 3: the over-riding of the Milverton moraine north of Listowel*

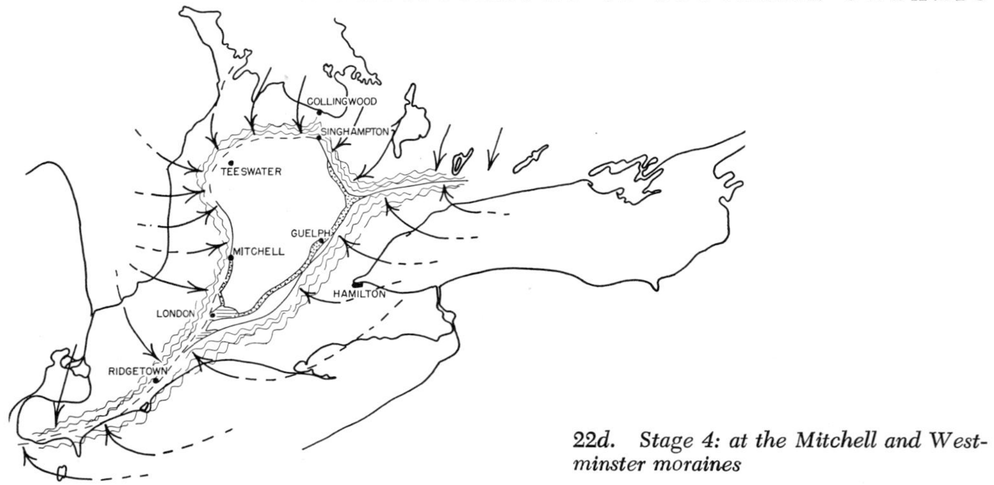

22d. Stage 4: at the Mitchell and Westminster moraines

water just prior to this period until their outlets gradually were deepened to drain them. One is marked by the marlbeds and stratified sand and clay south of Atwood; the other is in the Ellice swamp area in front of the Milverton moraine.

Figures 22*b* and *d* show the Ingersoll and Westminster moraines being built at the same time as the Milverton and Mitchell moraines, respectively. It should be clearly stated, however, that the moraines of the Huron and Erie lobes cannot be correlated with certainty before the Port Huron morainic system.

Because of their position and size in relation to their neighbours, the Seaforth and St. Thomas (Mt. Elgin) moraines are considered to be contemporaries (Figure 22*e*). Both are of similar brown, calcareous clay till and their southern extremities were built under water. After building the Seaforth moraine the Huron lobe made only two minor halts in front of the Wyoming moraine. The Erie lobe made two minor thrusts to push

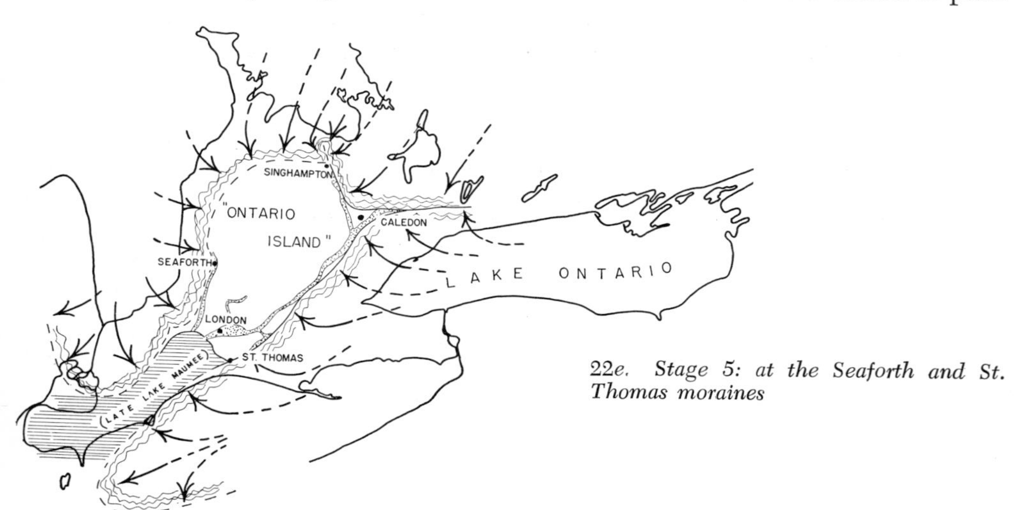

22e. Stage 5: at the Seaforth and St. Thomas moraines

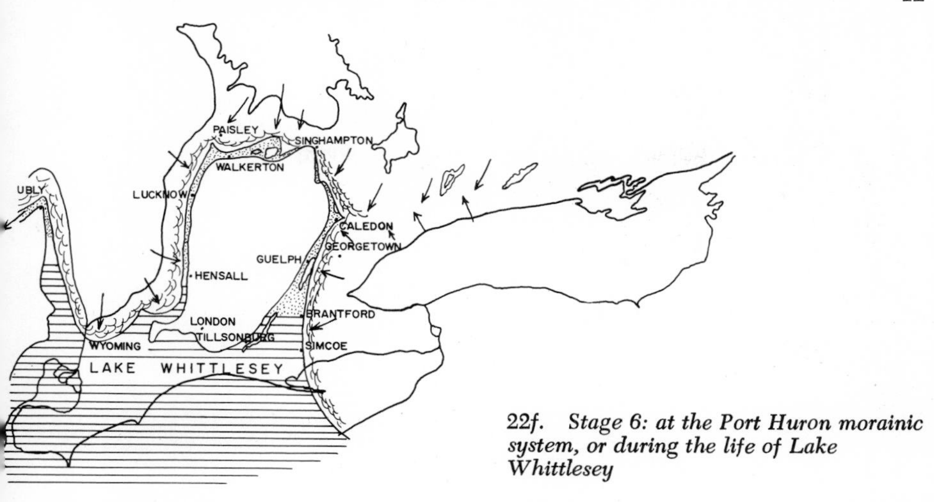

22f. Stage 6: at the Port Huron morainic system, or during the life of Lake Whittlesey

up the twin strands of the Norwich moraine before retiring past Tillsonburg.

Probably the glacier retreated a considerable distance behind the Port Huron moraine before it pressed forward again, and Lake Arkona existed during that interval of retirement. Lake Whittlesey, which stood about 30 feet higher than Lake Arkona, was contemporaneous with the Port Huron moraine. The strength of this moraine marks this as the longest halt experienced by the Wisconsin glacier up to this point in Southern Ontario (Figure 22*f*). The Huron lobe built a complex moraine of two or more strands. North of Goderich a spur of the moraine veering lakeward suggests that the ice lobe became truncated before the northern section was built. The Georgian Bay and Lake Simcoe lobe of the glacier advanced onto the highest points of the upland south of Collingwood and left a single-stranded moraine. From Singhampton the glacial drainage ran in either direction towards Walkerton and Caledon. It gathered volume and made progressively larger spillways as it advanced towards the points of contact with the Wyoming and Paris moraines. The Paris and Galt moraines of the Ontario–Erie lobe are both outstanding and the spillway before them is exceptionally large. The huge delta of Lake Whittlesey below Brantford must also be correlated with this splendid morainic system. The ice front at this stage has been traced through Michigan, Wisconsin, into Minnesota where the extension becomes indefinite (91).

The next stage map (Figure 22*g*) shows the Georgian Bay lobe of the glacier at the Gibraltar moraine while the Ontario lobe pressed against the Niagara escarpment and even over-topped it south of Milton building the Waterdown, the Vinemount, and the Niagara Falls moraines.

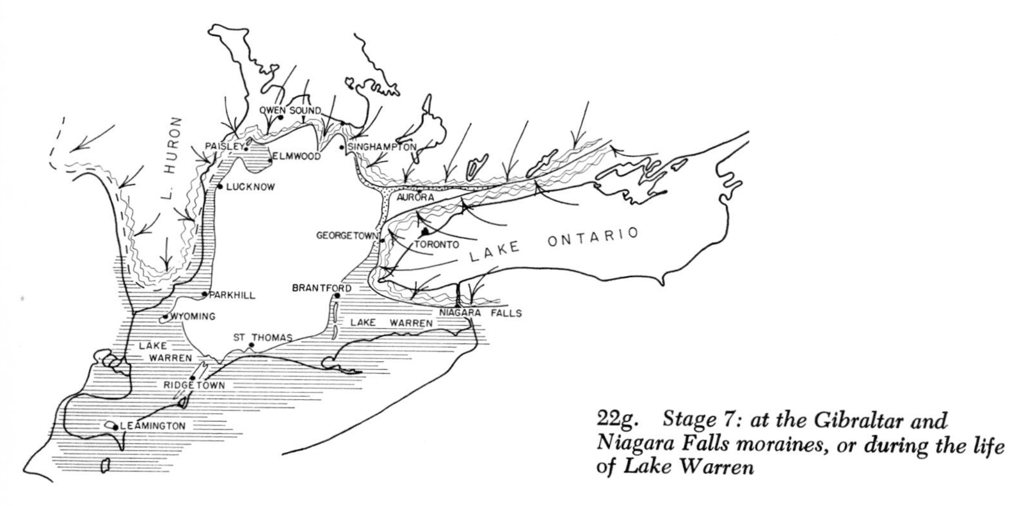

22g. *Stage 7: at the Gibraltar and Niagara Falls moraines, or during the life of Lake Warren*

By this time the Ontario lobe and what may be called the Lake Simcoe section had parted along the interlobate (Oak Ridges) moraine in Albion and King townships. The meltwater from both lobes piled sand on the crest of the moraine and then drained from Caledon East along the face of the Niagara escarpment past Georgetown to Campbellville leaving a disconnected train of gravel. At Campbellville it flowed over the top of the escarpment and soon met Lake Warren among the drumlins in Flamborough township. The sand and gravel beds in that township and the fine sand, silt, and clay between Dundas and Brantford were deposited at this time. Lake Warren in the Erie and Huron basins had its outlet westward across Michigan through the Grand River valley which allowed the water to drop 60 feet below the level of Lake Whittlesey. Also, there was a second stage ten to fifteen feet below the first so that Lake Warren is marked by twin beaches. This was an extensive lake that contacted the glacier north of Paisley and at Campbellville.

Leverett and Taylor (92) and others recognized two lakes following Lake Warren at 45 and 65 feet lower altitudes. Called Grassmere and Lundy, they did not leave fairly continuous shorelines similar to the Warren beaches in Ontario. The shallow gravel bar extending from near Leamington through Essex is at Grassmere levels. The beach at Lundy's Lane, Niagara Falls, for which the lower lake was named, is also a shallow gravel bar capping a faint moraine. Any extension of these beaches is missing in Ontario, indicating that these lakes were very short-lived. The Grand River built no deltas in them.

In Welland county several short gravel beaches occur below the Lundy water plane at altitudes of 625 to 600 feet above sea level. They appear to mark, rather indefinitely, the limits of a lake plain into which

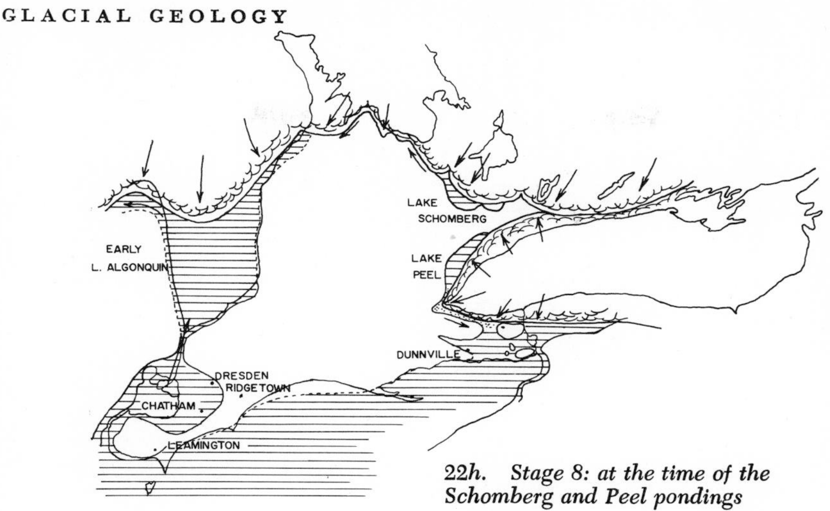

22h. Stage 8: at the time of the Schomberg and Peel pondings

the Grand River built a sandy delta at about 600 feet in the Dunnville area. This shoreline is no longer horizontal, but owing to warping of the earth's crust it rises gradually towards the north. This water plane, as shown on Figure 22*h*, was part of a lake in the Erie basin that also covered the flat plain around Lake St. Clair. A low but distinct bluff or thin beach strand at 605 feet marks this shore from Dresden to Sombra; this is at the level of Lake Algonquin. Following Hough's interpretation (75), it is suggested that the drainage at this stage flowed northward through the St. Clair river channel into early Lake Algonquin. Indeed this is the manner in which the St. Clair channel was cut down to the Algonquin level, a circumstance that is otherwise not well explained.

At this stage, as shown on Figure 22*h*, the Lake Ontario lobe stood at the Trafalgar moraine extending eastward from Nelson to Streetsville. It held a lake between the escarpment and the south slope of the Oak Ridges moraine known as the Peel ponding or, better, Lake Peel. The varved clays deposited in this lake are shallow and weathering has destroyed the stratifications except in the deeper beds occasionally found in depressions. The clay veneer is thin on the higher ground, but even here it is usually enough to produce a Peel clay loam profile instead of the Chinguacousy profile which develops on the underlying shaley, clay till. The Credit and Humber Rivers built small deltas at 700 and 725 feet which mark the upper limits of this lake.

The Lake Simcoe ice lobe dammed water ahead of it against the north flank of the interlobate moraine. This lake is clearly marked by beds of varved clays. The varves are unusually thick, the couplets in places being four to six inches thick. The silty layers are light gray and quite marly. At first this lake stood at high levels and emptied over the moraine south-

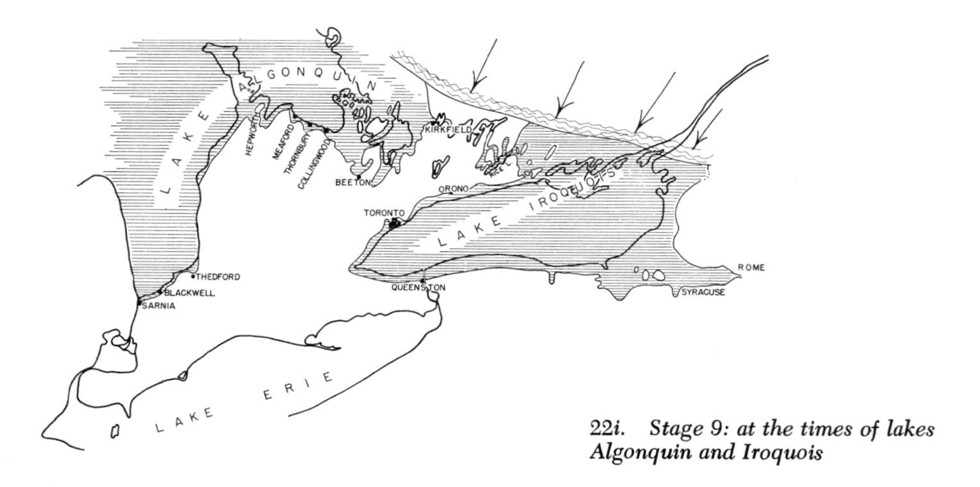

22*i*. *Stage 9: at the times of lakes Algonquin and Iroquois*

ward past Palgrave when the Lake Ontario lobe was still pressing against the south slope. Later the lake level was lower, the outlet probably being towards Owen Sound although it cannot be traced. The position of the ice front at this stage is shown on Figure 22*h* approaching the Oak Ridges moraine diagonally and overriding it east of Uxbridge. The extent of the retreat before this readvance is not known. The drumlins on the south slope, pointing southwestward, were built by the northern ice sheet. The limit of this advance is not clearly marked except near Trenton where a ridge of sand and gravel appears at the contact line between these drumlins and those of the Lake Ontario lobe. The Lake Ontario ice lobe in time withdrew to the eastern end of the basin uncovering the Mohawk valley at Rome, N.Y., which allowed the lake level to fall, thus initiating Lake Iroquois (121) (23). The northern lobe receded nearly to the edge of the Shield in Iroquois time as shown in Figure 22*i*. With the uncovering of Kirkfield and Fenelon Falls, Lake Algonquin found a new outlet down the Trent valley.

Judging by the strength of its shorecliffs and beaches, Lake Iroquois was much longer-lived than any of the earlier glacial lakes. However, the Dummer moraine which represents the work of the glacier during this period is no match for some of the earlier moraines. Moreover, it terminates abruptly just east of Tamworth and is not found in the Thousand Islands area. The Iroquois shoreline rises from 365 feet above sea level at Hamilton to 700 feet east of Stirling, which mostly is due to uplift since the downdraining of the lake. Lake Iroquois came to an end when the glacier withdrew from the slope at Covey Hill uncovering lower outlets into the Hudson valley. There are a few fragmentary beaches below the Iroquois beach, but nothing to equal the succession of Algonquin beaches

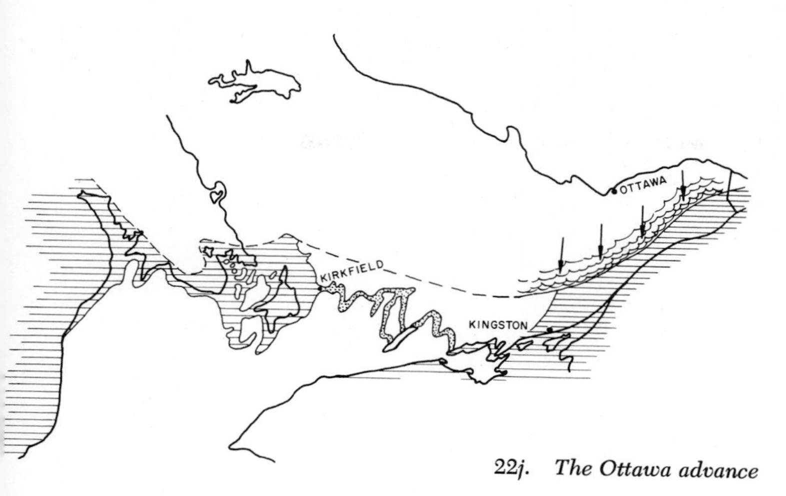

22j. The Ottawa advance

found in the area bordering Georgian Bay.

Lake Algonquin at first was confined to the southern part of the Huron basin, roughly south of Port Elgin. The outlet, following Hough's interpretation (75), was southwest from Chicago through the Illinois River valley toward the Mississippi River, and this entails an open passage between the Huron and Michigan basin. South of Point Clark the Algonquin beach is not in evidence, having been undercut by the present lake except around the Thedford marsh behind the sand dunes south of Grand Bend and near Sarnia where, if present, it cannot be separated from later beaches. As the glacier retreated, it soon fell away from the escarpment between Owen Sound and Collingwood and retired from the Lake Simcoe area. The shoreline may be traced continuously from Point Clark to near Wiarton, around the Georgian Bay to Collingwood then around the Nottawasaga flats and Lake Simcoe lowlands to Kirkfield. At Kirkfield the gravel beach leads to and continues for a mile or two into the outlet eastward toward Balsam Lake. This implies that the lake level at that stage coincided exactly with the level of the Kirkfield outlet —a rather unlikely coincidence.

It has been suggested that the earliest stage of Lake Algonquin stood 40 to 100 feet above the level of the Kirkfield outlet at 883 feet and that the water levels dropped when this outlet was reached (92) (78) (32). This was suggested in spite of the fact that there is no distinct, continuous shoreline in this area above the main Algonquin beach. On the slopes west of Lake Simcoe fragmentary beaches are found, especially about 100 feet above the Algonquin beach. We previously regarded them as belonging to a lower stage of Lake Schomberg. It was not possible to trace any such shoreline westward along the lower slopes of the Niagara escarpment past Collingwood toward Owen Sound, nor could a higher outlet channel eastward in the area south of Kirkfield be found. Thus the evidence for a higher early Lake Algonquin is vague and conflicting, but it is mentioned as a possibility because the alternative interpretation is hardly acceptable, as mentioned above.

Lake Algonquin maintained contact with the ice front as it receded north of Kirkfield. Although beaches are scarce, there are lacustrine clays, silts, and sands generally below levels of 910 feet at Uphill, 1,070 feet near Huntsville, about 1,245 feet east of Bernard Lake, 1,220 feet at South River, and 1,195 feet at Trout Creek (12). The falling-off of beach levels at South River and Trout Creek is due to crustal uplift rather than the uncovering of lower outlets. There are no outlets at these levels leading eastward towards the Ottawa valley. The first one to be found is east of Powassan, at Fossmill, leading down the valley followed by the Canadian National Railway and the Petawawa and Barron rivers emptying into standing water west of Petawawa.

It appears that the Algonquin lake levels dropped a maximum of 160 feet when the Fossmill outlet was opened. This would have uncovered both the lowlands around Lake Simcoe and the Nottawasaga flats. All the discharge was now through the Fossmill channel and a large delta of coarse sand was built in the Petawawa area. Uplift proceeded rapidly as the ice front retreated north of Fossmill. With its outlet now in the north the lake rose differentially toward the south, resubmerging the lowlands mentioned above. The fine, silty sands overlying coarser sand beds in the Algonquin plain around Alliston and north of Beeton are also attributed to the uncovering and resubmergence of this area. Eventually, in the south the lake reached the level of the Chicago outlet and the St. Clair River and these outlets took over. The St. Clair channel, being in clay rather than rock, deepened more rapidly than the Chicago outlet and so it eventually captured the whole discharge. With the outlet in the south, further uplift produced a succession of falling lake levels converging towards the south. This succession of rising and then falling water levels produced a series of beaches on the hillsides of the Penetang peninsula. Stanley's observation that the lower Algonquin beaches are parallel and his conclusion that these beaches must be due to separate outlets (128) (126) is not accepted; having made a thorough search we have not found outlets south of and above the Fossmill channel.

Lake Algonquin came to an end when the ice front retired past the Mattawa valley allowing the lake level to drop to the level of that valley thus initiating the Nipissing Great Lakes.

In the St. Lawrence lowland the retreat of the glacier from Covey Hill,

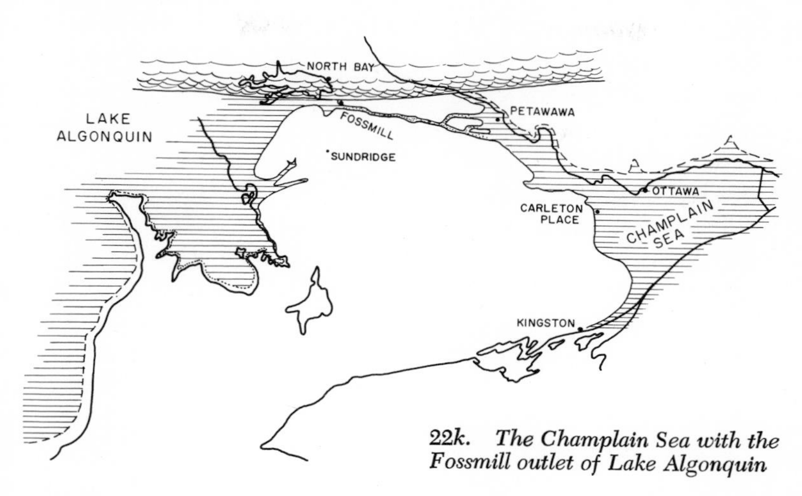

22k. The Champlain Sea with the Fossmill outlet of Lake Algonquin

N.Y., first uncovered a col lower than the Iroquois water plane and this served as an outlet for a lake called Frontenac. It emptied into the Champlain valley. Further retreat from Covey Hill allowed the water in the Ontario basin to drop to the level of the outlet in the Champlain valley, the lake at this level being called Lake Fort Ann. Neither of these lakes had any important effect on the upper drift in Ontario. Lake Fort Ann persisted until the wasting glacier left the lower St. Lawrence valley allowing free access to the sea. There is abundant evidence of the invasion of salt water into the St. Lawrence–Ottawa lowland in the form of marine shells embedded in the upper clay, sand, and gravel deposits. The bones of whales have also been found (19). In the upper St. Lawrence valley marine shells have not been found west of Brockville (97), but since salt water stood at Brockville the same water body almost surely extended into the Ontario basin. The lack of marine shells may simply reflect the brackish nature of the water at that time. The Champlain Sea covered all of eastern Ontario as shown on Figure 22*j*. The shoreline in the west is rather indefinite as few beaches and no bluffs are found on that rocky slope. Figure 22*j* shows the ice barrier north of Petawawa, with the Fossmill outlet of Lake Algonquin in operation emptying into the Champlain Sea. The Petawawa sands were deposited as deltas at that time.

Eastern Ontario bears the marks of three different ice invasions. The first was from the northeast and left the Malone till (98), known mainly from excavations for the St. Lawrence Seaway. The drumlin-like hills near the St. Lawrence River east of Cornwall point in a northeast-southwest direction parallel to the river. We suggest that they may be drumlins of Malone till overridden later by the Fort Covington advance. The

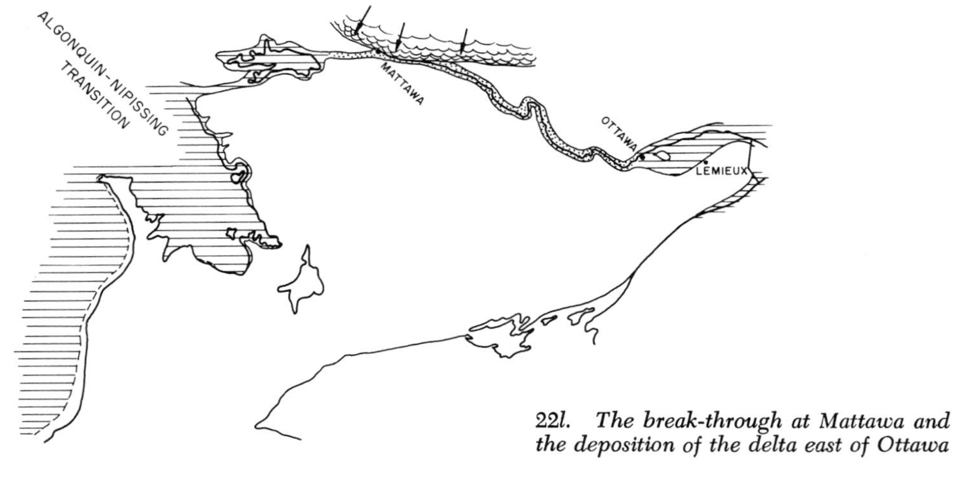

22l. The break-through at Mattawa and the deposition of the delta east of Ottawa

Fort Covington till lies above the Malone till and was left by ice advancing from the northwest. The till ridges in Glengarry and the southern parts of Stormont, Dundas, and Grenville counties lie roughly at right angles to this thrust and are interpreted as moraines built under water by this Fort Covington advance. The third movement was from the north. It produced drumlins in the area north of the low till ridges just mentioned. The drumlins all have a north-south alignment while in this same area a few fragmentary, sandy moraines occur running roughly east and west. These deposits, spread out by wave action, account for the sand in southern Grenville and Dundas counties, near Winchester, Moose Creek, in the Merivale vicinity south of Ottawa, and many other places.

Retreat of the ice front from Fossmill to Mattawa was accompanied by rapid uplift of the land. Consequently, the level of Lake Algonquin dropped prior to the uncovering of comparable levels on this slope. There was no drainage across this slope above 850 feet, or until the Ottawa River was reached. The break-through occurred two to three miles east

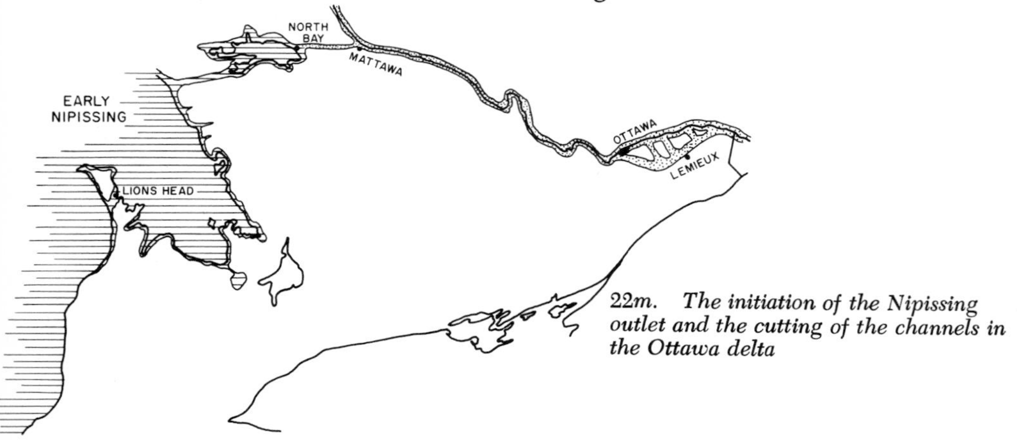

22m. The initiation of the Nipissing outlet and the cutting of the channels in the Ottawa delta

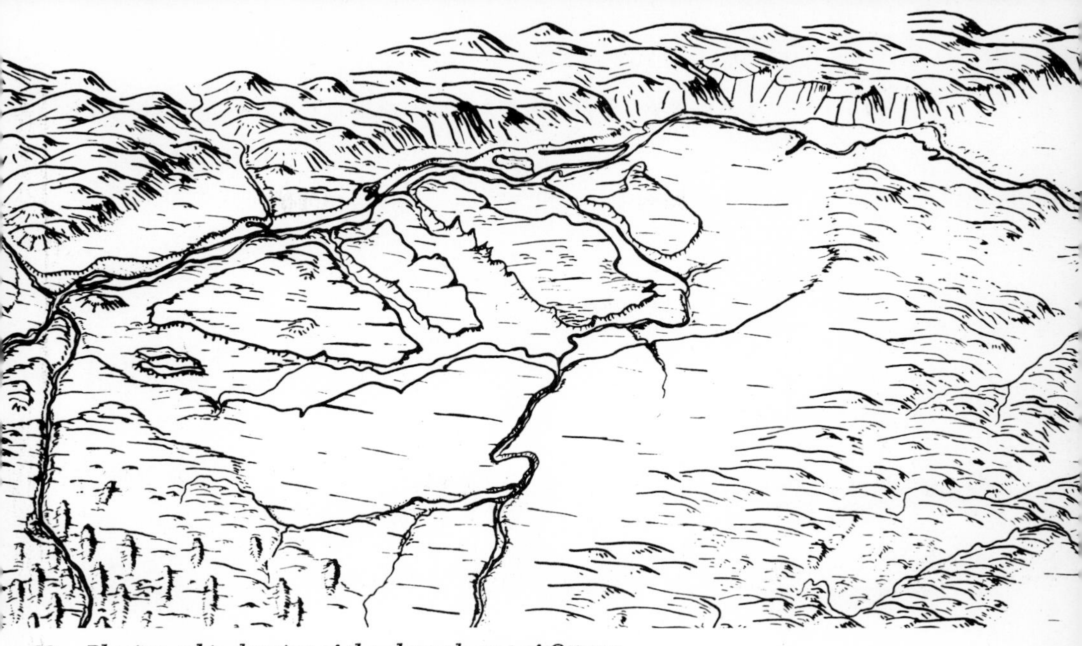

23. *Physiographic drawing of the channels east of Ottawa*

of Mattawa. By this time all the land above about 240 feet in the Ottawa valley was uncovered and only a bay east of Ottawa remained submerged. The break-through east of Mattawa allowed Lake Algonquin to drop from about 850 feet to the level of the Mattawa valley. The amount of lowering probably amounted to between 100 and 150 feet. We suggest that this sudden discharge cut the great channels through the Petawawa sand beds and redeposited the sand as another delta in the bay below 240 feet east of Ottawa (Figure 22*l*).

The delta sands east of Ottawa are now traversed by great flat-bottomed channels 20 to 30 feet deep and two to six miles wide. There are two main channels with several cross-channels cutting through the sand beds into the clay below. They are the work of the Ottawa River when this land first rose above sea level and they resemble tidal estuaries. The river now occupies the north channel. The southern one is drained in part by Bear Brook and the South Nation River while in two large undrained sections the Mer Bleue and Alfred peat bogs have developed.

The Nipissing Great Lakes came into being when the glacier retreated beyond the Mattawa valley uncovering a new lower outlet for the upper Great Lakes. The sill was just east of North Bay, but the marks of a great river are less impressive here than downstream toward Mattawa where scoured rocks and boulder pavement appear in various sections of the abandoned channel. Initially, the level of the outlet was over 100 feet below the St. Clair River at Sarnia, and some of the floor of Lake Huron was exposed in the south. Uplift in the north tilted the basin until the

24. *Stream-cut bluff near Vinette, Russell county. The foreground is the bottom of a channel of the Ottawa delta.*

St. Clair outlet again became active and it was at this two-outlet stage that the strong Nipissing bluffs and beaches were built. Except where undercut by the present lake this Nipissing shoreline is invariably present not far from the present shore. The rising water levels not only facilitated the cutting of cliffs and the accumulation of gravel beaches, but resulted in planing the lake floor to a remarkably flat surface. Lake Nipissing came to an end when North Bay rose above Sarnia, the outlet towards Mattawa dried up, and the St. Clair River gained the whole discharge as at present. Progressive uplift then uncovered the Nipissing plain, often producing a series of beaches between the Nipissing shoreline and the present shore. This was the last land in southern Ontario to be uncovered in the retreat of the Wisconsin glacier.

It was during the recession, before vegetation covered the ground, that the thin layer of silt was deposited on the surface of the till plains. It is regarded as wind-blown dust similar to the deeper loess deposits of the Mississippi valley.

POSTGLACIAL DEVELOPMENTS

Following the retreat of the glacier two processes have been operating to produce the landscapes as they appear today in Ontario. One is erosional and the other depositional. The first and the more important of the two is the cutting of river valleys, which includes the main valleys with their branches right out to the smallest gullies. These valleys and their relation to the Pleistocene deposits within the various watersheds will be dealt with in some detail later under the heading of River Valleys.

Erosion by wind should be mentioned here also. Blow-outs and sand dunes are common in deep dry deposits of coarse sand, but especially in kame-moraines. Some of the dunes are freshly made and are actively forming although this has been greatly controlled by reforestation. Most of them are old fixed dunes, long since covered by forests, with a soil profile developed at the surface.

In the wet, undrained depressions bog plants flourished and accumulated eventually to produce peat bogs. The shallower lakes are mostly filled with peat, but a central lake with peat encroaching upon it from the sides is also fairly common; the accumulation is continuing at the present time.

A peculiar and distinctive type of relief is found occasionally along the sides of the abandoned channels of the Ottawa River. It is due to mud flows (90, p. 200). The marine clays of this area are sensitive, which means they are very soft when wet and, under saturated conditions, landslides and mud flows sometimes occur. In the latter case the clay beds slump into a muddy "porridge" which flows to lower ground. The upper few feet of clay are more stable than lower beds, and blocks of this crust drop into the mud as it gives way beneath them and are floated away from the bank. The end result is a semi-circular crater within which there are irregular concentric ribs formed by the blocks of crust tilted at various angles. In one case noted on the eastern outskirts of Ottawa and described by Crawford (90, p. 200) the mud came to rest on the floor of the old stream channel nearby. However, in most cases the material is not present on the channel floors so the flows no doubt occurred while the early Ottawa River still used the channels. The craters of these mud flows occupy several hundred acres in Russell and Prescott counties. The choppy relief is unique and cultivation of these lands is extremely difficult. Moreover, the instability of these clays close to any valley constitutes a serious danger to buildings and even to life.

SURFACE 3 FEATURES

In an earlier section a few points about the distribution and formation of moraines, drumlins, eskers, and shorelines were given to aid in the interpretation of glacial history. It is now in order to refer to the map and to describe these surface features in some detail.

MORAINES

In this section both the moraines and associated spillways are described. The first of these features dealt with are the Orangeville and Waterloo moraines. They represent the first land to appear when the ice lobes separated west of the Caledon Hills. The Blenheim and Leamington moraines are included here. Next, the group of clay till moraines north of London formed by the Huron lobe will be described. Then, the comparable groups south of London formed by the Erie ice lobe will be dealt with. This leaves only the sandy Wawanosh moraine and the Saugeen kames, the work of the Georgian Bay lobe, before describing the Port Huron morainic system. It seems better to discuss moraines and spillways in these groups rather than in a strictly chronological order.

After dealing with the horseshoe-shaped set of Port Huron moraines, the Gibraltar and Banks moraines, the Tara Strands, and the series above the Niagara escarpment around the west end of Lake Ontario are described. The moraines of south-central Ontario, the Oak Ridges, Trafalgar, Trenton, Oro, Corn Hill, Edenvale, Mara, Dummer, and the fragmentary moraines of the St. Lawrence–Ottawa lowland are then discussed in that order.

In describing the moraines we built upon the well-known work of F. B. Taylor (131), which has been a standard reference since 1914.

The boundaries of the moraines on the map usually outline the higher, more distinct crests. In some cases lower slopes and low outlying ridges which might have been included are left with the adjoining till plains.

Typical materials in a kame at Glen
'*s on the Trent River. The cross-*
'ding and irregular grading are charac-
'stic.

26. *Surface topography of Southern Ontario*

INTERLOBATE MORAINES BUILT BETWEEN THE ONTARIO–ERIE AND THE GEORGIAN BAY AND HURON LOBES

Orangeville moraine. This moraine was first mapped and named by Taylor (131), but he did not regard it as an interlobate deposit formed in the first split of the ice lobes in Ontario. This conclusion was reached in 1949 (18) and the interpretation was further changed in 1963 to include the extension northward to Horning's Mills and beyond. This latter section was formerly mapped as part of the Singhampton moraine. It is clearly visible to the east of No. 10 Highway south of Primrose where it is over 100 feet high with an abrupt western face. It is flat-topped except for occasional till ridges and consists mainly of gravel and sand. Two spurs projecting westward are made of similar glacio-fluvial deposits and break the continuity of a train of shallow sandy outwash along the west flank of the moraine. The gravel in this section, as in the section south of Orangeville, includes a good deal of dull, rusty brown siltstone, while calcite is abundant in the sand fraction. In the Hillsburgh area the main component is fine sand.

At the head of the Hockley valley there is a cluster of kames which extend into the First Concession of Amaranth. This is a usual development associated with valleys indenting the Niagara escarpment. It is convenient to include it with this moraine, but the main extension of the moraine is two miles east of Orangeville and crosses Highway No. 10 one to two miles south. One very big gravel hill stands up conspicuously to the west of the highway a mile south of the town. The spillway on the eastern edge of town is a later development associated with the Singhampton moraine.

South of Orangeville, Caledon Lake is held in position by a barrier of

27. *Spillway at Hillsburgh*

morainic hills to the south. This barrier appears to break the continuity of a broad channel with a gravelly and swampy floor extending through the moraine to Hillsburgh and Cedar valley. This spillway cuts from the centre of the moraine at Caledon Lake to the southern border at Hillsburgh. The dry terraces have been cleared and cultivated, while white cedars and tamaracks remain in the swamps.

West of Belwood the moraine is reduced in size, fading to a clay till ridge near Alma particularly, but a mile or two south of that village a spillway reappears soon to be flanked by sandhills as in the Hillsburgh area. This group of kames is on the eastern side of the spillway which stops short of Floradale. The train of stream gravels continues from Floradale to Wallenstein, beyond which it disappears. A low till moraine lies east of it, meeting the Conestoga River at Hawkesville.

Waterloo moraine. The Waterloo hills occupy an oblong tract west of Kitchener extending from St. Clements to Ayr. Some of the hills are kames and most are of sandy till. Sandy outwash prevails in the hollows. The topography is less rugged than on most kame-moraines and the abundance of fine sand and sandy till is unusual. In these traits it resembles the Orangeville moraine, but it lacks the accompanying spillway. It also contains more calcite than dolomite in the sand and siltstone in the gravel.

The origin of these sandhills is puzzling. They lie near the interlobate line but extend southward between the Guelph and Woodstock drumlin fields. The modest topography and sandy till have resulted from a readvance of the Ontario–Erie lobe overriding the stratified sand and gravel.

Beyond the Nith in North Easthope township, the sandhills are cut through by a spillway from Lisbon through Amulree. In taking this course it cuts through high ground. The kames east of the channel stand up 100 feet or more above the adjacent plains. The higher knobs rise above 1,350 feet, and the floor of the channel is at approximately 1,175 feet above sea level.

Between Stratford and Tavistock three faint moraines of heavy till cross the smooth till plain. South of the Perth county border along the Trout Creek and around Lakeside is another fragment like that in Easthope township, with kames and till intermingled. The Cobble Hills on the Middlesex–Oxford line are an isolated group of kames in direct line with the other moraines. From here to the Blenheim moraine there are no other forms which could be interlobate in origin except possibly the highest sandhills in the vicinity of Mt. Brydges.

Blenheim moraine. This moraine was named by Taylor who wrote that "northeast of Blenheim it has a relief of 20 to 30 feet and is well defined." It is unusually wide, being nearly seven miles across at one point, and north of Duart splits into two strands. Very few kames are included but from a practical standpoint the dearth of gravel is offset by the beaches of glacial lakes. The twin Warren beaches encircle the highest section, then extend southwestward as a gravel bar on its crest through Cedar Springs to the bluff of Lake Erie where it disappears, having been undercut by the lake. It reappears a couple of miles west of Port Alma but fades out before reaching Wheatley. Without this gravel bar to set off the moraine, it would be hard to trace it south of Blenheim, which is surprising in view of its bulk east of Ridgetown. Regarding its extension westward to Leamington, Taylor said, "There is a low flat ridge hardly visible as a ridge."

Leamington moraine. In describing his Essex moraine Taylor included the "high knob west of Leamington." This stands about 100 feet above surrounding lake plain and the highest part, north of Ruthven, is of heavy till. The shoulders and sides are mantled by gravelly beaches of Lakes Whittlesey and Warren.

The Essex moraine was originally described as "a low broad ridge of till, very smooth and with such gentle side slopes as to be quite inconspicuous to the eye as a ridge" (131). Although the gravel bar along its crest was not mentioned, it was later mapped as a Grassmere storm beach (92). In fact it is the beach that provides most of the relief. Following the rule of mapping only moraines that can be distinguished in the field, we have omitted the Essex moraine, except the fragment at Leamington. The "occasional low stony knobs of till" of the Kingsville Boulder Belt are also omitted because they are so scattered.

MORAINES OF THE HURON LOBE NORTH OF LONDON

Milverton moraine. Taylor's description of the Milverton moraine as "a slender lightly built moraine, rather narrow but quite well defined, its relief being generally 20 to 30 feet, sometimes 50 feet" needs no amplification. The material in it is a pale brown calcareous clay till with only a light sprinkling of pebbles and boulders included. Moreover it contains very little sand. No doubt the colour is given by the brown limestone which underlies this area. Despite considerable variation in texture and stoniness, the foregoing describes all the silty clay till which is common to a broad belt east of Lake Huron.

From Fullarton village, where it contacts the Mitchell moraine, the Milverton moraine crosses a till plain of little relief ending east of Listowel. North of that point there is a smooth till plain scored by flutings trending southward. This type of plain extends south of Atwood while south of Monkton the faint knoll-and-sag type of relief appears instead. The fluted till is more stony and gritty than the silty clay till of the Huron lobe.

Drainage from the northern section of the Milverton moraine went down the Nith valley, while the remaining part flowed towards the Thames valley but left no definite channel. The moraine serves as a dam to prevent westward drainage of the Ellice swamp. Stratified silt, sand, and clay appear on the margin of this bog.

Mitchell moraine. The Mitchell moraine is a single strand of heavy till like the Milverton moraine. It is well defined from Walton to Brodhagen, Mitchell, Bannock, and Prospect Hill and is crossed by No. 7 Highway just east of Elginfield. A couple of miles beyond it comes to an abrupt end where it is overriden by the Lucan moraine. At the northern extremity it comes to a rather indefinite ending two or three miles north of Walton. Actually it appears to curve westward to Blyth, losing its identity in the moraine west of that village. The kames west of Brussels are not part of it as they are set on the drumlinized till of the Georgian Bay lobe.

From Mitchell southward a spillway, now followed by the Thames, runs along the front of the moraine ending near Thorndale. North of Brodhagen the flat and occasionally swampy nature of the terrain and the presence of stratified sediments indicate temporary ponding of water on the lower land east of the moraine.

Lucan moraine. The third in the series north of London, this moraine can be traced from the townline of Caradoc through Lobo, Ilderton, Lucan, and Woodham, north of which it is fragmentary. The ridge passing two miles east of Staffa, curving northwestward to Constance, may be its extension or the product of a slightly earlier thrust. Another possible correlative is the ridge north of Farquhar which fades out west of Cromarty.

When the ice front was at the Lucan moraine, drainage flowed past Elginfield and down the Medway River. Beginning near Cromarty two or three shallow troughs extend to Kirkton, the more westerly one being undrained in two separate sections. The most pronounced channel occurs between Granton and Elginfield where a broad valley over half a mile

28. *Clay moraine in South Oxford. The crest of the moraine is on the skyline.*

wide is covered with stratified sand and silt. It is rather odd that the present drainage does not continue down the spillway as an extended Medway River. The break through the Mitchell moraine near Prospect Hill and the capture of the upper reaches of the Medway valley by Fish Creek appear to be later developments.

Seaforth moraine. As Taylor expressed it, "the Seaforth moraine is somewhat stronger and bulkier than the three preceding it." South of Blyth it consists of brown clay till like the others. All these moraines have a general height of approximately 50 feet and the knob-and-kettle topography is not too rough to prevent the clearing and cultivation of the land. The southern reaches of the Seaforth moraine, beyond Watford where the ice front was under water, are too faint to be mapped. From Watford a broad ridge extends towards Arkona. Just south of that village it doubles sharply to skirt the south side of the Ausable valley, sweeping gradually around to the north until it is crossed by No. 4 provincial highway east of Clandeboye Corner. For a few miles south of Seaforth it is weak but elsewhere it is stronger. South of Blyth it ends amid a cluster of kames in the southern extension of the Wawanosh moraine.

Glacial drainage failed to produce a definite channel in front of the Seaforth moraine north of Hurondale. South of that hamlet a continuous shallow trough, now followed by tributaries of the Ausable, lies on its eastern flank. It is present in best form between Denfield and Lucan Crossing. There is some doubt about where this spillway met glacial

Lake Maumee. The delta at Coldstream and Poplar Hill appears to be the highest point where there is definite evidence of standing water. However, we examined the bedding and pebbles in a gravel pit near Denfield (lot 31, concession XV, London township), and found well-rounded, evenly sorted pebbles, like beach gravel. On similar evidence Taylor suggested that the lake extended as far north as Exeter.

MORAINES OF THE LAKE ERIE ICE LOBE SOUTH OF LONDON

Between London and Tillsonburg there are several moraines similar in form and composition to those north of London. Although they are distinct ridges they never attain the height or ruggedness of the more sandy and bouldery moraines between Paris and Caledon. They all fade out towards the southwest where the Erie lobe was under the water of glacial lakes. The greyish brown, silty clay in these ridges is similar to that found in the moraines north of London. The predominant pebbles are of pale brown or brownish-grey cherty limestone from the underlying formation while greenstones and conglomerates are common.

Ingersoll moraine. This moraine, the first south of the Thames River, does not touch the town of Ingersoll but swings away from the river along the boundary of West Oxford township. It skirts a group of drumlins, then turns northward to the river again east of Woodstock crossing No. 2 Highway three or four miles east of the city. West of Woodstock it is of silty clay till with a few kames while its extension, found in the disconnected morainic hills between Woodstock and the Nith River, consists of loose, loamy or sandy till.

The great Thames spillway carried the drainage when the glacier stood at the Ingersoll moraine and while at least two later moraines were being constructed. It is a broad channel cut to bedrock, a depth of 75 to 100 feet. The gravel terraces between Thamesford and London were left by the same spillway and are present in the bottom of the channel farther north. A few miles north of Woodstock the deep channel ends, but a wide train of sand and gravel can be followed over the low divide into the Grand River basin. A deep channel appears again at Plattsville continuing towards Ayr along the course of the Nith. From Ayr to Blair it is a swampy trough drained only by small brooks and at Blair it contacts the great gravel terraces in front of the Paris moraine.

Westminster moraine. The Westminster moraine is separated from the Ingersoll by Dingman Creek. Similar in form and composition, it is a clay ridge approximately 50 feet high and a mile and a half wide. In diminished form it can be distinguished where it crosses the Talbot Road midway between Lambeth and Talbotville Royal. From there eastward it grows larger to a point near Crampton where it meets the spillway that heads towards the Thames at Putnam, and beyond that it cannot be identified.

Dingman Creek from Derwent to Lambeth drains a broad valley floored with clay, silt, and sand or gravel. These sediments relate to a late stage of Lake Maumee.

St. Thomas moraine. The St. Thomas moraine is the strongest moraine of the series between London and Tillsonburg and extends for 60 miles from Wallacetown to Eastwood. South of St. Thomas it was built by a submerged ice front, but nevertheless it is prominent and includes more gravel than usual as far as Wallacetown, where it dwindles rapidly and disappears under stratified clay and silt. At St. Thomas a gap occurs in the ridge so that in a sense the selection of that name for the moraine is questionable. A happier choice would have been Mt. Elgin, after the little village that perches on the moraine where it is best developed. At Burgessville the moraine touches the Ingersoll moraine, parting from it again at Oxford Centre and running as a separate strand towards Drumbo where it becomes buried under outwash sands. Beyond the Grand River it shares the fate of the other members of this series in being merged with or overridden by the Paris moraine.

The upper reaches of Kettle Creek and Reynolds Creek follow glacial drainage channels in front of the St. Thomas moraine. These channels are marked by deposition rather than erosion, the broad valley being lined with silt or, less extensively, with sand and gravel.

Norwich moraine. Between the St. Thomas and Tillsonburg moraines is another which may be named after the village of Norwich. In the southwest it splits into two narrow strands separated by a branch of Catfish Creek. At Lyons village they lie within a mile and a quarter of each other. At Norwich and New Durham it is a distinct continuous ridge, while between Cathcart and Galt there are only disconnected segments separated by sand and gravel beds.

Tillsonburg moraine. The fifth moraine southeast of London passes through Tillsonburg and that name was applied to it by Taylor. When building this moraine the glacier was washed by the water of Lake Whittlesey south of Tillsonburg. The sand and gravel at Otterville were deposited by ice-border drainage. Southward, delta sands bury the moraine completely in several stretches between Tillsonburg and Sparta. For about five miles west of Sparta there appears, oddly enough, an isolated ridge over 100 feet high. Its only traces west of Kettle Creek are in the form of a few knobs of boulder clay between Union and Fingal.

From Tillsonburg the moraine can be traced northward to the outskirts of Ayr, beyond which its identity is lost in the gravelly hills west of Galt. This moraine is well marked by contours on the Tillsonburg, Brantford, and Galt map sheets. The Nith River cut a wide gap through it west of Paris.

SANDY MORAINES OF THE GEORGIAN BAY LOBE

Saugeen kame-moraines. Within the Saugeen drainage basin between Wingham and Flesherton the Georgian Bay lobe left scattered groups of kames. Quite often these sandhills are found along the shoulders of the river valleys. Associated with them are great quantities of coarse pitted outwash. The Holstein vicinity is almost covered by gravel terraces at several levels, which altogether occupy an extensive area. The Saugeen kames exhibit some of our roughest morainic topography and consist of unusually coarse material.

Wawanosh moraine. From Clinton northward through East and West Wawanosh townships and past Langside there is a morainic belt about five miles wide. It is a complex of low glacial till ridges and hills of sand and gravel in which the gravelly knobs invariably stand up as the highest peaks at 1,050 to 1,100 feet, while the less rugged parts of the moraine are generally of silty till. The kames are concentrated in the north and the south. In West Wawanosh township there is mostly clay till, as in the Wyoming moraine adjacent on the west. Along its eastern border this moraine merges almost imperceptibly with the drumlinized till plain. The gravel trains and sandy outwash common to the depressions of that plain continue right to the moraine.

THE PORT HURON MORAINIC SYSTEM

The result of a strong readvance following a poorly defined retreat, the Port Huron moraine is strongly developed in Ontario and the adjacent states. Thus, it serves as a marker at one important stage in the recession of the last glacier. In Ontario it consists of the Wyoming, Singhampton, Paris, and Galt moraines.

Wyoming moraine. The Wyoming moraine on the eastern side of Lake Huron is the correlative of the Port Huron moraine in Michigan. Consisting of boulder clay like the several moraines before it, it was originally described by Taylor as "one of the strongest in the series." His mapping was vague north of Lucknow but he extended it to Gibraltar on the edge of the Blue Mountain. In the present work, following a policy of separating the moraines of the Huron lobe from those of the Georgian Bay and Lake Simcoe section, the Wyoming moraine is terminated at the bend west of Walkerton. Incidentally, the moraine which proceeds from that point extends not to Gibraltar, but to Singhampton.

29. *Till moraine, Hensall: poor beans on eroded clay knolls in the Wyoming moraine*

From Ailsa Craig southward the Wyoming moraine was built under the water of Lake Whittlesey and forms a single broad ridge which fades out four miles west of Wyoming. From Ailsa Craig northward to the Bayfield River there are two strands lying close together with a big spillway to the east and a smaller one between them. The main spillway splits north of Zurich and passes on both sides of the modest ridge on which Hills Green is situated. This ridge and Taylor's Clinton moraine, which lies on the east side of the spillway from Brucefield to the Maitland River, might well be regarded as part of the Wyoming moraine, although doubtless they are the result of earlier thrusts of the glacier. From Goderich to Lucknow one main strand can be identified and it is broken by a gap at the Lucknow River. In addition the spillway south of Lucknow is divided by short morainic ridges in several places. On the western slope of the moraine, glacial drainage left small channels and intermittent gravel trains which provided some of the gravel for the beaches of Lake Warren.

North of Lucknow the moraine is broad with two crests usually in evidence. It lies west of the Greenock swamp and contacts the Singhampton moraine north of Glammis in a cluster of kames.

As indicated on our maps the southern part of Taylor's Goderich moraine is regarded as the continuation of the Wyoming moraine; therefore the former name is dropped. The section on the brow of the Blue Mountain and its extensions is given the name of Banks. Thus the Banks moraine becomes the third of a group including the Singhampton and Gibraltar moraines. We do not consider Taylor's Kincardine moraines definite enough to be mapped. Finally, there is no moraine between Port Elgin and Hepworth. In his cursory survey of this area Taylor apparently failed to recognize the drumlins. The two bits of moraines shown on his map east of Owen Sound are also drumlins.

Singhampton moraine. The northern limb of the Port Huron morainic system in Ontario runs through rough stony land and is made of loose stony loam in which dolomite predominates. It is a single-crested ridge running across the drumlinized plain. Kames appear at frequent intervals. Between Kinghurst and Berkeley the adjacent terrain is very rough, making it difficult to segregate the moraine. The kames west of Flesherton are not included in this moraine. These, along with the gravelly outwash, constitute the "Artemesia Gravels" which early writers (94) regarded as sea beaches.

From Singhampton to Caledon this moraine lies on the brow of the Niagara cuesta, although broken by the larger re-entrants in the escarpment. It is fragmentary from Singhampton to Maple Valley, then gains prominence past Honeywood. South of Hornings Mills this moraine shifts to the east and combines with the Gibraltar moraine to form the massive ridge east of the Violet Hill spillway. Between Orangeville and Caledon several drumlins of the Guelph group lie to the west of it; apparently that area had been moulded by the Ontario–Erie lobe before the moraine was built. The thrust which built the moraine advanced onto the sides of these drumlins.

Glacial drainage in front of the Singhampton moraine flowed both west and south from Singhampton, gaining volume as it progressed. Along the eastern arm a great gravel train appears south of the Pine valley. North of the Hockley road the spillway is called locally Eden valley. It provides level ground for the race track and fair grounds on the northern outskirts of Orangeville, and appears as a swampy depression across the

30. *Orangeville moraine as seen from Highway 10 near Primrose*

eastern approaches to that town. From there it extends down the path of the Credit River to Cataract. West of Singhampton the meltwaters left a small channel to Flesherton and south of that village entered shallow catch basins from which they escaped down the Saugeen valley. The big gravel pit at Durham is in this spillway, and gravel trains lead into it from the north. From Allan Park to Hanover the terrace is unusually broad, as shown on the map, and the gravel beds are fairly deep. Being more confined in the Walkerton area the drainage carved distinct valleys which now contain cedar swamps. Two main and two minor routes were used to reach the Greenock swamp, judging by the evidences of stream action. The Greenock swamp and the surrounding land is floored with silt and fine sand. No doubt it held standing water when this spillway was active.

Paris moraine. Taylor's original Paris moraine included an extension from Caledon to Gibraltar. In this account the name will be applied only to the section south of Caledon which was built by the Ontario–Erie ice lobe.

From Acton to Galt it is a high bouldery ridge. At or near Paris this moraine was waterlaid and it becomes steadily fainter towards Lake Erie. It is still a sharply defined ridge at Scotland but is weaker at Vanessa and quite faint east of Delhi. Farther south it is "scarcely perceptible as a ridge but exerts some control over minor drainage." Taylor's mapping projected this moraine east of Big Creek towards Port Burwell. However, west of that course the sandy till that comes to the surface between Langdon and Carholme may be a segment of the moraine which barely escaped burial by sand.

The Paris moraine differs in composition from those just described. South of Paris it is sandy and to the north it consists of loose bouldery loam including considerable coarse outwash, the material being mostly dolomite from the Niagara cuesta. Towards Caledon an admixture of red shale is visible. The prominence of the moraine is reduced south of Paris because it is partly buried by stratified sand. North of Delhi the waves of Lake Whittlesey built gravel bars on selected sites.

The work of the running water that drained along the ice front during this stage is no less impressive than the work of the glacier itself. The spillway from Caledon to Brantford is unsurpassed in southwestern Ontario. Broad gravel terraces, often at two or more levels with swampy stretches in the lowest one, can be traced all the way. From Erin to the brow of the escarpment the Credit River now occupies the old valley. Near Brisbane for a short distance there is no stream in it, then the Eramosa River takes possession until it meets the Speed at Guelph which in turn gives way to the Grand at Preston. From No. 7 Highway the spillway can be seen where it passes through the farm of the Ontario Reformatory at Guelph. Nearby on the Cutten Fields golf course the gravel terraces are well displayed and a stream-cut bluff diagonally truncates the end of College Hill.

Galt moraine. The Galt moraine lies just behind the Paris moraine and never far from it except in the extreme south. At the southern boundary of Brant county they are less than two miles apart while the distance between them from Simcoe to Delhi is ten miles. South and east of Guelph in several places the two come together producing a broad composite moraine. The Galt moraine passes south of Acton, crosses No. 7 Highway two miles east of that town, and extends in diminishing strength towards the escarpment, fading out near Rockside.

31. *Typical morainic topography: an oblique aerial view of the Paris moraine near Erin*

The Galt moraine is similar in form and composition to the Paris moraine. It is a rugged stony ridge of loose loamy till north of Brantford and a smoother sandier ridge from Brantford to Simcoe. South of Acton several drumlins appear in intimate association with the moraine. Kames occur quite frequently.

It may not be clear from the map that the glacial drainage broke through the Paris moraine at Eden Mills and followed the earlier spillway to Paris leaving only local drainage to pass along the Galt moraine. These spillways are marked by gravel terraces rather than by deep channels.

MORAINES BEHIND THE SINGHAMPTON AND GALT MORAINES

Gibraltar moraine. This moraine lies behind the Singhampton and is similar but slightly weaker. West of Chesley it was built by a submerged ice front and as a result is reduced in height and is relatively smooth. Its crest stands at about 900 feet above sea level. East of Chesley the moraine becomes rough and bouldery like the Singhampton moraine. As far east as Williamsford an abundance of coarse outwash appears along its southern flank and several shallow lakes or ponds appear amid this sand and gravel. In Holland and Euphrasia townships, where the overburden is thin, the drift is rocky and the surface is not well planed. Here, the separation of the moraines from the till plains is no easy matter. East of the Beaver valley, the intervening till plain is also rocky and uneven, while the moraine itself stands out between Lady Bank and Gibraltar. Precambrian boulders are numerous on the brow of the escarpment in this area. At Gibraltar the moraine lies on ground above 1,700 feet, while at Paisley the altitude is only 900 feet above sea level.

The eastern arm of the Gibraltar moraine is intimately associated with the main terraces of sand and gravel on the face of the escarpment. These may be seen in splendid form on the Dufferin county line at Lavender, between Whitfield and Perm, or on any sideroad in Mono township. They are particularly well displayed at Violet Hill where the photograph in Figure 32 was taken. The moraine itself includes a good deal of sand. The till is coarse in texture and is strongly influenced by red shale and sandstone derived nearby. Just north of Mono Mills the gravel terrace is above the escarpment but it falls below again from that hamlet to Sleswick.

Banks moraine. The Banks moraine is not quite so well developed as the Singhampton and Gibraltar moraines; otherwise it is similar. At the crossroads-village of Banks it lies on the very crest of the Blue Mountain, making a sharp-angled bend at that point. Three or four of the knobs rise over 1,700 feet above sea level or 1,120 feet above Georgian Bay. To the west this moraine holds a position on the edge of the escarpment to Walters Falls or Massie with a gap at the Beaver valley. The moraine is distinct from Banks to Duncan even though the adjoining plain is rocky and uneven. North of Kolapore and near Duncan the moraine splits, part of it dropping below the edge of the escarpment to form little valley moraines across re-entrants. West of the Beaver valley the whole promontory east of Walters Falls is best classed as a morainic complex; in fact the Banks and Gibraltar moraines coalesce through Holland and most of Sullivan townships. However, the little valley moraines east of Massie are looked upon as part of the Banks moraine. In Elderslie township the two are distinct, with the Snake River draining the depression between them. The Banks moraine serves as a natural southern border for the Arran drumlin field. In that section it is a smooth, single-crested strand.

From the angle at Banks the eastern arm of this moraine soon falls below the top of the escarpment. In the Pretty River valley it forms a striking semi-circular barrier with a gap cut through it by the stream. With several breaks it appears to pass west of Duntroon to Glenhuron and Dunedin. Clay from the reddish shale of the escarpment makes up the bulk of the material in this section. The moraine is missing on the face of the scarp near Creemore, but it reappears on the lower slope west of Banda and from there to the Hockley valley lies close behind the Gibraltar moraine. That stretch is predominantly sandy and it is intimately associated with outwash. The Banks moraine loses its identity south of Hockley where it merges with the sandy complex which is part of the interlobate moraine of central Ontario.

Tara strands. Across the Arran drumlin field, at right angles to the long axes of the drumlins, run several small moraines. The first appears near Tara and the last southeast of Wiarton. Usually less than 40 rods across, they are nevertheless distinct moraines of loamy till with an occasional kame. Gravelly terraces in two of these moraines north of Hepworth Lake were mistakenly referred to as beaches in the 1951 edition of this book.

Beaver valley moraines. In the lower Beaver valley there are two small distinctive moraines. The one north of Heathcote may be a contemporary of the Corn Hill moraine. Its distinctive trait consists of transverse ribs which may be seen from the county road near the Beaver River. The pattern shown on the aerial photograph may be duplicated in the Banks

32. *Spillway at Violet Hill. The dark areas in the broad channel are cedar swamps.*

moraine west of Chesley and east of Wasaga Beach in the Crossland area. These are all the work of submerged ice fronts and in form they are similar to Deane's ice block ridges (33).

On either side of Thornbury the Nipissing shorecliff is cut in a thin bouldery moraine which affects the stoniness and drainage of the land on several farms.

Vinemount, Niagara Falls, and Fort Erie moraines. Around the western end of Lake Ontario there are three moraines above the "Hamilton Mountain" named by Taylor: Vinemount, Niagara Falls, and Fort Erie. North of Dundas, extending to Waterdown and beyond, the three Waterdown moraines are distinct, apparently landlaid ridges of stony till with trains of gravel and sand in the hollows between them. South of the Dundas valley one definite ridge of boulder clay lies near the brow of the escarpment, while two broad low clay moraines lie farther back, fading entirely east of Smithville.

The Vinemount moraine is fairly continuous and is clearly the extension of the Barre moraine of New York State. It was probably built at the same time as the Trafalgar moraine. The second moraine, although discontinuous, may reasonably be related to the faint moraine west of Niagara Falls, which seems to be the extension of the Tonawanda moraine in New York. The Niagara Falls moraine is a mere swell in the clay plain east of the Welland canal, except in the Lundy's Lane section where it is topped by a gravel bar. The big kame at Fonthill is in line with this moraine and may be included with it although it is a valley-head feature of different composition. The kames and outwash around Ancaster and Copetown are similar. The correlation of the third moraine of the Hamilton area with the Fort Erie (Buffalo) moraine is doubtful since there is a complete gap between Ridgeway and Binbrook except for the tiny morainic ribs on the till plain discussed by Loken (95).

MORAINES OF SOUTH-CENTRAL ONTARIO

Oak Ridges moraine. Probably the best known moraine in the province is the one that forms the height of land north of Lake Ontario. It is also the bulkiest; in Albion township it buries all but the top of the Niagara escarpment, at Richmond Hill it is up to 800 feet deep and at Uxbridge it is seven miles across. It has the character of a typical interlobate moraine, that is, one built between two opposing lobes of the glacier. Much of it is till, but the crest is covered with sandhills and coarse outwash. Occasionally a knob or ridge of till projects above the fluvium.

33. *The Oak Ridges moraine in Albion township. The irregular rounded knobs and basins are typical. (Metropolitan Toronto and Region Conservation Authority photo)*

Fine sand and even silt appear in some of the depressions.

The Oak Ridges moraine contains a great deal of limestone derived from Trenton and Black River formations. As soil-building material the sand is fairly high in phosphorus and low in its potash content. The till is also high in lime. It is highly impervious to water and difficult to excavate, probably due to having been overridden.

This moraine is limited on the map to the sandy crest and the shoulders of the ridge. The northern and southern slopes are mapped with the till plains. While the southern border of the moraine is fairly straight, the northern one is scalloped, in keeping with a lobate ice front. The result is a moraine of variable width; the sandy crest pinches out near No. 27 and No. 7 Highways while north of Port Hope a complete gap appears in the moraine.

Trafalgar moraine. Taylor's Scarborough moraine took its name from Scarborough bluffs. It extended from the escarpment near Nelson to Scarborough, then swung northeastward towards Claremont. In his description Taylor remarked on the weak expression between Streetsville and Toronto and admitted not having found it above Davenport hill within the city itself. In an article written in 1943 we objected to the extension from Scarborough bluffs towards Claremont, because all of that area is drumlinized (16). Further study has shown that the surface at Scarborough bluffs is distinctly drumlinized as well as the area to the northeast. In view of this we have discarded the name Scarborough and substituted Trafalgar after the township where it is best represented. Subdued morainic relief appears west of the Humber in the Richview area and continues past Burnamthorpe to Streetsville. West of Streetsville the moraine is quite definite, standing in contrast to the smoothed till plain just above the Iroquois shorecliff and the clay plain to the north. The Trafalgar moraine is built of reddish boulder clay containing a great deal of Queenston shale.

Trenton moraine. The ridge of sand and gravel extending from Trenton to Smithfield and westward to the sand terrace north of Brighton is considered to be a moraine. Because it was submerged in Lake Iroquois it was smoothed by wave action, but the glacio-fluvial nature of the deposit is shown by the irregular bedding in the gravel pit near the railway north of Trenton. The sands in the terrace north of Brighton may have come from complete mutilation of its westward extension by wave action.

In the Bay of Quinte area there are two drumlin fields in which the alignment is different. Those formed by a Lake Ontario ice lobe lie parallel to the lake shore, approximately 70° west of south. The other group were then formed by the northern ice sheet and trend 30° west of south. The Trenton moraine is at the southern limit of these drumlins, where they contact the Lake Ontario group. It marks the limit of the advance of the northern ice lobe which overrode the Oak Ridges moraine east of Uxbridge.

Oro moraine. Northwest of Lake Simcoe a broad belt of sandhills extends from near Midhurst through the northern part of Oro township to Bass Lake. In structure it is similar to the Oak Ridges moraine and it forms the highest land in the vicinity, rising to 1,300 feet above sea level. The sands are largely derived from Precambrian rocks, and very little local limestone is present.

The explanation of a moraine in this position presents the same sort of problem as does the Waterloo moraine. The last thrust of the Wisconsin glacier in this area was from the northeast, as is duly shown by the moulding of the adjacent till plains. The Oro moraine does not lie at right angles to this thrust. It may be the result of a split between two minor lobes and as such would be an interlobate formation. The other possibility is that it is actually an earlier sandy moraine built by a Georgian Bay lobe and overridden from the northeast by the last advance of the glacier.

Corn Hill moraine. Around the southern end of Georgian Bay there are four short moraines. The foremost is a broad ridge running diagonally across Nottawasaga township. It lies against an outlying remnant of the Niagara cuesta. Two or three miles west of Sunnidale Station it disappears under delta sands and is not found again to the east. The boulder clay is a mixture in which local rocks such as red and grey shale or sandstone, limestone, and siliceous rocks from northern Ontario are well represented. Corn Hill is the name applied to this ridge by Hunter (58). There is no village of that name, nor is corn grown extensively.

Although they are not connected with the Corn Hill moraine, the two short ridges running along concessions V and VI of Sunnidale are the work of the same ice lobe. The red shale is missing but otherwise their composition is similar.

Edenvale moraine. Between the Minesing flats and Jack Lake there is a ridge of boulder clay with sand spread on the surface. Through it the Nottawasaga River has cut a gorge 90 feet deep. The ridge appears to extend only from Jack Lake P.O. to Phelpston. However a drainage divide extending from Crossland to Wyevale may be a continuation. The surface of this ridge is smooth and it lies more in line with the broad curving uplands of north Simcoe county than with the probable position of the receding front of the last ice sheet.

Mara moraine. Northeast of Lake Simcoe, extending across the limestone plain at the north side of the drumlin field in Mara and into Rama and Eldon townships there are two thin, discontinuous moraines. They mark a northwest-southwest trend of the ice-front. Some sections are surprisingly rough for moraines built in water but other sections are levelled by the waves of Lake Algonquin.

Dummer moraine. During the life of glacial Lake Iroquois the ice barrier crossed the St. Lawrence valley in the Thousand Islands vicinity. On the basis of scattered morainic knobs near the border of the Black River limestone, Coleman extended this ice barrier to the Kawartha Lakes. The limestone in this belt is thinly covered with till consisting mostly of coarse angular limestone fragments. The surface is littered with blocks of limestone and big Precambrian boulders. The finer matrix generally has a faint reddish colour from red shale and siltstone which occur just under and appear along the northern border of the Black River limestone. This extensive and loosely knit feature disappears east of Tamworth. The very small moraines which produce a washboard effect in the Napanee and other valleys (95) can hardly be considered as part of this moraine.

EASTERN ONTARIO MORAINES

In the St. Lawrence–Ottawa lowland east of Kingston the moraines are fragmentary and smoothed by wave action. The hills of sand and gravel near Seeleys Bay still have the form of kames. However, there are quite extensive areas of glacio-fluvial sands that have been largely reworked and contain marine shells. They are concentrated in a broad arc through Augusta, Edwardsburg, and Matilda townships to Maxville, but are also found just south of Ottawa and scattered throughout Renfrew county. A moraine crosses the Ottawa River at Petawawa. The stony ridges of Glengarry and southern Stormont and Dundas are mapped as till plains rather than moraines because of the smoothing of their surfaces.

DRUMLINS

Between six and seven thousand drumlins have been mapped in Southern Ontario in widely scattered groups. They are drawn to scale on the large-scale map so that their position, shape, and size are shown. Their height, form, and the steepness of their slopes are dealt with in the following descriptions, together with a few points on the composition of their tills.

Drumlins consist, almost invariably, of medium-textured till. While there is considerable variation in the proportion of sand and clay in these tills, heavy clay or light sand is seldom found. To illustrate this point the particle-size curves for several typical tills from drumlins are shown in Figure 16. The grouping of the curves is quite apparent. However, lens of gravel and sand or even stratified clay are sometimes found in drumlins.

Drumlins afford some very good soils, their chief limitations being steep slopes and stoniness. Wet spots due to seepage along their sides are quite common, but otherwise the soils of drumlins are well drained. The hollows between these hills are usually wet and often swampy. The soils on drumlins being medium in texture are easy to till, adaptable, and fairly durable. The sides of drumlins are regular slopes which may be worked on the contour and strip-cropped; in fact, here is the major application of cultivation in the province. It is already practised to a considerable extent where the lot lines happen to be in line with drumlins. Where the roads and fences run diagonally across the drumlins, as in Peterborough and Northumberland counties, they present a severe handicap to contour farming.

The drumlin fields present a pleasing landscape. They are not usually so highly cleared as the clay plains; woodlots of hardwoods with perhaps a few evergreens intermingled, and cedar swamps are left to grace the countryside. Even the stone piles along the fences and in the fields add variety to the landscape though they are otherwise objectionable to the farmer. Most of the farmsteads have attractive well-drained settings.

Drumlins of the Lake Ontario ice lobe. The Lake Ontario lobe left three or four groups of drumlins west of the Niagara escarpment and others north of Lake Ontario. Those in the Guelph area number about three hundred, most of them lying west of the Paris moraine in association with beds of stream gravel. A few more lie south of the Paris and Galt moraines near Moffat, Puslinch, and Freelton. The field around Woodstock is smaller and except along the Thames River there are no gravel beds betwen the drumlins. The small group near Caledonia lies partly buried in varved clay. East of the escarpment there are drumlins south of Georgetown and on the south slopes north of Lake Ontario, many of them running almost directly up the slope. The long thin drumlins around the Bay of Quinte and eastward are roughly in line with the shore of Lake Ontario and rest directly on a limestone floor.

The Guelph drumlins rest on Lockport and Guelph dolomite. The till is only moderately stony and consists mainly of dolomite, the fine sand fraction being nearly half dolomite. The Queenston shale content imparts a reddish cast. It is strong near the Niagara escarpment, diminishing westward. The drumlins are mild in form with moderate slopes, seldom more than 8 per cent. Although they vary in size there is a preponderance of drumlins that are one-half to three-quarters of a mile long and 80 to 120 rods wide, with a height of 60 to 70 feet.

Near Acton the Paris and Galt moraines lie close together while the drumlins lie immediately behind and in front of them and even between them. In several instances south of Acton drumlins half buried by the moraine were noted.

To the south of Freelton there are several peculiar drumlinized ridges. They run roughly east and west but are not regular in outline. Their crests approach a height of 100 feet above both the rocky tableland to the south and the Beverly swamp to the north and west. On some of them there are gravel bars at Lake Wittlesey levels. (82)

The Woodstock drumlins are underlain by Bois Blanc limestone and are farther away from the source of Queenston shale. The till is a little browner than that around Guelph; there is some dolomite in it but the pale brown limestone predominates. Boulders are not numerous. North of the Thames the drumlins are indistinct and fade out entirely north of Mud Creek to be replaced by a fluted till plain.

The small group in the Caledonia vicinity lies below the Onondaga escarpment. They are underlain by the shale, sandstone, and limestone of the Salina formation while dolomite lies adjacent to the east. Probably the local bedrock was drift-covered during the later glaciations. However, there is a good deal of grey and red shale mixed with dolomite and limestone in the local till. As usual there is a sprinkling of pebbles and hardheads imported from beyond the Palaeozoic border.

Immediately east of the Niagara escarpment in Halton county the till plain is fluted and several recognizable drumlins occur north of Milton. The till here is reddish due to Queenston shale and consequently is less calcareous than most of the Southern Ontario tills. Even the sand fraction contains an abundance of shale fragments.

Between the Trafalgar moraine and the Iroquois shoreline the till is shallow but it is fluted. The flutings score the plain from a few degrees south of east, and in places they are worn in the soft bedrock itself.

In Scarborough township the till is a light clay loam or silty loam in texture. It is derived mainly from the underlying grey and black shale and the Trenton limestone which borders it on the east. It is not particularly stony, but there are plenty of field stones to be picked on the farms and occasional large boulders to contend with in ploughing the land or digging excavations. The soil is highly abrasive to ploughshares. The drumlins are low in form and inclined to be quite long. On aerial photographs the depressions between them are very conspicuous near Wexford and Malvern. A few drumlins are found scattered over Markham and eastern Vaughan townships.

South of the Oak Ridges west of Orono the drumlins are related to the Lake Ontario ice lobe. They are quite widely scattered, about half of them being in the Iroquois lake plain where they rise above the clay beds. Near the ancient shore they were obliterated by wave action and only the boulders remain. West of Oshawa shale is plentiful, but to the east it disappears and the till is dominated by limestone.

Near the Lake Ontario shore between Cobourg and Brighton there are several drumlins which point east and west; it is not until Belleville is reached that any number of them appear. However, driving along No. 2 Highway the drumlins of the main central Ontario field, pointing south-west, can be seen on the slope to the north.

Prince Edward county has between fifty and sixty drumlins, most of them centrally located. They are of the long narrow type characteristic of the Quinte field. The southern end of Lennox and Addington has about 175 drumlins, which is none too many as between them the lime-stone is barely covered except where there is lacustrine clay. This area was all submerged at the time of Lake Iroquois and the crests of the drumlins are strewn with boulders as the result of wave action.

Drumlins of the Georgian Bay ice sheet. The drumlins around Marsville, the flutings and incipient drumlins of the Dundalk-Arthur plain, together with the extensive field to the northwest, are the work of ice advancing from Georgian Bay. Well over 1,500 distinct drumlins have been mapped; our previous estimate of 2,000 of these hills is a little too high (109). It is well to remember that in a drumlin field a varying proportion of the till is planed and moulded although not made into distinct drumlins; for instance the rolling till plain between Mount Forest and Walkerton has very few drumlins.

In form and composition the Marsville drumlins are similar to the Guelph drumlins. Those in the townships south of Teeswater are also similar except that instead of dolomite they contain a preponderance of pale brown or grey limestone. The outwash sands and gravels between them are reminiscent of Erin and Eramosa townships. North of Durham in Grey county there are some big drumlins with steep sides, composed of very bouldery till. As they are north of the Bois Blanc limestone, dolomite predominates. The Arran group is of similar till, but with a higher proportion of igneous rocks and a little red shale. A count of small pebbles from an exposure west of Owen Sound gave 25 per cent of dolomite and most of the remainder hard rocks of Precambrian age. Clay flats are associated with the drumlins in parts of this area while in the heart of the field the hills are packed close together so that it is not easy to map them separately. Many of the larger ones have flutings along their sides. In general they are long and have steep slopes, the largest being two miles long with a breadth of three-eighths of a mile.

Manitoulin Island has about 175 drumlins, mostly of the slender type. They were all submerged in Lake Algonquin and bear the mark of this experience in the form of boulder pavement on their crowns and sides. It is unfortunate that an area including so much shallow soil should find some of its deep soil rendered unfit for cultivation by excessive stoniness.

The composition of the till in the Manitoulin drumlins varies considerably. In the Tehkummah vicinity the pebbles in one typical drumlin consisted of 35 per cent dolomite, 10 per cent sandstone, and 55 per cent Precambrian rock in which quartzite, diabase, greenstones, and other dark basic rocks were well represented. Dolomite in the drumlins increases westward with red shale appearing in those around Ice Lake.

The rolling topography, smooth slopes, and generally prosperous farms of a drumlin field make a very pleasing landscape, and nowhere is this better illustrated than in the Bighead valley back of Meaford. The drumlins here stand up plainly, being higher than usual and nearly oval in outline. A normal member of this group would measure 60 rods by half a mile, with a height of 75 to 100 feet. Near Woodford several high ones appear. No drumlins remain in the Algonquin lake bed near Meaford; undoubtedly there were some, but they were unable to stand up to the waves. The local till is a pale greyish brown clay loam with a reddish cast. The red colour is imparted by Queenston shale and the illuviated layer of the soil has the buff colour which is a mark of shaly soil. However, limestone and siliceous rock are also well represented in the till. For some reason no comparable group of drumlins were formed in the Beaver valley a few miles to the east.

34. *Smoothly rounded and oval–drumlins in the Bighead valley*

35. *An island in Rice Lake with the typical drumlin form*

Drumlins of the Simcoe and Kawartha Lakes ice sheet. An area of about 2,000 square miles in central Ontario contains roughly 4,000 drumlins. Most of them are north of the Oak Ridges moraine but the group south of the ridge in Northumberland county is important. The field extends beyond the east end of the moraine across Hastings into Lennox and Addington.

Most of these drumlins are well formed, 60 to 80 rods wide and one-half to three-quarters of a mile long, with a height of about 75 feet. Some very low ones occur north of Lake Scugog. Others in Emily and Smith townships are short and nearly round. Those in Northumberland county are notable for their large size. East of Orillia and in parts of Douro and Dummer townships, both in areas of fairly shallow till, they tend to be elongated.

In the belt just south of the limestone border it might be supposed that the bulk of the material in the till would be from the Shield, but this is not so. In fact the amount of Precambrian material increases with the distance south of the Precambrian-Palaeozoic rock boundary (65). In much of this drumlin field local limestones make up 45 to 65 per cent of the till but the proportion decreases in the finer material of the matrix. Stoniness tends to decrease from the border of the Shield towards the Oak Ridges moraine while sand content increases.

Eastern Ontario field. Between the St. Lawrence and Ottawa rivers, across Carleton, Dundas, Stormont, and Glengarry counties, drumlins are scattered; there are also a few in Grenville. In Glengarry and Stormont the till is extensively planed and moulded into ridges but distinct drumlins are scarce. In Dundas drumlins are widely distributed and for the most part separated by clay plains. The most closely spaced are those in Osgoode and North Gower townships.

36. *Oblique aerial view of an esker, Jessopville. The serpentine humpy form is well shown.*

The till in Glengarry and Stormont is much like that in Peterborough and South Hastings except for a sprinkling of sandstone. In Dundas, grey or black shale is mixed with the limestone and siliceous rock, while in Carleton there is a little shale with a preponderance of sandy dolomite. All of it is stony and since wave action has concentrated the boulders on the surface in many places the soils cannot be other than stony. However, the sides of the drumlins in eastern Ontario are seldom as steep as they are in Northumberland, Peterborough, or Grey counties.

ESKERS

Eskers have been noted frequently in geological writings but it was not until 1943 that a map of them for Southern Ontario was published (109), and that was only on a small scale. The inclusion of these features on our master map improves this situation materially. No doubt more detailed surveys will discover other small eskers but our map is fairly complete.

Eskers are of interest to agriculture because they afford poor soil. Mostly they are cleared and kept in pasture although the dry sand and gravel ridges produce scanty herbage. However, they are valuable sources of gravel, especially for road building. Extending across till plains they are usually crossed by a road every mile and a quarter and thus are convenient sources of road metal. Small pits have been opened on almost every road and in places whole sections have been worked out. The gravel is often coarse and crushing may be required. The deposits are also quite uneven and may include fine sands. Because eskers were deposited last and rest on the till there is no overburden and standing water is not a problem in these ridges.

The first eskers to be formed were those around Guelph, chief among which is the one that begins south of the Ontario Agricultural College farm on the outskirts of Guelph and ends at West Montrose. On the O.A.C. farm it is barely recognizable north of the dairy barn where some gravel has been removed. There is another small gravel pit where the esker crosses College Avenue. The spillway swept away a section of it at the Speed River, but it reappears west of the city and proceeds as a modest single ridge with a few breaks to West Montrose.

In 1909 F. B. Taylor reported having traced an esker ten miles southeast from Mount Forest, saying that he did not reach either end (129). It extends five or six miles farther in either direction and is one of the five lengthy eskers that cross the Dundalk till plain. The Egerton esker rises in Glenelg township and ends in the centre of the Luther bog. Four or five miles to the west lies the Brice Hill esker, while another shorter one crosses the corner of Melancthon township southeast from Riverview. Figure 36 is an aerial view along the latter not far from Jessopville. A much bulkier ridge follows a swampy path through the centre of the township splitting into single eskers north of the Osprey town line. Southwest of Mount Forest conditions were equally favourable for eskers. but they are all less than ten miles in length.

In McKillop township two good eskers are encountered running at right angles to those just described. They are related to the ice lobe which built the Mitchell moraine, but from a more practical point of view they provide the only gravel in the township excepting the kame four miles north of Seaforth.

The longest esker mapped was traced from a point near Codrington west of the Trent River to a point on the edge of the Shield beyond Beaver Lake. Thus its course is out of line with the drumlins. Mirynech's recent survey of the Trenton map-area (103) has traced this esker to Biddy Lake near Colborne. It crosses Hastings county from Frankford to Marlbank with short gaps in the Moira River flats near Foxboro. It almost

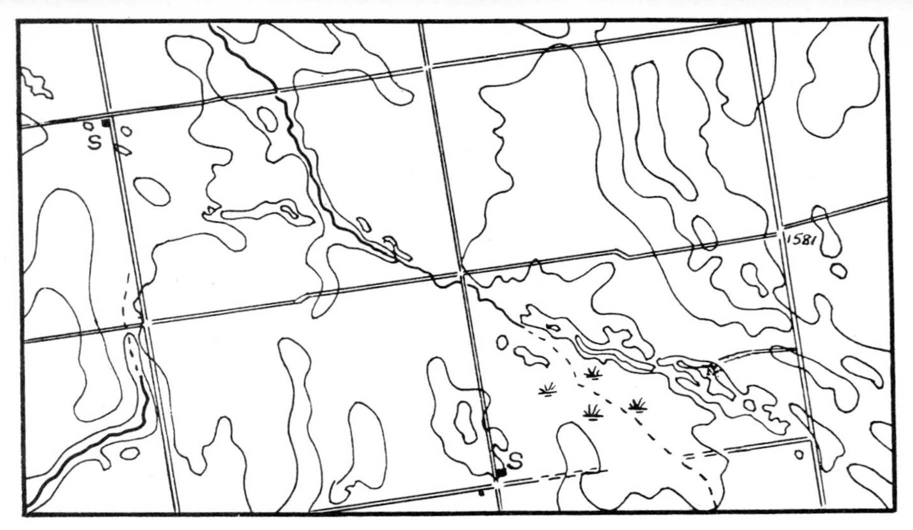

37. *A section of the Palmerston map sheet on which the contours intermittently outline an esker running from upper left to lower right*

divides Beaver Lake in two. The road from Erinsville to Lime Lake follows it, running along its crest for several miles. A tributary some 18 miles in length extends to Tweed past Phillipston and Thomasburg, and a shorter branch is mapped south of Tamworth.

The stream that laid down this gravel flowed to the west crossing the limestone border at Beaver Lake. On examining the gravel in the esker on either side of this little lake we were surprised to learn that, although not one limestone pebble could be found north of it, the gravel a scant mile or two over the boundary was approximately three-quarters limestone.

The biggest esker encountered lies between Norwood and Westwood north of Rice Lake. On the topographic map it takes five contours along the highest part. Here it exhibits a fairly common trait in having an extra ridge flanking the main one. The Omemee esker which crosses No. 7 Highway just west of that village is also a big one, especially in the Pigeon River flats and it contains very coarse, cobbly materials.

The slope from the north shore of Lake Ontario to the Oak Ridges moraine is fairly steep and neither here nor on comparable slopes on the north side of the moraine are there any good eskers. What appears to be a short esker half-buried by clay a mile or two north of Brampton bears mention because the town water supply comes from it. The Kawartha Lakes region has a series of these remarkable gravel ridges comprising mostly limestone gravel. Cameron Lake is bordered by one on the east and south, while towards Cannington quite a group of them appears. The Omemee and Norwod eskers have already been mentioned and there are others in the Peterborough area.

There are no typical eskers in Ontario east of Kingston owing to the fact that the receding ice front was always under water. However four of them, now badly reworked by wave action, have been recognized. All of them have lost their knobbiness and are reduced in height, with the extra sand spread alongside. One runs north from Kars to the border of Nepean township. Gravel pits in this ridge expose esker structure at its base and regularly stratified beach gravel, containing marine shells, above. Most common among the shells are those of *Leda glacialis* (*Portlandia artica*, Gray), and *Saxicava*. The second esker can be traced from a point one mile east of Kemptville to Reids Mills and dicontinuously to West Osgoode and Greely. Some sections have been entirely reworked, for example at Reids Mills, where the gravel was apparently thrown into beach ridges. The third esker runs south of Finch for three or four miles. The fourth is north of Bearbrook. All four eskers run roughly north and south in line with the stripe of the drumlins.

ABANDONED SHORELINES AND LACUSTRINE SEDIMENTS

More than half of the total area of Southern Ontario was submerged for a time during or immediately after the retreat of the last glacier. Water-laid sediments are practically free of stones, this feature and the level topography of the beds being the surface earmarks of a lake plain. The sediments themselves are stratified. Some glacial forms often remain to break the near-levelness of the lacustrine and marine plains but never without being modified to some extent. Wave action worked to produce a boulder pavement on their surfaces.

An outline of the several glacial lakes which successively inundated lands at various levels was given in the previous section. In a report of 1909 F. B. Taylor (129) concluded that the plain between Stratford and Milverton was a till plain even if it were "as flat as the ocean in time of calm." From this it is evident that he had looked for stratified clay. However, he did not mention any of the tracts where stratified sediments do appear. One is found in the Conestoga valley. Near the marlbeds south of Atwood in concession VIII of Elma, a freshly dug ditch in 1941 showed stratified clay and sand several feet deep over the till. The marl required shallow water for its accumulation, therefore there is no doubt that shallow water stood in the basin until a tributary of the Maitland emptied it. Two other tracts of lacustrine clay, without the marl, occur north of Brodhagen in front of the Mitchell moraine and southeast of Tavistock. Both lie on divides between two watersheds and drain in two directions leaving central areas where no draining courses are established.

The London correlative of Lake Maumee. The ponding east of London has already been discussed as a correlative of glacial Lake Maumee; while it is not bounded by a distinct shoreline the limits of the stratified beds are reached just under the 900-foot contour. The finer sediments give way to stream gravels along either branch of the Thames River. The central area is level and it is no accident that a block of it near Crumlin was selected for an airport. The beds of uniform silt exposed on the campus of the University of Western Ontario, north of the stadium, are Maumee sediments. They also underlie the gravels along the North Thames below Thorndale.

Two beaches above the level of Lake Whittlesey appear near Lambeth. The upper one is followed by the road northwest of the village for a mile. It is a broad gravel beach just above the 850-foot contour. The other is 20 feet lower and extends from Scottsville towards the first bend in No. 2 Highway west of Lambeth where it gives way to a low bluff. At these levels lacustrine deposits, when present, consist of shallow pockets of silt.

Lake Whittlesey. In Ontario, Lake Whittlesey is notable for its deltas while its shoreline is but faintly developed. The best stretch of beach is that between Arkona and Watford. It consists of shaly gravel. The continuation of the shoreline around the end of the Seaforth moraine is a distinct bluff, except for a bit of a beach near Sutorville Siding. Continuing towards Kerwood the shorecliff diminishes in form, fading out before reaching the vicinity of the village. Two short gravel beaches appear on either side of the railroad west of Kerwood. Neither beaches nor bluffs appear within five miles of Strathroy to set a sharp boundary between the sands of Caradoc and the Seaforth moraine. North of the railroad tracks, west of Komoka, a shorecliff at about 825 feet appears on the end of the Lucan moraine. Although it seems to be a little too high for Whittlesey levels, the exact tilt of that water plane is not known and leaves this point in doubt. However, it is at the edge of the Caradoc delta.

Between the Thames River and St. Thomas no shore features are found near the 800-foot contour. Three miles west of that city on a ridge where the power line intersects the Michigan Central Railway, two disconnected gravel beaches lie close to Lake Whittlesey's level. Beaches appear in a few places east of St. Thomas so that the levels are established even though the shoreline is indefinite for most of the way. A gravel beach crosses the Edgeware Road north of Yarmouth Centre. South of Mapleton a short beach and a low bluff east of it cross the extreme southwestern corner of South Dorchester township. The water level at these points was about 780 feet above sea level. The next beach to be seen lies on the side of the Tillsonburg moraine; one bit occurs at Tillsonburg Junction, while a longer one begins a mile and a half southwest of that point and runs south of No. 3 Highway for four miles. These beaches lie just under 800 feet above sea level. The moraine south of Tillsonburg, lying between the Canadian Pacific Railway and the Little Otter Creek, also has a fragmentary beach encircling it at the same level and several more short gravel beaches occur south of Otterville against two prominences of the Tillsonburg moraine. Near Windham Centre on the Paris moraine at 825 feet, or slightly higher, is another gravel beach, which probably is the work of Lake Whittlesey. North of this the best evidence found was a gravel beach at about 875 feet lying west of the road two miles north of Scotland. The delta sands continue north of Windham Centre towards Paris; the point where they give place to the stream terraces is not definite, but above Burford gravel predominates in the beds.

On the Paris and Galt moraines, north of Windham Centre and Waterford respectively, gravel bars near the 850-foot contours are too high for Lake Whittlesey and must, therefore, be pre-Whittlesey in age. Karrow's more detailed surveys of the area east of Galt has found and mapped several short beaches at Whittlesey levels (82) (83).

The two major deltas of Lake Whittlesey in Ontario are well known as the Norfolk and Caradoc sand plains. The first is at the mouth of the Grand spillway in front of the Paris and Galt moraines and the second in the Thames valley west of London. Oddly, there is no good delta in front of the Wyoming moraine in the Ailsa Craig vicinity.

The Norfolk sands lack the typical deltaic gradation from coarse to fine sands. Medium-to-coarse sand is widespread. Also, a peculiar "island" of silt appears north of Simcoe above Lake Warren levels. The Whittlesey and Warren deposits are continuous in Norfolk and Elgin counties, the Warren deposits being regarded mainly as reworked Whittlesey sediments. There are no distinct deltas of Big Creek, Otter Creek, and Kettle Creek in Lake Warren.

Recent studies of the composition of these sands (13) have found that they contain more calcite than dolomite on the average. The main source of the calcite is the limestones of the Lake Simcoe area beyond the Niagara escarpment.

Stratified silt deposited in Lake Whittlesey occurs east and west of

St. Thomas and as a silty "island" surrounded by sand north of Simcoe. The small tract of silt near Kerwood is not connected with the Caradoc delta, the sediments probably being derived from nearby till in which shale as well as limestone is abundant. West of St. Thomas through Southwold, Adelaide, Ekfrid, Metcalfe, and Brooke townships to Watford, Whittlesey clays bury the till and smooth the surface.

Lake Arkona. On the map, a beach is placed between the Whittlesey and Warren beaches in the Arkona-Alvinston vicinity. This is the beach of Lake Arkona which existed prior to Lake Whittlesey when the glacier held a position of retreat before its readvance to the Port Huron moraine. This lake built three beaches on the west side of Lake Huron northwest of Port Huron in the Black River area. They were submerged under the higher waters of Lake Whittlesey and more or less scattered by wave action.

The only place in Ontario where the Arkona beach is definite is between Arkona and Alvinston. It is best expressed from Arkona to Kingscourt Junction to Armstrong Station where the shoreline skirts the end of a moraine and where conditions are favourable for the cutting of a bluff rather than the accumulation of a beach ridge. From Armstrong Station to Alvinston the abandoned Canadian National Railway line separates it from the twin beaches of Lake Warren, which are finely formed in that section. It extends through Alvinston and through the cemetery south of the village to the Sydenham River where it ends.

Other places where beaches are found at Lake Arkona levels are at Glencoe, on the Blenheim moraine near Clachan and Duart, on the hump west of Leamington, and around the Sparta ridge south of St. Thomas. The one south of Glencoe is the only well-formed beach. At Clachan there is a gravel bar capping high ground, which is not specific evidence of a shoreline. Near Leamington the gravels interpreted as the Arkona beach lie on the north and east sides of the clay ridge, and extend towards Leamington as a spit with its crest above the 725-foot contour. Although the distinct form of a beach may be wanting, except near Glencoe, these are massive gravel deposits, much heavier than the beaches of Lake Whittlesey.

Lake Warren. Glacial Lake Warren at its highest stood about 60 feet below the level of Lake Whittlesey and left its mark on a much larger area in Ontario. It covered a broad belt north of Lake Erie and a narrower

38. *Ripple pattern in Lake Warren plain northeast of Goderich. The dark streaks are due to wet soil in hollows between successive sand bars. The two Warren beaches cross the lower right corner. (R.C.A.F. photo)*

border east of Lake Huron. In the east it was confined by an ice barrier near the top of the Niagara escarpment, while the crest of the interlobate moraine may have been showing at this time. The other ice barrier stood just north of Paisley at the position of the Gibraltar moraine, although it eased away from that moraine towards the end of the lake's existence. The abandoned shoreline is now tilted so that it rises from near 700 feet in the south to 925 feet at Paisley. East of Ridgetown it rises to 775 feet at Simcoe and 835 feet above sea level several miles north of Brantford. There are two beaches, the second being ten to fifteen feet below the first, usually nearby and seldom more than half a mile away.

On the slope east of Lake Huron the Warren beaches are feeble but fairly continuous. East of St. Thomas the shoreline runs through sandy land where the shore features are ephemeral, being subject to wind erosion. The two strands can be seen and traced on aerial photographs north of Goderich. The northernmost occurrence is a well-developed shoreline around a morainic ridge three miles north of Paisley. A few gravel bars west of Pinkerton are the only evidence of the Warren beach in the Saugeen embayment. Bruce township does not include a good example of these beaches, although the sand and gravel south of Willow Creek is in line with them where they end south of the Kincardine town-line. The two beaches, a quarter of a mile apart, appear in lots 13 and 14, concession XI of Kincardine, and can be traced southward in spite of some discontinuity right to Wyoming. The beaches generally consist of two or three feet of gravel. Deeper accumulations are eagerly sought as sources of gravel, which is scarce. One of these occurs west of Lucknow near Lothian. West of Dungannon village the steam gravels along the Lucknow River provided material for the waves to work with. Outside Goderich deeper gravels form the sites of two cemeteries and several gravel pits. The shoreline is clear-cut south of Goderich; where beaches fade out a small bluff is usually present instead. The two beaches are close together at Blake. They pass through Dashwood and are fairly strong as they cross the road respectively one half-mile and one mile west of the corner. Some old dunes remain near Mount Carmel along the higher Warren shoreline, and an unusually strong shorecliff and bouldery terrace extends for six or seven miles from Parkhill Creek to the Ausable River. On all this slope the Warren beach stands at the boundary between the rolling lands above and the level plain below. Under the beach or the shorecliff for a variable distance, seepage from the higher ground is apt to collect, creating swampy conditions.

East of Lake Huron the gravel in the beaches is characterized by limestone without shale. South of Thedford brown and blackish shale is found mixed with the limestone and the crystalline rocks from the north. The two beaches, close together, continue with but few breaks through Forest to a point four miles west of Wyoming. In that section, with the benefit of Lake Huron to the north, the warm gravelly soils of the beach are planted to peach and apple orchards in spite of a tendency to drought. The Warren shoreline surmounts the Wyoming moraine west of Wyoming, cutting a small bluff where it passes the crest and building a gravel spit a mile northwest of Copleston. On the southern slope of the Wyoming moraine the beaches are thin, fading out shortly after crossing the Warwick township boundary. According to the levels, a bay of Lake Warren extended to Warwick village, but in that protected cove no beaches were left. The silt and fine sand around that village probably are Warren sediments but the beds continue northward along Bear Creek, rising to 725 feet—slightly above the level of the shoreline. The beach gravels along the 725-foot contour east of Kingscourt are thought to be of Arkona age; below that level in the Kingscourt and Sutorville areas the Warren shoreline is very faint or missing. It reappears near Armstrong Station as a single thin beach of shaly gravel. Within two miles of Alvinston the two beaches are again present in fine form, continuing past Alvinston to the bank of the Sydenham River east of Tancred.

Warren beaches are lacking between the Sydenham and Thames rivers; however, the edge of the delta extends from Shields Station to Wardsville. Around West Lorne and Rodney the land at 700 to 710 feet is close to the level of the Warren waters. Here, too, in the absence of beaches or bluffs the upper shoreline is placed at the edge of the sands. The Blenheim moraine was separated from the mainland by three miles of shallow water. The Leamington moraine also stood out as an island in this lake.

On the Blenheim moraine the Warren shoreline is generally represented by beaches on its southeastern flank and a shorecliff on the opposite side. From Blenheim through Ridley, Morpeth, and Palmyra a pair of beaches run continuously and the gravel ridges are held in high favour for building sites on the farms through which they pass. In tracing them it was discovered that the contours on the Ridgetown map sheet between Morpeth and Palmyra are in error, the 700-foot contour swinging too far lakeward. The beach crosses this contour as drawn for three miles west of Palmyra. East of Duart the beaches have given place to a ten-foot bluff and bouldery terrace. The bluff continues past the county boundary but a beach reappears near Taylor. North of Kintyre a low bluff cut in sand or gravel beds crosses over the 700-foot saddle. The Blenheim moraine has two crests in the Kintyre and Clachan section. On both crests the Arkona beach appears in strength, making the task of identifying the Warren beach difficult. The top of the moraine at Clachan is covered by beach gravel. A faint shorecliff crosses the road a few rods south of the corner at Clachan, while a beach crosses north of the corner. These are considered to be the two Warren shorelines. The low bluff extends northeast towards the beach at the schoolhouse on the next side road, the elevation at its base being close to 710 feet above sea level. Between Clachan and Ridgetown the upper Warren shoreline is a bluff, with a few exceptions, but a beach runs from Ridgetown to Blenheim and Cedar Springs. This beach has an abrupt northern face with clay appearing almost immediately below.

The dome west of Leamington is almost encircled by a gravel beach between the 675-foot and 700-foot contours. MacLachlan (96) puts the elevation of the upper Warren beach near Ruthven at 680 to 685 feet above sea level.

There is no shorecliff on the end of the St. Thomas moraine near Wallacetown, and there are no beaches in the vicinity. In fact the one unmistakable occurrence of the twin Warren beaches east of Wallacetown is at Simcoe (Woodhouse concessions V and VI, lots 3 and 4). This shoreline is tentatively drawn through a line of sand dunes through the townships of Malahide, Bayham, and Walsingham. Two sandy beaches cross the townline a mile and a half south of Glen Meyer but they have poor continuity. Dunes extend north of Langton, and low bluffs appear south of Wyecombe and extend towards Lynedock. East of Big Creek a line of dunes can be traced from Blaney to Simcoe, although some of the high ground west of Simcoe was above the level of Lake Warren. The upper beach east of Simcoe stands at 775 feet above sea level, thence the shoreline runs along the eastern side of the Galt moraine with bluffs being more common than beaches. However, some land west of Galt moraine is just at Warren levels, which allows this to be considered as part of the lake. The dunes at Atherton, Nixon, and Lynnville join with the low bluffs a quarter of a mile east of Lynnville and with another which crosses the townline west of Waterford to suggest the shoreline of a shallow bay or lagoon north and west of Simcoe.

North of Simcoe the shoreline on the eastern side of the Galt moraine has good continuity to Waterford. On the northern outskirts of Waterdown a short but distinct gravel bar above the 775-foot contour is a Warren beach. North of Waterford the Warren shoreline cannot be followed continuously. Good beaches appear between Paris and St. George, but at 830 to 840 feet above sea level, which is somewhat higher than expected for a Warren beach. The best one is southeast of Blue Lake in lot 18, concession II of South Dumfries crossing the road three-quarters of a mile south of the corner where it soon gives way to a bluff. Other short gravel beaches appear two miles east of the corner of Highway No. 5 west of St. George. A beach near Branchton in lot 1, concession VIII, of North Dumfries and several other short beaches on drumlins in Beverly township were built by Lake Warren (83).

Gravel beaches on the big kame at Fonthill are doubtless of Warren age, as they have about the right altitude of 850 feet; in other words, a small isolated island appeared in glacial Lake Warren.

The Norfolk sands have already been discussed as deposits of Lake Whittlesey in the delta of that lake, although much of the sand is below Warren level. The sands are continuous and there are no distinct Warren deltas. The beds of silt, sand, and fine sand between Brantford and Campbellville are considered to be deltaic in Lake Warren. The Thames River brought down enough sand to cover 600 square miles below Newbury, most of it spread thinly on the clay floor. The Ausable, Bayfield, and Maitland rivers built small deltas while in the Saugeen embayment there are four square miles of deep sand beds north of Walkerton.

While silt is valued as the parent material of good soils it is the bane of the road builders. Most of the silts in southwestern Ontario are deposits of Lake Warren. Deep varved silt occurs around Brantford and eastward towards Dundas. Similar deposits are found near Fingal, south of Chatham, and between the Blenheim Ridge and Rondeau Harbour. The county soil maps show it is Haldimand loam or Beverly loam and fine sandy loam.

Deep beds of clay, some of it interstratified with silt, make up the superficial beds in the Niagara peninsula, particularly south of the Grand River. More appears back of Port Rowan. The clay between Paisley and Elmwood is 60 feet deep in places, these beds being notable because the brown and grey varves are so distinct and regular. Shallow deposits on the clay till of the slope east of Lake Huron, and in Lambton, Kent, and Essex counties tend to smooth these plains. This includes Pelee Island where the surface is quite flat.

In the Niagara peninsula north of the Grand River, exposures consistently show crumpled clay strata within a few feet of the surface, overlain by a thin sheet of heavy till and sometimes topped by more stratified clay. The crumpled strata are indicative of overriding by the glacier. Aerial photographs of this area show the "marbling" that is usually associated with heavy till. This boulder clay is unusually free of pebbles and boulders no doubt consisting of material reworked from the lacustrine beds below. Similar clay till, often overlying crumpled clay strata, occurs on the borders of Lake Huron.

An odd feature of the Warren lake plain east of Lake Huron is a belt of sand running roughly parallel to the beach and usually separated from it by a strip of clay. Occasionally this sand contacts the beach but it also occurs over 100 feet below the beach. It is therefore not a lower shoreline even though frequent wave-built bars appear. The deposits are of shallow, stratified sand, two to four feet being the prevailing depth. The surface exhibits peculiar form in the shape of wavy swells running in the same direction as the beach. On aerial photographs they produce a striking ripple pattern as shown in Figure 38. We regard this sandy strip as a complex of off-shore sand bars in Lake Warren. In the north towards Kincardine only a few scattered gravel and sand bars are found.

Below the Warren levels in the Niagara peninsula, there are fragmentary shorelines, indicative of temporary water levels, which must be mentioned. The gravel bar at Lundy's Lane, on the crest of the Niagara Falls moraine, stands at 675 feet above sea level, or about 175 feet below the highest shoreline of Lake Warren which appears on the big kame near Fonthill. Several gravel beaches appear in Welland county at elevations varying from 600 to 650 feet above sea level, and with too much local variation for them to be regarded as the work of a single water plane. Among the highest is the gravel bar at Ridgeway, on which an elevation of 657 feet was established. This is almost at the same level as the wave-washed limestone mesa near Ridgemount. Another fairly continuous line of beaches appears at 625–630 feet in the Ridgemount area. Beaches at similar levels appear on the low upland in the southern part of Crowland township. Low bluffs and beaches also appear just above the 600-foot contour in Wainfleet township, a few miles west of Port Colborne, while a distinct shore cliff and thin beaches are to be seen at about 595 feet on the boundary between Moulton and Sherbrooke townships, not far from the Lake Erie bluffs at Lapp Point. The strongest and most conspicuous beach of the group, however, seems to be the one followed by the county road from Crystal Beach to Sherkston. An elevation of 618 feet was established on its crest in Crystal Beach, which corresponds very well with the upper limit of the deltaic sands near Dunnville on the Grand River.

At the western end of the Lake Erie basin, the best formed beach below Warren level is the Essex ridge, a well-known gravel strand followed through Essex by a road between Windsor and Leamington. Between Essex and Cottam the elevation of its crest is just over 650 feet. Taylor (68) related it to the Grassmere beach in Michigan but it seems rather high to be a true correlative unless MacLachlan's (96) theory of special local deformation can be shown to apply. Two lower beaches run parallel, a few miles to the west. The upper one at 630 feet has about the same altitude as the heavy storm beach beside Highway 3 from Romney to Port Alma, which may correspond to a Lundy water plane. Strangely enough, this ridge now forms the water divide between the streams flowing north into Lake St. Clair, and drainage into Lake Erie which is less than one half-mile to the south. The crest of the bar near Ouvry is about at the Grassmere level, while others to the east trend upward to Warren levels. At Ravenswood and Camlachie, just south of Lake Huron, there are beach deposits, just on or above 650 feet, which may be correlated with the Essex water plane.

Lower than the Warren beaches, the shorelines of these lakes are all rather unobtrusive features of the present landscape. At the Warren, Grassmere, and Lundy water planes, the lakes in the Huron, Erie, and St. Clair basins were continuous, with progressively narrower and shallower straits across the saddles at Sarnia and Windsor. The island north of Leamington stood progressively a little higher and became a little larger with each recession of the water level. By Lundy time a considerable body of land had emerged in the western part of Essex county but there was a large submerged area to the east. Shallow open water must have persisted in the area from Tilbury to Wheatley almost until early Algonquin time.

Once the isthmus near Wheatley became dry land, an almost completely enclosed water body in the Lake St. Clair basin came into existence. It received the drainage of the Thames River and other smaller streams, and it was free to drain either to the north or to the south, depending upon the direction in which the ultimate outlet was established. The best evidence of an established water plane is the shoreline near Dresden which was correlated with the Elkton by Taylor (92), although obviously not at the same level. We shall therefore refer to this stage in the St. Clair basin as Lake Dresden.

Early Lake Algonquin and its correlatives. The low flat plain which borders Lake St. Clair is one of the more significant features of southwestern Ontario. It is lowest in Dover township where large areas, actually below the level of the lake, have been reclaimed by dyking, ditching and pumping. The upper limit of this plain is marked by the Dresden shoreline. It consists of low bluffs and thin gravel beaches which can be followed from Dresden to Sombra, at or close to a contour of 605 feet above sea level. East of Dresden the shoreline of the Thames embayment of this water level is somewhat indistinct. However, the soil patterns over a wide area from Kent Bridge to Lake St. Clair strongly suggest the deposits of a delta. Between the Sydenham River and the Thames River, the surface deposits are sandy and the sands have been subject to wind action both above and below the 605-foot contour. There has been little recent dune formation, but old blow-outs, built by westerly winds, are evidence of a shore zone although no distinct beach remains.

To the south of the Thames, from Kent Bridge to Chatham, the edge of the deltaic sand is found just above 600 feet and is bounded in places by indistinct shoreline features such as thin beach strands and very low bluffs.

Near Kent Bridge, the early Thames River split into two major distributaries, one of which followed the path of Big Creek while the other followed the course of the present Thames River through Chatham. Both distributaries meandered widely, throwing off minor distributaries and, although the abandoned channels are shallow, they are quite distinct and show up clearly on aerial photographs. The sands which border and lie between them must be regarded as the delta of the Thames in this old lake. These formations have greatly influenced the soils of the area and are clearly reflected on the soil map of Kent county.

From Chatham to Windsor the shoreline of this ancient water plane is somewhat vague but the land below the 600-foot contour is very flat and bears a deeper deposit of lacustrine clay than the clay plain to the south. In the city of Windsor there is evidence of stream cutting below the 600-foot contour while, south of Windsor at corresponding levels, shallow, gritty sand deposits are found on the clay. South of Amherstburg, the thin beach through Malden Centre stands just above 600 feet and no doubt marks this water plane at the western end of Lake Erie.

In the previous edition of this book we stated our belief that both the earlier, higher level of Lake St. Clair and the higher level of Lake Erie, which invaded the low area around Dunnville, were contemporaneous with Lake Algonquin. We did not at that time dispute the interpretation of Leverett and Taylor (92), who, in turn, had accepted the conclusion of Fairchild (49) concerning the drainage of Lake Dana (hypo-Warren or hyper-Iroquois) waters toward the Rome outlet. In doing so, however, they were at some pains to discredit Spencer (92, p. 407) who had earlier contended that Lake Algonquin did not drain southward.

In more recent studies, Hough (75) has traced a connection between early Lake Algonquin and Lake Chicago, thus demonstrating that the water level in the Huron Basin was controlled by the bedrock sill at Chicago. He gives this as the reason why the Algonquin beach was formed at 605 feet at Sarnia, instead of at a lower level which must surely have resulted if the drainage had been toward the south. The fact that Fairchild did not actually trace the continuity of his suggested drainage connections and, in addition, the fact that neither Fairchild nor Leverett and Taylor successfully demonstrated that they were dealing with the same water planes in the eastern and western ends of the Lake Erie basin, both strongly support Hough's contention. Since we have no doubt from our own observations that the Algonquin shoreline at 605 feet and the Dresden shoreline at 605 feet are representative of one and the same water plane, continuous through the Sarnia strait, we endorse Hough's hypotheses of northward drainage at Sarnia and control by the Chicago outlet over the water level at the time of the early Algonguin beach. It follows, of course, that we accept Hough's idea of the early Algonquin stage in the Lake Erie basin because we had already come to a similar conclusion. However Hough's portrayal of the situation along the north shore of Lake Erie (74, Fig. 67A) is, even on such a small-scale map, completely unacceptable.

Although the Dresden water level can be traced past Windsor and Amherstburg into the western end of the Lake Erie basin, it cannot unfortunately be followed to the eastern end. The north shore of Lake Erie, for a long distance, is marked by bluffs which in some places are more than one hundred feet in height, while an extended area of old lacustrine formations, including nearly everything below Warren levels, has been eroded away (14). Thus the nearest land, low enough to record the features of Early Algonquin time must be at Long Point and, much more extensive and important, the area from around Dunnville eastward to the Niagara River.

The "hinge line" or isobase of zero uplift for the 605-foot water plane as portrayed by both Leverett and Taylor (92) and Hough (74) should pass somewhere between Long Point and Dunnville, hence the beach toward the east will be up-warped and quite probably split. This seems to have been the case.

A considerable area of the Erie plain in the Niagara peninsula lies below 600 feet and it is all remarkably flat. On the western end of this flat plain there are shallow sand deposits over the clay, no doubt representing deltaic formations of the Grand River. In searching for a shoreline to surround this low ground one finds the fragments in the Humberstone, Crowland, and Crystal Beach areas which have already been mentioned. Of these the strongest and most likely correlative of the Dresden shoreline is the strand from Sherkston to Crystal Beach with its benchmark of 618 feet above sea level.

Here, unfortunately, there is no neat coincidence of water planes at

640, 620, and 605 feet to correspond with the Glenwood, Calumet, and Toleston stages of the Chicago outlet. This is beyond the "hinge line" and neither the intervals nor the elevations correspond. One cannot be sure that the original Lundy at Lundy's Lane represents the same water plane as does the Lundy in Michigan which is found at 620 feet above sea level. Possibly uplift produced more than one representative. Here, also, one may be reasonably sure that some of the lower fragments are true "early Erie" beaches, related to downcutting of the Fort Erie outlet at a slightly later time.

There are two features east of the Niagara escarpment, Lake Peel and Lake Schomberg, which were contemporaneous with early Lake Algonquin as shown in Figure 22*n*.

The Peel plain extends northeastward from the Niagara escarpment, through Halton, Peel, and York counties. The evidence for submergence is not as strongly marked as in some lake plains, nor is it always unequivocally present. There are widespread beds of stoneless clay over the till, while in some areas the beds are deep enough to accommodate the soil profile and to preserve some depth of unweathered stratified clay below. An extensive area of this sort was exposed during the building of Malton airport. Another pocket with 8 to 10 feet of stratified clay near Drumquin was sectioned in 1964 during the rebuilding of the Oakville Seventh Line road. Flat areas with some depth of stratified clay are found near Ashgrove, Brampton, Concord, and Ambler. Near Unionville, an extensive flat area has surface deposits of stratified sand over clay. The shoreline is not anywhere marked by continuous beaches or shorecliffs, but small sandy areas in the Credit valley near Norval, and in the Humber valley near Nashville, appear to have been deltas in this lake. The chief effects of submergence, however, appear to have been the bevelling or smoothing of the till surface and the deposition of just enough clay to produce a Peel clay loam, instead of the Chinguacousy clay loam of the adjacent till plain.

To the north of the interlobate moraine, and apparently contemporaneous with Lake Peel, there existed another body of water to which we have given the name Lake Schomberg. Because it is superimposed upon a strongly drumlinized till plain the Schomberg plain is not so nearly level as the Peel plain but it is characterized by more extensive, and deeper, beds of varved clay. In many parts of this plain the varves are unusually thick, three- or four-inch couplets being not uncommon. In colour, the clay layers are brown, while the silty layers are pale gray. The latter layers also are quite marly in appearance and texture. Samples of the clay deposits from some areas showed, upon analysis, that the beds

39. *Varved clay over till, north of Streetsville on the Peel plain. The base of the two-foot rule rests on till.*

contained about 50 per cent of calcium carbonate, mainly rock flour, ground up by the glacier as it overrode the edges of the Palaeozoic limestone formations to the northeast. In some places, as for instance to the east of Newmarket, the material is more largely composed of silt-sized particles, but again in the Lake Scugog area there is less silt and more clay.

The physiographic characteristics of these three areas are somewhat different. The Schomberg area may be described as "drumlins and clay flats," the Newmarket area as a dissected clay plain, while the area north of Lake Scugog comes close to being a "bevelled till plain" like the Peel area, but is more nearly flat and also more poorly drained. The Scugog area is joined to the Schomberg area by a series of narrow valleys through the intervening upland. For the most part they have swampy floors, but along their edges it is possible to trace the same type of calcareous, varved clay that is found in the main areas.

Since these water bodies did not leave well-formed continuous shorelines, it is obvious that they were rather temporary phenomena and, probably, had fluctuating water levels. The identification of outlets and, therefore, correlation with recognized stages in the Great Lakes basins is also difficult. At a higher level there is good evidence that a spillway across the interlobate moraine at a present level of 950 feet was in operation at Palgrave leading past Caledon East and Georgetown. There is a fairly good set of beaches to be found along the western side of Lake Simcoe, indicating a shoreline at about 90 to 100 feet above the main Algonquin beach in this area. Such a water plane, and in fact any water plane below the Palgrave channel, must have flowed to the west until the Kirkfield outlet was opened. This could only have been during early Algonquin time.

The outlet of Lake Peel must at first have been over the Niagara escarpment where a series of gaps are found at different levels. One of the more spectacular of these is the Lake Medad channel between 800 and 825 feet above sea level. However, there had to be later and lower outlets as the glacier shrank away from the escarpment. A few scattered areas of sand near Mount Nemo and Rock Chapel give indication of water action, but perhaps most of the water escaped through a marginal area of stagnant ice.

Lake Algonquin and Lake Nipissing. Lake Algonquin and Lake Iroquois were the longest-lived of the glacial lakes of southern Ontario. Their shorelines are about as mature as those of Lakes Huron and Ontario, and they are far better developed than any of the earlier shorelines. In fact several of the stages of Lake Algonquin are better marked than the shoreline of Lake Whittlesey, and are nearly as prominent as the Warren shoreline. Lake Nipissing produced beaches or bluffs second in size to those of Lake Algonquin in the Huron basin. The Algonquin shoreline consists of a bluff over most of the 240 miles from Sarnia to Collingwood, beaches appearing near the streams and occasionally elsewhere.

Sarnia to Tobermory. Between Sarnia and Grand Bend, where the Algonquin and Nipissing shorelines are both horizontal, they occur at approximately 605 and 595 feet, respectively, above sea level. They cannot always be separated. However, the gravel beaches between Sarnia and Vyner lie just above the 600-foot contour and are regarded as Algonquin beaches. The sand dunes between Point Edward and Blackwell may be attributed to both Algonquin and Nipissing. This also applies to the larger belt of dunes between Kettle Point and Grand Bend. Between these belts of dunes the ancient shorecliffs have been undercut by the modern lake except north of Gustin Grove where a 50-foot shorecliff departs from the present shore making a small coastal plain where a few fruit farms have developed.

Behind the two belts of dunes mentioned above are silty flats containing marlbeds, black muck, and shallow lakes. Lake Wawanosh in the Blackwell area has been drained and its bed reclaimed for special cash crops. Occasionally in the spring, ice along the shore of Lake Huron blocks the outlet of the canal, temporarily re-establishing the lake. Part of the larger lagoon in the lower Ausable valley has been drained and the land reclaimed. It is well known as the Thedford celery marsh. North of Thedford the edge of the lagoon is sharply marked by a bluff which is probably an Algonquin feature.

The map does not show the shorecliff between Grand Bend and Clark Point, which suggests that the Algonquin and Nipissing bluffs have all been undercut by the present lake, although they may have had much to do with the initiation of the high bluffs along this shore. At its best near Goderich the bluff is over 100 feet high and it is seldom less than 50 feet high. Taylor reported seeing one bit of beach four miles south of Bayfield that appeared to be an Algonquin feature. From Clark Point northward to Port Elgin the Algonquin bluff runs some distance back from the shore of Lake Huron. The terrace below it carries a boulder pavement modified by numerous gravel ridges and sand dunes. The Nipissing shoreline is usually present as a smaller bluff, a line of dunes, or a beach ridge.

Another lagoon in the Saugeen valley was confined by a massive barrier beach built from the south side of the valley through the outskirts of Port Elgin. The bar is mainly of gravel with some dunes added. It causes the Saugeen River to deflect five miles northward before heading out to the lake. On the floor of the lagoon are level beds of silt and fine sand which, like all the soil materials along the eastern shore of Lake Huron, are highly calcareous.

Where it encroached on the drumlins north of Port Elgin, Lake Algonquin cut into their sides, no doubt obliterating some before it retired. Between the drumlins the waves constructed the beaches which impound Marysville and Gould Lakes. In that area there is a heavy storm beach consisting of several strands of nicely graded gravel. From Gould Lake a sand and gravel spit projects towards Hepworth, while behind it a succession of strands fan out around the northeastern end of the lake. From a point east of Park Head Junction a sandy beach swings around Hepworth separating Shallow Lake and McNabb Lake, both now drained, from the body of the lake plain. The familiar pattern of wave-cut drumlins connected by beaches reappears east of Clavering. The swampy depression around Hepworth Lake is below Algonquin levels, but that lake is held on the west by drumlins rather than a baymouth bar. Another Algonquin beach lies on the side of a drumlin three miles west of that corner at the border of a scoured, boulder-strewn floor of dolomite. Along Colpoys Bay the dolomitic scarp served as the shore. A well-preserved beach of coarse shaly gravel at approximately 810 feet is built across a shallow re-entrant in the scarp west of North Keppel and it runs southeastward in diminished strength.

In the Port Elgin–Southampton area the Saugeen River debouched a large amount of sand into this lake. It appears that sand from this source was swept northward alongshore towards Sauble Beach and Hepworth where it merged with sand brought down by the Sauble River. This sand plain has a full complement of dunes which have become active since the land was cleared. The peculiar form of these blow-outs is well illustrated on the aerial photograph in Figure 40. Near French Bay the bouldery remnants of a drumlin may be seen.

On much of the limestone plains of the Bruce peninsula it is common to find lines of rubble running roughly parallel to the length of the peninsula. As seen on the ground the bigger ones look like little moraines. They are visible from aloft and show up clearly on aerial photographs as illustrated in Figure 41.

North of Wiarton the hills above 850 feet, mostly on Cape Croker, were islands in Lake Algonquin, but only fragmentary beaches are present. Lower Algonquin beaches of sandstone shingle are common in the Cape Croker Reserve. The big gravel bar south of Lion's Head, which is conspicuous from the main road into the village, is one of the lower Algonquin beaches. Several gravel beaches on the road to Tobermory have been used up in highway construction.

The Nipissing beach was mentioned in connection with the dunes in front of the Blackwell and Thedford marshes. From Grand Bend to Clark Point the Nipissing bluff has been entirely undercut by Lake Huron. North of Kincardine this shoreline is near the present shore to beyond Inverhuron, taking the form of gravel strands and sand dunes. At Inverhuron, where the Little Ausable River empties, there is a fine sand beach on the shore backed by a belt of dunes which in turn is delimited by a small bluff. The low shorecliff prevails with very few gaps to Sauble Beach where it gives way to sand dunes that continue for nearly fifteen miles to Red Bay. As a rule the dunes rise to a height of 50 or 60 feet. A series of small shallow lakes and swamps lie behind the dunes while in front of them a wave-washed floor of limestone is littered with a splendid array of beach strands.

40. *Crescentic pattern of blow-outs on the sand plain near Hepworth. The river is the Sauble. (R.C.A.F. photo)*

The most important effect of Lake Nipissing on the Bruce peninsula was to leave the marly silt in the flat plain of Eastnor township. At this point the lake overtopped the peninsula leaving the section to the north an island. The beach at Lion's Head on the border of the silt plain is too high to belong to Lake Nipissing. However, on the Lake Huron side curving beaches west of Spry Corner and near Pike Bay at 640 feet probably belong to this lake. The shoreline north of Stokes Bay and Lion's Head is hardly accessible and we have very few records of it. The Nipissing beach is the upper one of the series of shingle beaches at Hope Bay and Dyer Bay. The related beaches at Dunks Bay are more sandy.

Wiarton to Collingwood. From Colpoys Bay to Owen Sound the highest stage of Lake Algonquin was confined by the sheer bluff which constitutes the Niagara escarpment in this sector. The gravel ridge at 735 feet

41. *Near Shallow Lake. The succession of wavy lines are little moraines. (R.C.A.F. photo)*

crossing the road at Lindenwood is not the highest beach. Two finely formed beaches can be seen along the road south of Hogg. The higher one curves around the drumlin-shaped outlier of the escarpment one mile south of East Linton; the lower one continues a mile farther then hooks to the base of the escarpment. Both consist of coarse, shingly gravel and are finely formed.

The Algonquin bed around Colpoys Bay and Owen Sound consists of wave-swept shelves of sandstone and shale. Red shale appears in the Colpoys Range, elsewhere it is not important. The shallow soil is interrupted west of Kemble by a deeper scattering of rocky debris which is either a moraine or an ice rampart.

Between Owen Sound and Annan several drumlins bore the brunt of the Algonquin waves and are surfaced with boulder pavement. A barrier beach lies across Telfer Creek valley. For two miles on either side of Annan the shoreline consists of a bluff which is most prominent on the sides of drumlins. From the end of this bluff a barrier beach with several crests bridged the strait south of Coffin Hill. The clay and shale of the adjacent lake bed is badly gullied near the Nipissing bluff from Leith to Coffin Hill. From Coffin Hill to the mainland on the east the waves failed to close off the little bay. The main force of the waves came from the northeast judging by the gravel spit projecting southward from the northeastern shore of the bay. There is another composite bar which crosses the corner at Balaclava, its crest taking a 775-foot contour on the Owen Sound map sheet. Around Cape Rich the Algonquin bluff is cut in Queenston shale. It has been undercut subsequently at the Claybanks by Lake Nipissing and Georgian Bay. Beaches were thrown across either end of the crooked valley in the bedrock now identified by Sucker Creek and Mountain Lake; Mountain Lake is held up by the baymouth bar at its eastern end. At three lower elevations near Cape Rich fairly strong Algonquin beaches occur, while another fainter strand a few rods west of Cape Rich corner lies half buried under silt. These beaches are described by Stanley (127) who studied the Algonquin shorelines in considerable detail from here eastward.

Between Cape Rich and Collingwood the Algonquin shorecliff coincides with the base of the Niagara escarpment (Blue Mountain) along the three former headlands separated by the Bighead and Beaver river valleys. Across the mouths of these valleys, back of Meaford and Thornbury, great curving barrier beaches appear. Comprising a dozen or more gravel strands, they are splendid examples of their kind, the highest crests being

42. *This R.C.A.F. photo of the country just west of Collingwood shows: top (north), Georgian Bay with waves near the shore; upper right, multiple beach strands in the Nipissing lake plain; from upper left corner southeastwards, the Nipissing shorecliff with gullies extending into the Algonquin clay beds above; lower left, the wooded lower slopes of the Blue Mountain ending at the Algonquin shorecliff, which is not distinct at this place.*

built up well over 25 feet. South of Meaford there is a shorecliff with a series of beaches left on the terrace below.

In front of the barrier beaches just described the Bighead and Beaver rivers have built deltas of sand and silt. The delta in the Beaver valley grades typically from coarse sand to fine sand to silt, the silt appearing from Thornbury to the eighth line of Collingwood township. At the mouth of the Bighead River there is a much larger sandy area. Some sandy ridges, running north and south back of Meaford stand up above the level of the sand plain. They are regarded as glacio-fluvial deposits and no doubt account for the extra sand in the Bighead valley.

The Nipissing bluff is fairly continuous south of Colpoys Bay not far from the shore. It is higher around Cape Commodore and a series of beaches lie below it. The bluff continues to Owen Sound, usually accompanied by a boulder pavement and shingle beaches, with a spit built from the south occurring at Gravelly Bay. The bluff continues in good form to Collingwood. A barrier beach appears at Leith across Telfer and Keefer Creeks. The Nipissing bluff has been undercut opposite Coffin Hill, at the Claybanks, and again east of Meaford. The Claybanks, it should be noted, are worn in shale, as are the other high bluffs opposite the promontories of the Blue Mountain. Near Meaford and Thornbury where they are cut in clay and sand or till, the bluffs are fairly small, which is fortunate as they run through both towns. On the bouldery terrace below the bluff a particularly numerous set of sand and gravel strands occurs between Craigleith and Collingwood. Figure 42 shows them as displayed on an aerial photograph.

The Nottawasaga basin. Between Collingwood and the boundary of the Precambrian rocks north of Orillia the bed of Lake Algonquin has an area of 1,200 square miles, exclusive of Lake Simcoe. The shoreline is estimated to be 400 miles long. It is highly irregular, with great bays and islands. Not only one but several beaches or bluffs are present. No finer assemblage of abandoned shorelines is to be found in the province. The map shows the upper shoreline, the stronger lower beaches, and the bouldery terraces. Where several beaches lie close together it is impossible to map them all on the scale of four miles to one inch; for instance the broad gravel bar shown at Colwell actually has several strands.

In the Stayner area the Algonquin shoreline lies against the base of the Corn Hill moraine with a broad, wave-swept, bouldery belt before it. Where it rounds the end of this moraine five miles east of Creemore, a big gravel spit comprising seven distinct strands was built across the valley of Coates Creek. The line of sand dunes south of New Lowell is due to lower stages. In the sandy area north of Everett the Algonquin sands join other sand beds above Algonquin levels and the shoreline is indefinite. The lake south of Alliston was shallow and somewhat marly with poor shore features.

The sediments in the arm running from Beeton past Cookstown on the east are of calcareous silt with some fine sand and clay interbedded. Around Alliston and northward to Angus the sediments are sandy having come from the terraces on the escarpment down the Boyne, Pine, and Mad Rivers. Three or four feet of fine sand over coarser sand is the common sequence. The Nottawasaga River is well entrenched in Essa township giving good drainage to the sand plain adjacent. Away from the rivers the water table is higher, even producing bogs on the Camp Borden Reserve and in the col east of Colwell. A glance at a contoured map will reveal that the railways made use of the level grades provided by the Algonquin flats.

It would be a fruitless task at this time to describe all the Algonquin shorelines in Simcoe county. However, some of the really magnificent features must not be missed; beyond that the map will have to suffice. The big multiple-stranded spit at Colwell has already been described and some other good gravel beaches appear along the base of the bluff south of Colwell. The valley-sides west of Allandale were considerably steepened by undercutting of the ancient lake. The Canadian Pacific Railway follows the wave-cut terrace from Colwell to Midhurst and beyond. In this section the bluffs stand over 200 feet high at the maximum. The lake managed to throw a barrier beach across the mouth of the Little Lake depression, and great quantities of gravel have been removed from it by the Canadian Pacific Railway. West of both Hillsdale and Waverley bold bluffs have been cut in the sides of the uplands. Within the long valleys of the Coldwater and Sturgeon Rivers, themselves floored with sand, the shore features are at their weakest. The northern ends of the North Simcoe islands have bold upper shorecliffs with boulder pavement and in places a series of beaches lower on the slopes. The Penetang peninsula exhibits a great deal of boulder pavement, the boulders invariably being hardheads from north of the limestone border. The sandhills north of Lafontaine hill are old fixed dunes, some of them having started to shift again. Around Lafontaine village deep beds of fine sand and silt underlie a group of good farms.

The Lake Simcoe basin. In the Orillia area the detailed account of the Pleistocene geology by R. E. Deane deals largely, on the map and in the report (32), with the features of Lake Algonquin. His map shows a corner of the level clay plain in North Orillia township. The island at Ardtrea is encircled with a startling series of shore features. The crescentric spit which confines Bass Lake and the barrier beach two miles east are both massive beaches. The north-facing promontory between is

43. *Algonquin bluff, Waverley, Simcoe county*

truncated to provide a terrace and a bluff which rises 70 to 80 feet. The waves completely decapitated the hill on which the Ontario Hospital is situated giving it a "pancake" top. Wave action produced some boulder pavement on the drumlins east of Orillia while clay was accumulating in the hollows.

On the west of Lake Simcoe from Orillia to Bradford the Algonquin shoreline is definite and continuous, although not so strongly developed as on the exposures facing Georgian Bay. Beaches are scarce between Orillia and Barrie, at least at the upper level, but they appear at frequent intervals south of Barrie. The shoreline is faint around the Holland Marsh, reappearing in strength as a sandy beach west of No. 11 Highway near Holland Landing. The sediment in the adjacent plain is sandy; perhaps it may be regarded as a Holland River delta. East of Cook Bay the shoreline is irregular in outline, with several islands appearing. In the wet and sandy depressions that lead southward, the absence of definite beaches makes it hard to distinguish between Algonquin sediments and those at higher levels which are older than main Algonquin. None of these broad valleys are completely blocked by a barrier beach, but the long spit north of the railway and two to four miles east of Egypt nearly closed one of the entrances. From the point of a drumlin at Wilfrid a strong gravel beach with two main crests extends northward. Six gravel pits are found on this beach within the same number of miles. Another big pit appears on a spit a mile or more south of Beaverton. Beaches take precedence over bluffs in the area east of Lake Simcoe, the most pronounced bluff occurring along a three-mile stretch directly east of Beaverton. North of here in an area where the limestone often comes to the surface many fragmentary beaches appear in offshore situations mostly on the slightly elevated tracts of limestone. Such beaches may be seen between Kirkfield and Carden. From Balsover, a beach runs with good continuity to the main outlet of Lake Algonquin at Kirkfield.

Lake Nipissing east of Collingwood. Lake Nipissing inundated a narrow but continuous border around Georgian Bay east of Collingwood, its abandoned shore being well marked at 50 to 55 feet above the water of Georgian Bay. The lagoon behind the sand dunes at Wasaga Beach is part of the lake plain, while the flats around the Minesing swamp are an annex at the same level south of the moraine at Edenvale. It is an uninviting strip of land that Lake Nipissing left around Georgian Bay as the soil is usually stony or wet. The deeper beds of sediments which give rise to good soil lie submerged in the bay.

Immediately west of Collingwood the 15-foot bluff of Lake Nipissing is easily identified. A group of good farms is situated above it. On the southern approaches to that town a beach divides the Nipissing and Algonquin plains, and several lower beaches appear in the vicinity of the golf course. East of Collingwood a low bluff extends to the point of contact with the Wasaga dunes. In the Jack Lake vicinity two or three of the lower Algonquin beaches are close to or even in contact with the bluff and beach at Lake Nipissing, while in the lagoon two other gravel bars lie partly buried by marl. On the eastern border of the lagoon, also, a beach above the Nipissing beach should be identified with Lake Algonquin, along with a lower beach which is mostly buried in the marl. The latter is the one which Stanley discusses as the Payette beach (128).

The barrier across the mouth of the Jack Lake lagoon takes the form of a belt of sand dunes. They are associated with the famous Wasaga Beach. Reflecting the prevailing winds the crests of the dunes trend 30 degrees to 40 degrees south of east. The majority of them rise to about 100 feet above the bay while the highest crests are 175 feet above.

The Nipissing bluff is bold all around the Penetang peninsula and on the headlands between Midland and Coldwater.

Manitoulin Island. All of Manitoulin Island was under the water of Lake Algonquin except the highest part of the tableland east of West Bay. We must report that no beach was found on that promontory to mark the shoreline, but it has an elevation of over 1,000 feet above sea level and the upper Algonquin beach should occur at approximately 950 feet (92). Fragmentary beaches are found on the island and boulder pavement is prevalent on drumlins. Algonquin sediments provide the best soils on Manitoulin. They are mostly silty but vary from silty clay to fine sand.

They occupy the lowlands north and south of Lake Mindemoya, around Wolsey Lake, in Gordon township, and in smaller patches elsewhere. The sediments are mildly calcareous and are derived mainly from local till. In fact the scanty till has been washed off extensive areas of limestone on this island.

The Nipissing shoreline shows up as low bluffs or sandy beaches and dunes at most points along the south shore of Manitoulin Island, but it is sometimes missing in areas of bare limestone. Low bluffs are cut in the clay south of Manitowaning. A bluff, with dunes on the limestone floor below it, appears at Providence Bay and Burpee. A series of sandy beaches is found near Murray Point and Carrolwood Bay. Along the precipitous north shore of the island the Nipissing beach often appears at the heads of the bays. A low bluff extends across the settled part of Barrie Island. A small bluff appears five to seven miles west of Little Current, while a distinct shorecliff and bouldery terrace crosses the road just south of Sheguiandah. We have not visited the Duck Islands but what appears to be a Nipissing shoreline is conspicuous on the aerial photographs of Great Duck Island.

Cockburn and St. Joseph Islands. Nowhere is there a better display of Algonquin and Nipissing shorelines than on Cockburn and St. Joseph islands. Both islands have morainic hills or uplands which were high enough to stand up as islands in Lake Algonquin and are encircled by high shorecliffs, except on the west side of the "mountain" on St. Joseph where there are some beaches. At least half of these islands is covered with boulder pavement, while most of the remainder consists of sand plains. The northwestern third of St. Joseph Island contains a good deal of lacustrine clay, interbanded pink and grey in colour, and fine in texture. Lake Nipissing is responsible for particularly bouldery terraces on the borders of these islands, bounded inwardly by bold shorelines at about 640 feet above sea level. The main beaches are marked on the master map and may be referred to easily.

On Cockburn Island the beaches which extend towards the northwest from the central hill are a series of massive gravel strands on the shoulder of high ground. The beaches south of the group of drumlins are also well formed. The boulder pavements are sandy and may be wet or dry, but the water table is generally near the surface. Practically all the stones and boulders are of igneous rocks from the north. Southeast and south of the central hill some sand dunes appear. The Nipissing shoreline consists of bluffs on the west and north, and a beach, mostly on the east and south side of the island.

On St. Joseph Island the upper Algonquin beaches which impound the little Mountain Lake actually form a terrace over a quarter of a mile wide and is a tremendous deposit of coarse gravel. The upper Algonquin beach, at about 900 feet, also crosses the road between Carterton and Otter Lake. A good series of lower Algonquin beaches appear on the slope west of Otter Lake. Sand dunes occur in a belt extending westward from the northern corner of the "mountain". Another good set of lower Algonquin beaches is found between 700 and 800 feet on the slope west of Hilton Beach. The Nipissing shoreline mostly takes the form of a bluff. In the northwest it is cut in the clay of St. Joseph township. It is continuous from Otter Lake to Rains Lake in the southern end of the island and rises to a height of 50 feet. In the latter vicinity there is a great number of beach strands within Nipissing levels. North of Hilton Beach the Nipissing bluff sometimes coincides with a limestone scarp, but an extremely bouldery terrace is invariably present below it.

Lake Iroquois. From the Niagara River to Trenton the Iroquois beach lies two to eight miles from Lake Ontario except at Scarborough bluffs where it swings in to the present shore. East of Trenton a much wider belt was flooded. The Iroquois shore features are comparable to those of Lake Algonquin; that is, they are unsurpassed in boldness and are about as well developed as those of Lake Ontario.

From Queenston to Toronto the beaches consist of shingle of calcareous siltstone that comes from the local shales, together with limestone gravel and the inevitable sprinkling of granite, quartzite, etc., from the Canadian Shield. Beaches are scarce in the Niagara fruit belt, occupying only seven or eight of the 45 miles of shoreline. No. 8 Highway is built on the beach through St. Catharines and eastward to Homer. It also touches part of the massive beach west of Stony Creek where it passes through the cherry orchards that have found the gravelly soil to their liking. In fact No. 8 Highway closely follows the Iroquois bluff from Queenston to Hamilton. At Hamilton there is the magnificent bar across the lower Dundas valley which carries the highway, railroad, and the famous rock gardens. The early truck crops for which the Aldershot vicinity is noted are grown on the Iroquois beach. From here to Clarkson the shorecliff is cut in red shale and beaches are limited to a few short strands. The gravel bar across the mouth of the Credit valley is 75 feet thick at its maximum. Built from the southwest it forces the river northward below Erindale and has been the most important source of gravel in that area. From Cooksville through the Dixie district the beach has been covered with orchards and lately has begun to be used for building lots. In the Islington area the brow of the bluff has special value as real estate.

Like the Credit, the Humber and Don rivers are deflected by sand and gravel bars, although these are built from the east rather than from the west. Both bars have been used as sources of sand and gravel. In Toronto the Davenport Hill is well known as the Iroquois shorecliff while the McLennan Hill is the same bluff east of Yonge Street. In weak form because of the protected location the bluff crosses O'Connor Drive, west of Dawes Road.

East of Scarborough bluffs the ancient lake encroached upon a rolling till plain so that beaches and bluffs follow each other in close succession. The sandy belt mapped under this shoreline is an off-shore terrace which includes a good deal of wet boulder pavement. Much of the sand is less than six feet deep; deep sand occurs in the Ganaraska valley and is frequent east of Port Hope. While some of the latter sand is deltaic, most of it is derived from local till by wave action. The first heavy beach east of Scarborough bluffs is the one on which the gravel pits are seen two miles east of Malvern. It deflects the Rouge River eastward. There were eight barrier beaches across the mouths of small valleys in Pickering, West Whitby, and East Whitby townships, most of them now removed for gravel. Between them the shorecliff is definite but not high. North of Bowmanville the beach swings lakeward around several drumlins and at this point the Canadian National Railway, which follows the shoreline to the west and east, takes the direct route through the enclosed lagoon. No. 2 Highway touches a strong beach three and a half miles east of Newcastle and runs on its crest for two miles east of Newtonville. This beach is built around glacial hills after the manner of the beach north of Bowmanville. Gravel pits have been in operation on two hills northwest of Port Hope, working beach gravels lying below the altitude of the Iroquois beach.

Through Northumberland county there are big drumlins pointing southwestward. The waves levelled them near the Iroquois shore and cut high bluffs along the shoreline itself. Thirty feet below the beach the drumlins again appear intact except for a concentration of stones on the surface.

Probably the best beach between Port Hope and Trenton is at Little Lake, called Biddy Lake on the maps. It comprises several strands built from the gravel of an esker. Not only does it impound Little Lake but it serves as a drainage divide forcing the overflow from the lake to run north on a round-about route *via* the Trent River to Lake Ontario. From Biddy Lake to the Murray hills there are only short bits of beaches, the remainder of the distance is marked by a high bluff and a broad dry sandy terrace which reaches its best development around the Murray hills. At its best the shorecliff on this shoulder of land is 200 feet high.

From the Murray hills to Hastings the Iroquois shoreline is irregular and the shore features are weakly developed. Cramahe Hill has a distinct terrace cut in its eastern flank, and so has the smaller island lying east of it. The islands north of Campbellford also have distinct bluffs and beaches on their eastern exposures. Wave-built features are absent in the Rice Lake area.

East of the Trent River the hills around Oak Lake stood as an island in Lake Iroquois. Dr. Coleman described it vividly (23). Oak Lake is confined by a beach on the north and thus remains as it was at the time when its mother lake was in existence. There are gravel bars on either end of the island. The southeastern shore consisted of bluffs well up to 100 feet high.

Northeast of Oak Lake near West Huntingdon another hill just managed to break water in Lake Iroquois. Aptly, it is called Pancake hill, because it is flat-topped. The gravel bars at the northwest corner represent the most easterly of Iroquois beaches. The remainder of the surface is a boulder pavement. Four more of these decapitated hills may be seen north and east of Frankford complete with gravel bars on their northwest sections.

The gravel in all the beaches east of Oshawa is mainly of limestone, contains little or no shale, and makes good concrete aggregate.

The sediments in the bed of Lake Iroquois will be discussed later in the descriptions of physiographic regions.

44. *Boulder beach produced by wave action on a till ridge near Greenfield*

45. *Marine clay, Carlsbad Springs*

As the glacier receded from Covey Hill and uncovered outlets lower than the Mohawk valley the water levels dropped in the Ontario basin. However, none of these water planes produced well-marked beaches along the north shore of Lake Ontario. The Champlain Sea came into existence when the glacier receded from the lower St. Lawrence valley far enough to allow access to the sea. Marine fossils are not found west of Brockville but, since marine clay was deposited there in still, deep water and there is no barrier to the west, the sea probably extended into the Ontario basin until uplift of the land brought the Thousand Islands area above sea level.

The Champlain Sea. The first fact to record about the Champlain Sea is that, because of the rocky nature of the slope, it did not have shore features along its western shore. Two short gravel beaches near Westport, just above the 525-foot contour, are probably Champlain Sea beaches. Since the Champlain Sea lacks a definite shoreline we can only set its limits at the limits of the stratified clay, bearing in mind the beach at Kingsmere, north of Ottawa, at 690 feet above sea level.

There is, however, an abundance of beaches in eastern Ontario; as the land finally rose above sea level the waves threw up beaches or bars on prominences throughout the area. In the west, on the outcrops of Beekmantown dolomite, the beaches are usually built of shingle. Chazy shale also provided shingle for beaches, chiefly in the vicinity of Metcalfe. The drift yielded rounded gravel, while erosion of till on the drumlins or elsewhere created boulder pavements. This is the cause of so much excessively stony land in these eastern counties. Many of the boulder pavements were cleared when the land was brought under cultivation but the heaviest of them still remain.

The extensive sand beds in the Ottawa valley around Petawawa are deltaic, brought in by the Petawawa, Barron, and Ottawa rivers of the time. East of Ottawa similar sands represent the last major deposition of sediments in eastern Ontario. The fairly extensive sand beds in Grenville, Dundas, and Stormont counties originated as glacio-fluvial deposits.

The marine clays of eastern Ontario must be separated into at least three classes each of which has given rise to specific soil types. There is the grey calcareous clay common to Grenville, Dundas, Stormont, and southern Grenville counties. Other grey clay containing much less lime is found in Leeds, Lanark, and Renfrew counties. East of Ottawa the pink and grey clay is still lower in lime and no doubt from a different source. The latter two types were largely brought in by the early Ottawa River and by streams from the adjacent Laurentian uplands. In some places these clay beds are very deep, with depths of 180 feet, 186 feet, and 220 feet being reported by well-drillers in the Ottawa area. On the other hand there are innumerable small tracts of shallow clay on limestone.

LOESS (AEOLIAN SILT)

In our three articles on the physiography of the south-central, eastern, and southwestern parts of Southern Ontario (107) (14) (17), no mention was made of loess, or wind-deposited silt. Deep beds of loess like those of the upper Mississippi valley are not found in Ontario. However, there are shallow, discontinuous deposits of uniform silt on the till plains which we are now disposed to look upon as aeolian material or loess.

The identification of loess where the deposit is shallow is not simple because it must be distinguished from the leached horizons of the soil. Moreover, the loamy calcareous tills so common in Southern Ontario present a special difficulty. Upon disintegration the limestone pebbles break down to a silt, and it is not surprising that a lively controversy should develop as to the origin of the superficial silt. While one would not deny that silt is produced by weathering in the A horizons of these Grey-Brown Podzolic soils, the variable depth of the silt can hardly be attributed to weathering. Also, the pebbles quite often come right to the surface, probably due to an absence of loess at these points. It must be noted that the B horizon of the soil almost always occurs at the bottom of the silt even if it is six feet deep, but in a limited number of cases the profile was found, formed entirely in the silt, with unweathered silt below.

The Dundalk till plain commonly has a foot or two of silt on the surface, more or less. Deeper deposits of two to six feet occur uniformly in the Redickville-Honeywood area. The gravel beds of the spillways, so numerous between Palmerston and Wingham, often have an upper blanket of silt. Similar deposits on the gravel terraces in the Grand valley give rise to the soil type called Burford loam. Fifteen to thirty inches of silt is general on the surface of the Woodstock drumlins. In Middlesex, Huron, Perth, Waterloo, and Oxford counties the Huron silt loam of the soil maps outlines the distribution of loess on the surface of the boulder clay of that region, mostly on the Lucan, Mitchell, and Milverton moraines. It is also found on low ground in areas of Perth and Brookston silt loam.

In central Ontario the slopes bordering the interlobate moraine carry fine sand and silt generally, on the surface. It is a little more abundant

46. *Wind-blown silt above a brown horizon, with glacial till below, in Cramahe township*

on the southern side, being as deep as seven or eight feet in spots. In Northumberland county the deepest of it is related to Dundonald fine sandy loam, less deep deposits to Bondhead fine sandy loam, and shallow deposits up to two feet to Bondhead loam. In York county, over heavier boulder clay there is silt on the surface which has given rise to King silt loam.

The drumlins of eastern Ontario do not have a silt covering, nor does it appear on any of the marine sands and clays.

From the standpoint of the soils and their cultivation this stone-free silt on the surface is an obvious boon as they would be much more stony without it. Post-holes and drains would also be harder to dig and all tillage more difficult.

On clayey tills the covering of silt creates late soils with perched water tables in wet periods. On sands and gravel the blanket of silt produces more productive and durable soils. Thus the over-all effect is one of great benefit to the farmer.

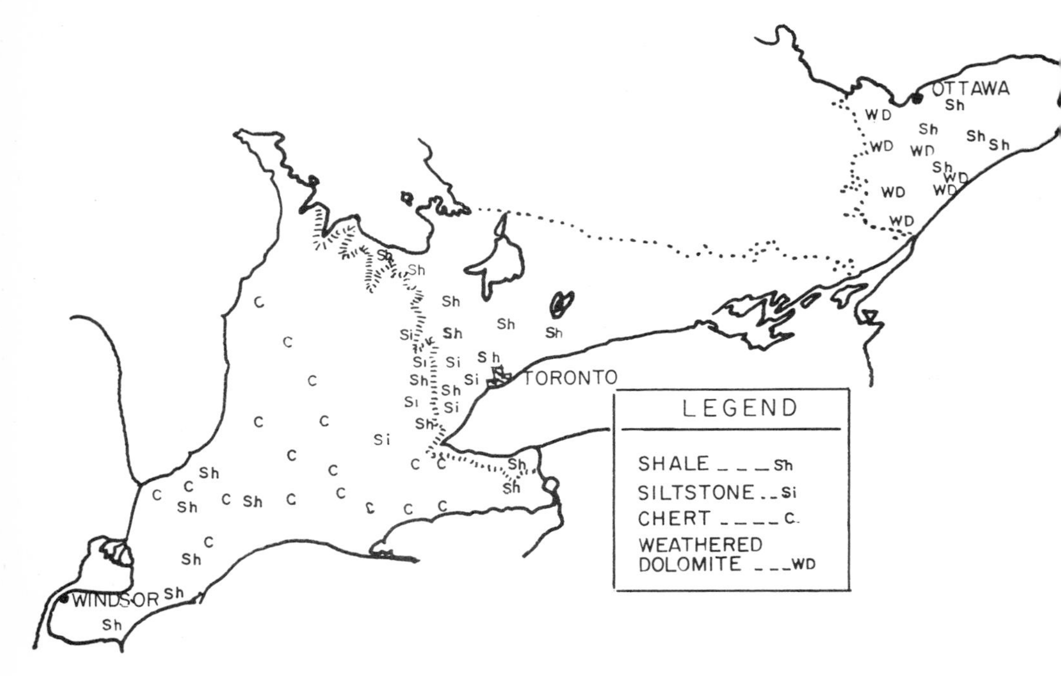

47. *The distribution of deleterious components in gravel in Southern Ontario*

GRAVEL AND SAND DEPOSITS

Among the Pleistocene formations are many gravel and sand deposits. The use of gravel and sand for road construction and concrete aggregate has increased rapidly in recent years in Ontario and in 1960 amounted to 75 million tons (69). Its value is increasing even more rapidly especially near the larger cities and where reserves are limited. In these latter areas the search for new gravel deposits is an urgent matter.

In his report on gravel (69), Hewitt includes a small map of southern Ontario showing only gravel-bearing deposits. On a larger scale our master map shows the formations where gravel may be found and is useful in locating gravel and sand. While there may be some difficulty in sorting out the gravel deposits on this map it has the advantage of identifying the other landforms in which gravel is not likely to be found. Soil maps may also be used for this purpose and are on a larger scale, as are other maps of Pleistocene deposits covering several selected map areas. However, none of these maps are detailed enough to include isolated little pockets and hills which may yield useful amounts of gravel. Neither will they show gravel deposits buried under a sheet of till. Aerial photographs are invariably used in modern surveys and these often give indications of the presence and the extent of granular materials.

The largest concentration of gravel in southern Ontario is found in the outwash associated with the Horseshoe moraines, particularly in the Orangeville to Brantford area. Beds of fairly uniform gravel 20 to 25 feet deep cover broad areas and 40 feet of material is not uncommon. The ancient beaches, the bulkiest being those of the long-lived Iroquois, Algonquin, and Nipissing lakes, also provide good deposits of fairly uniform material. An added feature of beach deposits is the good blend of grain sizes in the sand fraction. The main limitation of beaches usually is the restricted depth and extent of the deposits; mostly they consist of narrow ridges less than 20 feet deep. Unusually deep and wide bars have been built across the mouths of some of the larger stream valleys, or where a large deposit of glacio-fluvial gravel has been reworked by the waves, as at Colborne. The material contained in eskers is much more variable than that in river terraces and beaches and coarse, cobbly and even bouldery material is to be expected. Kames and related glacio-fluvial deposits consist of variable material and may be loaded with coarse gravel and cobbles or fine sands, usually the latter. Nevertheless, some of the most valuable pits are in deposits of this sort. The Oak Ridges moraine north and northeast of Toronto is dotted with pits and many prospective sites have been tested, some of them more than once.

Gravel and sand for concrete aggregate must be almost free of clay, iron oxides, and organic matter. There are definite upper limits of tolerance for shale, chert, weathered silty dolomites, calcareous siltstone, and cemented gravels. In southern Ontario the most widespread offender is shale. It not only disintegrates on freezing and thawing, but may react with the alkalies of cement and eventually cause cracking of the concrete. The deleterious materials in gravel are mapped in a generalized manner in Figure 47. Siltstone is associated with shale, coming from the harder strata in the local shale formations and varies from soft siltstone towards fairly stable arenaceous limestone. It is apparent that there was forward transport of this siltstone from areas of shale onto adjacent limestone and dolomite in eastern Ontario and over the Niagara escarpment. The spillway extending from Orangeville and Caledon to Brantford contains some siltstone, the amount diminishing towards Brantford. There is also some limestone mixed with the dolomite which is the main component. Soft weathered dolomite is troublesome in much of the St. Lawrence–Ottawa lowland, the amount in some places amounting to more than 15 per cent.

Chert is the dense, siliceous, flinty material from which arrowheads were made. That found in local limestones is white, cream, brown, or bluish-grey in colour. The pieces are irregularly blocky in shape with conchoidal fractures. Chert is prone to shatter on freezing and thawing in the presence of water and also sometimes produces alkali reactions in concrete. It is thought to vary in alkali reactivity and brittleness, but at present reactive and non-reactive chert cannot be separated on the basis of appearance, so it is all classed as deleterious.

Gravel of good quality is generally found in the region between Georgian Bay and Lake Simcoe and Kingston. Weathered granite that crumbles readily to sand sizes is common in that area, but since most of it is broken up during crushing it does not constitute a serious problem. There are never more than small amounts of shale and soft, silty stones. Limestone is the main component.

The other section of southern Ontario having very little deleterious material is in the area from Brantford to Durham, and northwestward as shown on Figure 47. Here, sound dolomite is the major component. It is fortunate that good quality and very great reserves of gravel are found together within reach of Hamilton and Toronto.

RIVER VALLEYS

River valleys are conspicuous features to which we have so far given little attention apart from a discussion of the preglacial valleys in the bedrock and the glacial spillways. Only the latter appear on the physiographic map. The valleys of the present streams are being cut in the drift and occasionally into the bedrock below. The peculiar relationships between their courses and the bedrock and glacial formations is of geomorphic interest. The volume of the streams, their gradients, and the type of drift determine the size of the valleys and affect the rate of erosion.

River valleys present a contrast to the glacial and lacustrine formations which they dissect. As a rule they have steep slopes that terminate abruptly, while glacial forms are generally rounded. Moreover, branching as they do towards the source, they form a dendritic pattern in contrast to the irregular arrangements of morainic knobs and basins, or the regular parallel alignment of drumlins. Very often, the stream valleys constitute the only important relief of till plains, terraces, and lake plains. Where dissection is well advanced there is an intimate association of level land and steep slopes. Dissected terraces are very common. The rolling and hilly topography of moraines and drumlins may also be modified and accentuated by stream valleys or gullies. Incidentally, the formation of gullies is often complemented by deposition in hollows farther down the slope.

The streams carry off slightly more than one third of the precipitation in Southern Ontario, which is normally 28 to 40 inches per annum, with most of the area receiving about 33 inches. The well-cleared clay plains in the warmest sections have several inches less run-off than the rocky lands farther north where woods and swamps are more plentiful. For example, the two upper branches of the Thames carry away 11 to 13 inches of water from their basins while the Beaver and Rocky Saugeen remove about 19 inches annually, the precipitation being 37 to 38 inches per year in both areas.

The watersheds of Southern Ontario contain few lakes outside the area of the Canadian Shield, with such notable exceptions as the Kawartha Lakes and Rice Lake in the Trent watershed and in the adjoining Severn drainage basin. Lake Simcoe, with an area of 284 square miles, is the largest inland lake in Southern Ontario.

Southern Ontario may be divided into several major drainage systems and the streams will be discussed under the following headings: (1) Georgian Bay Drainage; (2) Lake Huron Drainage; (3) Lake St. Clair Drainage; (4) Lake Erie Drainage; (5) Lake Ontario Drainage; (6) Ottawa River Drainage.

GEORGIAN BAY DRAINAGE

The larger streams draining into Georgian Bay from the south—the Severn, the Nottawasaga, the Beaver, and the Bighead—are discussed in some detail. A number of smaller streams are also considered worthy of mention.

Coldwater Creek, Sturgeon River and other streams. Five sluggish streams flow into Georgian Bay from the northern part of Simcoe county. The North River drains the clay flats of Orillia township. Four others, Coldwater Creek, Sturgeon River, Hog Creek, and Wye River occupy the flat-floored valleys which separate the upland masses of the Penetang peninsula. They appear to be fed largely by springs along the valley sides, the uplands being practically devoid of streams because of the vertical drainage in the sand till. These rivers have cut shallow channels, and parts of the broad valleys are still quite swampy. It is probably that these streams were once even more sluggish, since the water level of Georgian Bay has fallen nine feet in the last 300 years, as is indicated by investigations at old Fort Ste Marie on the Wye River near Midland.

Severn River. The Severn River, falling into the southeastern arm of Georgian Bay, drains an area of 2,000 square miles, two-thirds of which

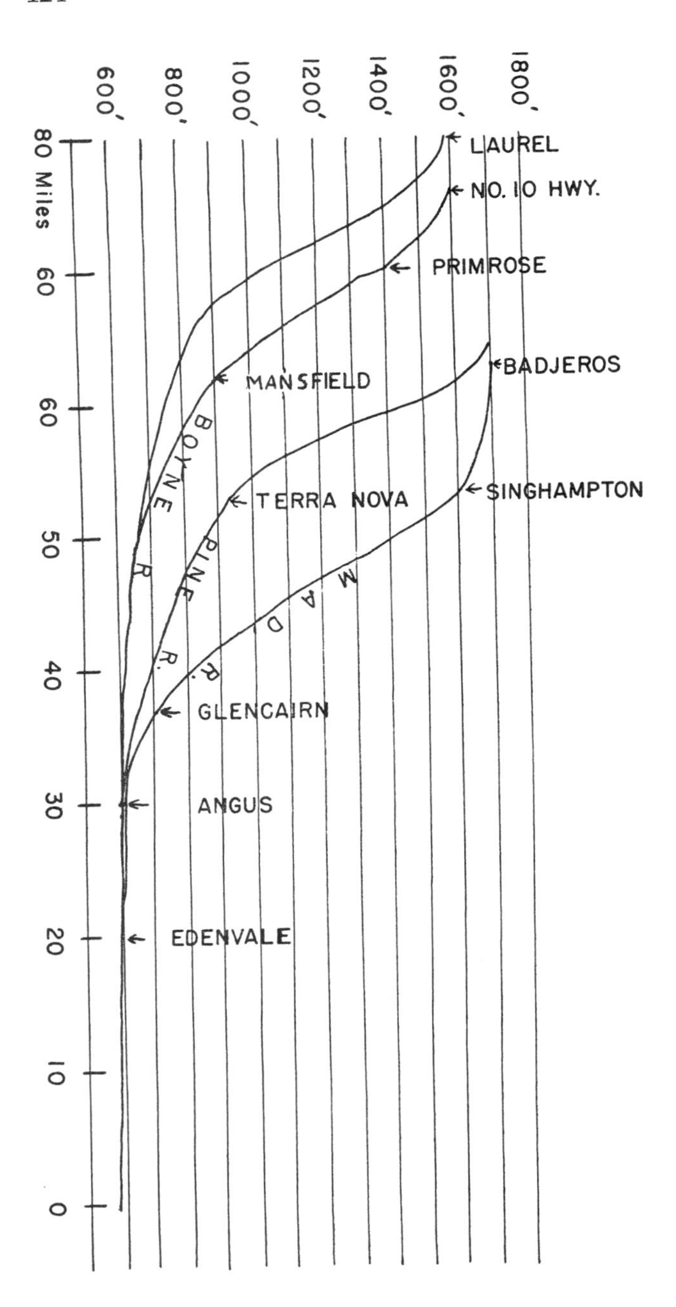

48. *Profile of elevations, Nottawasaga River*

is south of the Canadian Shield. Because its basin includes the great reservoir of Lake Simcoe (284 square miles in area) the flow of this river is fairly uniform and three hydroelectric plants are located upon it. The annual run-off amounts to about 15 inches.

The southern shore of Lake Simcoe is low and marshy and most of the streams entering from the south have low gradients and are bordered by marshland for many miles upstream. Most notable is the Holland Marsh along the Holland and Schomberg rivers, where, in recent years, 7,000 acres of land have been reclaimed and devoted to the production of vegetables. The valleys of these rivers probably originated in the preglacial drainage system, although now deeply mantled by glacial drift. There is little evidence of postglacial erosion except about the headwaters which drain from the Oak Ridges moraine.

In the early days of settlement, Lake Simcoe and the Severn River system were important transportation routes. They form a link in the Trent Canal system by which small boats may pass from Lake Ontario to Georgian Bay.

Nottawasaga River. The Nottawasaga River system drains an area of 1,145 square miles. The main branch is about 90 miles in length as it follows a round-about course from a point a few miles south of Shelburne to its mouth at Wasaga Beach. Its two longest tributaries, Mad River and Pine River, also rise in the high plain west of the Niagara escarpment. The border of the cuesta, served by these streams, is much better drained than the main part of the Dundalk plain. All three streams, and a number of smaller tributaries as well, flow down the escarpment in deeply cut rock-valleys several hundred feet in depth, with gradients often more than 100 feet per mile. This portion of the escarpment presents some of the roughest topography in Southern Ontario. It is mantled by glacial deposits, and sand or gravel terraces which have been deeply dissected by these rapid streams. The Mad River flows through Devil's Glen, a well-known spot of scenic beauty near Singhampton, while the upper branch of the Nottawasaga makes its way from the escarpment through the Hockley valley, also noted for its rugged scenery and challenging ski trails.

These rapid streams built great sandy deltas in Lake Algonquin and today they traverse them in flat-floored valleys 25 to 50 feet in depth. The Nottawasaga has not yet established complete drainage of its basin and a number of boggy areas still remain. Most notable of these is the Minesing swamp depression which is periodically inundated by flood waters. Here the Nottawasaga has built noticeable natural levees along

49. *Erosion in the Nottawasaga basin (Conservation Authorities Branch, Ontario Department of Energy and Resources Management)*

its banks so that its tributary, the Mad River, is forced to follow a parallel course for several miles before finding an entry. Willow Creek behaves in a similar fashion, crossing the central portion of the Minesing swamp.

Below the Minesing swamp the river cuts through the Edenvale moraine in a canyon over 100 feet deep, a quarter of a mile in width and about four miles in length. The lowland around Jack Lake, to the north of the moraine, is subject to flooding. Formerly it was a lagoon cut off from Lake Nipissing by a barrier beach and a line of sand dunes. The river has breached this barrier at the extreme southwestern end and is then forced to double back and pass between the dunes and Wasaga Beach for about four miles before emptying into the bay. About a mile upstream from the mouth of the river is Nancy Island Park, an alluvial island developed around the hull of the schooner *Nancy* which was sunk there during the war of 1812.

Beaver River. The Beaver River follows a deep preglacial valley which has been somewhat rounded out by ice action, and provides drainage for an area of about 225 square miles. This valley indents the Niagara escarpment for a distance of about 25 miles, from the shore of Georgian Bay to Flesherton. The largest branch of the river, however, enters the valley from the east at Eugenia Falls, about four miles below Flesherton, the upper part of the deep valley being occupied by a small stream known as the Boyne River.

The eastern branch of the river is utilized for the development of hydroelectric power. It drains an area of about 100 square miles of rocky land and swamp between the Singhampton and Gibraltar moraines. The channel is shallow in most of this area because bedrock is near the surface, but for two or three miles west of Feversham the river valley is 50 to 75 feet deep. A dam on the brow of the escarpment creates Eugenia Lake, a reservoir of 1,400 acres at an elevation of 1,422 feet. A flume carries the water down the side of the valley to the powerhouse located at about 900 feet, which generates 7,500 horsepower.

The upper part of the Beaver River drains a rocky, well-wooded country and maintains a strong summer flow, the best in the whole area in this regard. Its waters run clear even after heavy summer rains. Farther downstream the river is often muddy, charged with reddish-brown clay from the lower slopes of the valley.

From Kimberley to Heathcote the floor of the valley is almost perfectly flat and composed of sand and silt. This area is imperfectly drained and has been largely left in forest. Between Heathcote and Clarksburg the river passes a double barrier consisting of a trans-valley moraine and the great Algonquin beach which, near Clarksburg, exhibits fourteen separate strands. Throughout this section the river flows in a narrow winding inner valley about 75 feet in depth at its maximum but becoming shallower as it crosses the old delta to the present shore at Thornbury. At both Clarksburg and Thornbury there are power sites, while the river also provides the domestic water supply for Thornbury. Recently it has come into use as a source of water for irrigating apple orchards on the dry gravelly soil of the old beaches.

Bighead River. The Bighead River which enters Georgian Bay at Meaford is somewhat smaller than the Beaver, draining an area of about 120 square miles. It also occupies a broad re-entrant in the Niagara escarpment and drains a considerable area of wooded upland. Unlike the Beaver valley, however, the valley of the Bighead has no long extension inland. Above Bognor the river is not large, but many streams come down the drumlin-strewn sides of the valley and it grows rapidly between Bognor and Oxmead. This part of its course is crooked as it winds between drumlins, cutting a narrow inner valley in the inter-drumlin beds of clay.

Near Oxmead the river is diverted by the barrier beach of Lake Algonquin and makes a hairpin turn, swinging southward around several

50. *The Sydenham River falls over the Niagara escarpment at Inglis Falls, Owen Sound*

more drumlins before breaking through the old beach. For the last mile the river has cut through the sand and clay of Lake Algonquin to Meaford. The river mouth serves as a harbour for small boats and is protected by long concrete piers built out into Georgian Bay.

Sydenham and Pottawatomi rivers. Two small rivers, the Sydenham and Pottawatomi, empty at Owen Sound. They drain about 110 square miles of rocky or swampy country to the south and west, much of which is wooded. The Sydenham River rises in Williams Lake, a good-sized kettle-lake near the source of the Bighead. It flows west past Williamsford, separated from the North Saugeen by the Gibraltar (Chesley) moraine, then turns north to Sullivan Mills. From there to Inglis Falls it meanders through a swampy outwash plain, cutting a very shallow channel. At Inglis Falls it plunges over the Niagara escarpment into the beautiful glen in which Harrison Park is situated. The Pottawatomi also enters the lowland at Owen Sound by way of a small scenic waterfall after emerging from the swamp west of the city.

The cutting of deep valleys by these rivers has been prevented by the presence of bedrock near the surface, and the numerous swamps of the watershed are the results of the arrested valley development.

Disappearing streams. The boundaries of the watersheds along the Niagara cuesta must always be in doubt because of the number of disappearing streams in the area. West of the Beaver valley in Euphrasia township there are several examples. Wodehouse Creek goes underground near Wodehouse. Another brook disappears near Harkaway while two more find underground channels just north of Goring.

The rocky plains of Keppel township slope westward, but no stream of any consequence crosses the line into Amabel township. Hepworth Creek which drains Hepworth Lake, meanders for about six miles and finally disappears into the rock about a mile east of Hepworth village.

Other sinks are known in the Niagara peninsula above the escarpment near Smithville where certain small surface streams find underground exits.

LAKE HURON DRAINAGE

The last five rivers described rise above the Niagara escarpment and drain the border of the cuesta. However, the main slope of that high plain is towards the west and south, or towards Lakes Huron and Erie. The drainage of the highest portion is towards Lake Erie, chiefly by the Grand River. The westward slope is drained by the Saugeen, Maitland, and

51. *A disappearing stream: Hepworth Creek where it enters cracks in the dolomite bedrock east of Hepworth*

Thames rivers and their tributaries and by smaller rivers such as the Sauble, Lucknow, Bayfield, and Ausable.

These rivers drain an inner plain 1,000 to 1,700 feet above sea level, which is fluted or drumlinized and crossed by a network of abandoned glacial drainage channels. Surrounding this is a morainic belt consisting of several ridges and a big spillway. The spillway in several cases deflects the rivers southward giving them a trellised effect. Between the moraines and the lake is a sloping clay plain, the lower part of which is modified by the deposits of Lake Warren. North of Inverhuron, Lake Algonquin was responsible for a level sandy plain. Along the shore of Lake Huron, particularly in the central part, the waves have cut bluffs about 100 feet high and the numerous small streams entering the lake have cut deep gullies as they approach the shorecliff. Some of the larger creeks have dissected the slope farther inland, for example the clay plain near Kincardine is deeply trenched by the Penetangore River and its branches, while a few miles to the south the Pine and Eighteen Mile rivers have accomplished similar dissection. These are streams that dry up in summer and are subject to flash floods at any time of the year.

Except for a few areas near the Niagara escarpment and in the upper Saugeen basin, these watersheds are deeply covered with drift, and bedrock is not often exposed by the streams.

Sauble River. The Sauble River including its northern tributary, Rankin River, drains an area of over 300 square miles. The actual area is hard to estimate because of the sinks and disappearing streams in Keppel township.

The main branch of the river rises near Desboro in Sullivan township, Grey county, but the greater part of its course is in Arran and Amabel townships of Bruce county. It takes an extremely crooked course through two contrasting landscape types, the upper part being among the hills of the Arran drumlin field while the lower portion is across an Algonquin sand plain. The upper portions of its valley are purely accidental, winding through the inter-drumlin swamps in an aimless fashion.

A singular phenomenon occurs about a mile and a half southeast of Tara, in lots 32 and 33, concession IV, of Arran township, where the river passes completely under a drumlin nearly 50 feet high. At low water the stream disappears into cracks in the limestone in lot 33, and reappears as a huge spring about 80 rods westward in lot 32. During high-water stages the overflow takes a course around the northern end of the drumlin, adding an extra mile to the length of the river and necessitating two highway bridges.

The inter-drumlin areas contain several important lakes, the largest, Arran Lake, being over three miles in length with an area of 800 acres. Gould Lake, much smaller, is shaped somewhat like a horseshoe around a large drumlin. Chesley Lake, which has two small drumlin islands, finds a direct outlet to Lake Huron by way of Stoney Creek. The rock plain of Keppel township also has several lakes. Two of them, Shallow Lake and McNab Lake, have been drained in order to get at the deposits of marl with which they are floored. Hepworth Lake overflows into Hepworth Creek, whose disappearance underground has already been mentioned. The only visible outlet for these lakes is by way of Parkhead Creek, a small intermittent stream which flows into the Sauble River.

In the lower part of its course the Sauble traverses the sand plains laid down by Lake Algonquin. At first, apparently, it meandered quite widely but has now become entrenched. The deepest part of the valley is in the region of Parkhead and Hepworth, but even here it is only about 30 feet in depth.

Saugeen River. The Saugeen River drains an area of 1,565 square miles including some of the highest land in Southwestern Ontario. The main branch rises near Dundalk at an elevation of 1,700 feet above sea level and falls about 1,100 feet in its course of 115 miles to Lake Huron at Southampton. Its chief tributaries are the North Branch, the Rocky Saugeen, the Beatty River, and the Teeswater River. This system maintains a strong summer flow, the average for August being 75 cubic feet per second for each hundred square miles, or five times the rate of flow in the upper Thames. The headwater areas contain much rough and rocky land, and the farms have between 30 and 35 per cent in woods and swamp. In addition, a large part of the cleared land is used as permanent pasture. The soil is loamy or gravelly and the streams run clear even after heavy summer showers. The Teeswater drains the great Greenock swamp which serves to give it a uniform summer flow.

Above Walkerton the several branches of the Saugeen flow in the old glacial spillways which are associated with the Horseshoe morainic system. These old valleys with their broad gravel terraces have a general trend southwestward towards the Greenock swamp. At Walkerton the Saugeen River changes its course abruptly to flow northward through the moraine in a valley half a mile wide and 150 feet deep. In view of the well-developed valleys leading westward, this apparent case of capture can only be explained by the presence of an original low place in the moraine. North of the moraine, there is a sand plain of considerable extent in the neighbourhood of Eden Grove, representing a delta built in Lake Warren. Here the present valley is over 75 feet in depth, and reaches a maximum of two miles in width. Within the valley are numerous sand and gravel terraces, remnants of former flood plains. Noteworthy, too, are several isolated hills interpreted as cut-off spurs or meander cores, and there are swamps in the abandoned channels. Upon the present valley floor the river meanders widely, shifting its course from time to time with costly damage to highways and farm property. Below Paisley there is another deep passage through a ridge of clay till, and about two miles below Dunblane the river enters the former lagoon which was enclosed by the heavy Algonquin beaches at Port Elgin. From here on, its valley is not more than 50 feet deep except where it swings westward to break through the old beach to enter Lake Huron at Southampton.

While not serious along the headwaters, there is considerable dissection of the valley from Walkerton to a few miles north of Paisley. This is also true of the lower courses of the North Branch and the Teeswater which join the main stream near Paisley. There is a definite soil erosion problem in this area. While it is true that much of this sloping land is used for pasture and is thus not so unstable as it would be under cultivated crops, nevertheless one sees considerable gullying and slope-washing even in pastures.

To a limited extent the Saugeen was used as a transportation route in the early days of settlement and an excursion steamship once ran from Port Elgin to Walkerton, a distance of 30 miles. More important have been the power sites which permitted the establishment of grist mills, sawmills, and woodworking plants. Many of the original power sites are still in use, a tribute to the uniform summer flow. Another not-to-be-forgotten fact is that the branches of the Saugeen are known to be among the best trout streams in Southern Ontario.

Maitland River. The Maitland River which enters Lake Huron at Goderich drains an area of 981 square miles lying south of the Saugeen and Lucknow watersheds. With its several branches ramifying through gravel-floored spillways in the till plain above the morainic belt and the lower stretch of the river deeply entrenched in a clay plain, it is a sister of the Saugeen River. The proportion of woodland, swamp, and rough stony land is less in the Maitland drainage area than in that of the Saugeen; nevertheless a strong summer flow is maintained. The diversion of the river by the great spillway in front of the Wyoming moraine gives it a trellis pattern.

The main branch of the Maitland is about 90 miles in length, rising at 1,350 feet above sea level on the plain near Harriston and falling 770 feet to Lake Huron. Its headwaters are almost intertwined with those of the Conestoga and South Saugeen in a maze of shallow spillways, and in places the dividing line between watersheds is not definite. At Harriston the Maitland valley is flat, shallow and swampy, but just west of the town it enters the Teeswater drumlin field and at Newbridge the valley is 50 feet in depth, with remnants of old gravel terraces along the sides. This deeper valley persists for several miles, having a gradient of 100 feet in the next ten miles which provides power sites at Fordwich, Gorrie, and Wroxeter. From thence to Wingham the valley is insignificant as it winds its way through the gravel trains and swamps between the drumlins. The Little Maitland and Middle Maitland which come from the south to join the main stream at Wingham also have shallow channels. In this area these streams are little influenced by the original direction of flow of glacial drainage, often reversing it. Also it seems peculiar that the Middle Maitland should continue northward instead of using the direct westward route *via* Belgrave Creek. In fact, Belgrave Creek may yet capture the Middle Maitland. It is just as odd that the headwaters of Teeswater River flow northward from swamps close to the Maitland in the Wingham area.

From Wingham the river swings southwestward through the morainic hills of Wawanosh township. There is a valley up to 100 feet deep in this area, much of it caused by the present stream rather than glacial waters. At Auburn a good exposure of local drift is provided where the river has cut a 75-foot channel without striking bedrock. That exposure exhibits stratified clay, silt, and sand except for the boulder clay in the top few feet.

It is less than ten miles as the crow flies from Auburn to Goderich, but the river is diverted by the moraines so that it takes 30 miles to traverse this distance. In particular, it fails to take advantage of the cross-channel followed by the Canadian Pacific Railway southwest of Auburn and runs as far south as Holmesville before entering the main spillway along the Wyoming moraine. It makes a hairpin turn near Holmesville and flows north in the spillway to Benmiller, thus reversing the original direction of flow. Near Benmiller the valley exhibits splendid examples of incised meanders with successions of terraces on long slip-off slopes. About two miles below Benmiller an abandoned ox-bow channel is preserved just under 700 feet above sea level. At Goderich and within a mile of Lake Huron the valley is about 150 feet in depth with steep banks cut in unconsolidated material, except for the bottom six to ten feet, where limestone is exposed.

Tributaries entering the lower stretch of the river are all small except the South Maitland, and they enter the valley through narrow V-shaped gulches. The rejuvenation of these streams has, however, proceeded for only a short distance upstream. For instance, Sharp Creek, which drains a 12-mile stretch of the spillway north of Benmiller, has a fall of about 100 feet in the last mile. Even the South Maitland has been unable to maintain its cutting at the same rate as the main stream and shows a gradient of 25 feet per mile in the lower four miles of its channel.

The headwaters of the South Maitland arise on the highly cleared clay plain north and east of Seaforth. As with the Middle Maitland and Little Maitland, the upper branches sometimes consist of municipal ditches cut through flat wet land. The South Maitland flows along the eastern face of the Seaforth moraine for 12 miles before finding a gap towards the west.

The Maitland has had a notable influence on settlement. Its mouth provides the only good harbour for Great Lakes shipping along the whole west coast of Lake Huron. Goderich is a grain port and milling point, and many freighters take shelter in the harbour when navigation is suspended in winter. The river itself has not been much used for transportation, but numerous power sites have contributed to the location and growth of such towns as Wingham, Harriston, Brussels, Fordwich, Bluevale, Blyth, Wroxeter, Gorrie, Auburn, and other smaller points.

Lucknow River. The Lucknow River drains an area of about 100 square miles, most of which is composed of moraine and spillway with considerable swamp and woodlot, and therefore it maintains a permanent flow. Its headwaters all flow in the great Lucknow spillway and it breaks through the Wyoming moraine north of Dungannon, after which it turns towards the west and entrenches itself in the plain which slopes towards Lake Huron. This part of the valley is about 75 feet deep and a quarter of a mile in width. Waterpower has been developed at Lucknow, Mafeking, Dungannon, and Port Albert.

Bayfield River. Bayfield River enters Lake Huron at Bayfield, 12 miles south of Goderich. It drains an area of 200 square miles lying between the drainage areas of the Maitland and the Ausable. The characteristics and variations in this area are similar to those of the Maitland and the Lucknow to the north, except that the moraines are smoother and less elevated and the spillways are broader. The same trellis pattern of tributaries prevails in the upper part of the area and the same entrenchment has taken place between the moraines and the lake. The lower Bayfield valley is about 100 feet deep and half a mile in width, and is sufficient of a barrier to cause a complete break in the road system. High level terraces, old ox-bows, and even isolated meander cores are found in the valley, giving it some resemblance to that of the Maitland.

Ausable River. The Ausable River takes its name from the sand dunes south of Grand Bend through which it made its way before entering Lake Huron. Its mouth is at Port Franks but no doubt the outlet has changed many times because of the shifting sand. The total area of its drainage basin is estimated to be 160 square miles, and its headwaters gather in the swamps and gravel terraces along the front of the Wyoming moraine and in the clay plain north and east of Exeter. A great number of these small headwater streams now run in dredged ditches and are completely dry during the summer months. The Ausable river follows the depression in front of the moraine as far as Arkona. There, it is entrenched in a deep and crooked valley which cuts completely through the unconsolidated drift and penetrates many feet into the bedrock, exposing the highly fossiliferous Widder beds. The valley is about 100 feet in depth. From Arkona the valley runs almost directly north, cutting through the Wyoming moraine in a gorge 125 feet deep to enter the Thedford marsh. This was a shallow bay of both Lakes Algonquin and Nipissing in which a great deal of marl was deposited and which, in part, now contains peat beds. Crossing this depression to its northern corner the river entered the sand dunes, making a hairpin turn to flow southwestward between the lines of dunes for twelve miles to Port Franks.

In 1875 a canal was cut through the dunes, draining the swamp and providing a direct route to the lake. Between the head of the cut and the mouth of Mud (Parkhill) Creek, the Ausable is dead. The water from

Mud Creek finds its way directly to the lake through the short canal at Grand Bend. Of the disconnected channel traversing the dunes between Grand Bend and Port Franks we can do no better than to quote W. Sherwood Fox, " 't'aint runnin' no more'—at least, so it seems to dull incurious eyes. The truth is that though standing still it flows. After all what are the dunes but a watershed of a kind, a mother of springs and streamlets."

Mud Creek rises east of Dashwood and follows a trough between ridges of the moraine, making the same sort of sweep to the west as the larger Ausable. For a few miles east of Parkhill it is entrenched to a depth of about 75 feet and it is this section which will serve as a reservoir when the Parkhill conservation dam is built.

The largest tributaries of the east bank, Nairn Creek and the Little Ausable, drain the depression lying east of the Seaforth moraine. Between Lucan and Clandeboye, where the latter stream is crossed by highway No. 4, it flows in a valley 75 feet deep, only a small part of which can be the result of its own cutting. In many recent summers this stream has practically dried up.

The Ausable has a much more irregular regime than the rivers to the north. Low summer flow and spring flood make up its regular programme. The damage from floods is mainly experienced in the marsh area which, now drained, is utilized for the intensive production of vegetables and field crops. The river seems not to have been utilized to any extent for power, although a dam and small powerhouse were installed for a time near Arkona. The Morrison flood-control dam is on one of the upper tributaries near Exeter.

LAKE ST. CLAIR DRAINAGE

The drainage into Lake St. Clair includes the St. Clair River and two other important drainage systems, the Sydenham and the Thames.

The St. Clair River enters the lake through a number of distributaries which are building a typical bird's-foot delta. The newer part of the delta is low and marshy, but the older portion was built into a slightly higher water level and the water table is now appreciably below the surface. Sufficient depth for navigation by large lake freighters is maintained in the main channel by dredging. It is one of the busiest waterways in the world.

Sydenham River. The Sydenham River drains most of Lambton county and the adjacent part of Middlesex, an area of about 1,000 square miles. This is mostly a clay plain of little relief in which poor drainage prevails in the interfluves. The north branch, named Bear Creek above Wilkesport, rises near Arkona between the Wyoming and Seaforth moraines. It drains the south slope of the Wyoming moraine but turns southward at Brigden across the clay plain towards Wallaceburg. In this latter section its gradient is less than one foot to the mile and the valley is nowhere incised more than 30 feet in depth. The main or east branch is about 100 miles long. It rises between the Lucan and Seaforth moraines near Ilderton at an elevation of 925 feet and it falls 350 feet to Lake St. Clair. The grades are steeper in the upper section, the last 30 miles having a fall of only 20 feet.

The valley is best developed in the Alvinston area where it is over 50 feet deep and a number of tributaries enter it in narrow V-shaped gullies. These and tributary gullies have quite intricately dissected the adjacent clay plain.

The shallow valleys and low gradient has meant that this river has not been of much use for power, and floods sometimes occur. However, it is navigable up to Wallaceburg giving this growing industrial town access to the Great Lakes.

Thames River. The Thames River drains an area of 2,200 square miles and is second only to the Grand among the drainage systems of Southwestern Ontario. Its total length, from the source of the North Branch near Brodhagen to Lake St. Clair, is about 125 miles.

For purposes of discussion the river may be divided at London. Below this point it traverses fairly level lake plains in a small valley of its own making, and its gradient is low, the total fall from London to its mouth being about 175 feet or less than two feet per mile. Above London two branches drain over 1,000 square miles of till plains and moraines. The North Branch, which is the longest, falls about 400 feet at the rate of about 10 feet per mile and the gradient of the South Branch is almost as great. The city of London, located at the confluence, often experiences flood conditions.

The North Branch of the Thames originates from the union of several small creeks draining large flat areas in Logan township and at Mitchell it assumes the characteristics of a river in a well-defined but shallow valley. Below Mitchell the valley deepens rapidly to a depth of more than 75 feet at St. Mary's and over 100 feet at Porters Mills. Then the depth and gradient of the valley diminish from this point to London, a distance of 15 miles.

Between Mitchell and St. Mary's the North Thames receives several fair-sized tributaries from the east, but only one, Flat Creek, from the west. The Avon River, which flows through the city of Stratford, arises on the edge of the Easthope moraine in two swampy valleys or depressions which connect with the Nith drainage system to the east. In 25 miles it falls about 175 feet. Its lower valley is much deeper and narrower than the broad swampy valleys in which it rises. Trout Creek, from the east, joins the North Thames at St. Mary's, flowing in a valley much too large for it. This is the work of a glacial stream, described elsewhere, formerly connecting the drainage of the upper Grand River with the Thames. It contains a long swamp which now serves as the source of both Trout Creek and the South Branch of the Thames. At St. Mary's the course of the river is deflected westward for about five miles until, joined by Fish Creek from the north, it breaks through to the south of the moraine which controlled it. The North Thames has no further important tributaries until joined by the Medway also from the north, on the outskirts of London.

The south branch of the Thames rises in the long swamp west of Tavistock and flows east as though to join the Grand River system, then swings south and eventually finds the spillway through Woodstock and Ingersoll. This valley is about a mile in width and over one hundred feet in depth, being cut to bedrock, thus exposing the limestones which are being worked in the Beachville quarries. Cedar Creek, at Woodstock, is the only important tributary entering this section of the river. It drains an area of inter-morainal swamp to the south of the city and at times creates minor flood conditions. Near the small village of Putnam, about 13 miles below Woodstock, the river is joined by the Middle Thames on its north bank and by Reynolds Creek from the south.

The Middle Thames is also a misfit in a deep spillway which at one time carried the upper Grand River drainage, but it is not quite so large as its southern counterpart. The upper portion of this drainage, known as Mud Creek, is, as its name implies, a very sluggish stream. At Embro it is joined by a somewhat more swiftly flowing branch from the north. The deeply cut spillway ends at Thamesford to be replaced by broad gravel terraces extending towards London. It appears that the Middle Thames may at first have held a course westward as marked by a shallow channel, but it has been diverted towards the south.

Reynolds Creek follows a very circuitous route in the spillway fronting the St. Thomas moraine and has performed very little recent dissection.

From Putnam to Lake St. Clair the valley of the Thames may be considered to be of its own making with exceptions of the upper, broader portion of the gorge between the moraines near Hyde Park, which undoubtedly carried glacial drainage. The course of the river in Caradoc and Delaware townships and even as far as the eastern limits of Kent county is particularly noteworthy. Here, following the retreat of the Lake Warren waters, the newly constituted river wandered widely and has since entrenched itself to an average depth of about 75 feet below the old lake plain. The valley, in places, is more than a mile in width and exhibits many excellent examples of slip-off slopes, undercut bluffs, and abandoned channels. Here, too, the tributary streams are cutting down to the base level of the master stream, creating steep-sided narrow gulches along their lower courses. Some of this difficult terrain has been set aside in the Muncey Indian Reserve.

Downstream, the valley is smaller and the river is sluggish, especially below Thamesville. Periodic overflows in Kent county are the bane of farms and villages near the river. The alluvial clay from these floods has a weak reddish-brown colour in contrast to the greyish-brown clay of the region. It is given distinction as the Thames clay loam on the soil map of Kent county. Near Lake St. Clair the level of the river is above that of the adjoining territory which is protected by a system of levees. The Thames is navigable by barges, pleasure cruisers, and small steamships as far as Chatham, a distance of 16 miles from Lake St. Clair.

Flood conditions are most critical in the city of London, which has been partly built upon land which, physiographically, belongs to the river. The total rainfall of the region is nearly 38 inches per year. Gauging stations on the north and south branches of the Thames record a run-off of 12.2 and 12.7 inches, about six inches less than that of the Beaver and the Saugeen rivers. The flow, however, is highly irregular. In midsummer 1953 the North Branch was reduced to one cubic foot per second while in flood times it may carry over 20,000 c.f.s. A flood control dam was built at Fanshawe in 1952 capable of holding back 30,000 acre feet of water. The regulatory work of this dam is now supplemented by smaller ones at Mitchell, Stratford, and St. Mary's. A dam on the south branch is soon to be built at Woodstock. It shows similar, if not so great, variations. Floods occur on the Saugeen River, also, but it is the scanty summer flow of the Thames which accounts for the difference in run-off. Summer rainfall, for the most part, is lost by evaporation rather than run-off in this river basin.

LAKE ERIE DRAINAGE

Although the north shore of Lake Erie is over 250 miles long, not counting Long Point, it receives no large rivers except the Grand. The clay plain of Essex county drains almost entirely towards Lake St. Clair or the Detroit River, and only short small creeks enter Lake Erie. Long stretches of this shore, particularly in the great bight between Rondeau and Long Point, consist of high bluffs cut in unconsolidated drift. Near the shore, as along the Lake Huron shore, many small streams are cutting deep, steep-sided gullies. Since this shoreline is receding, these gullies are constantly growing headward under the influence of new gradients. Larger creeks such as Kettle Creek, Catfish Creek, Otter Creek, and Big Creek extend right beyond. Several brooks have accomplished some remarkable dissection between Turkey Point and Port Dover. The bluffs are low between Port Dover and the Grand River and the overburden on the limestone is usually less than 25 feet deep; consequently the creeks which drain the clay plain to the north have shallow valleys. East of Port Maitland, as in Essex county, the height of land on the adjacent plain is near the Lake Erie shore and southward drainage may be disregarded.

Kettle Creek. Kettle Creek enters Lake Erie at Port Stanley where the bluff is about 125 feet high; consequently the valley is deep and the bordering sand plain cut by gullies. It is quite obvious that some branches have been separated from the truncated river system by the encroaching shoreline which has here receded at least ten miles from its original position. Although very steep-sided, the valley is flat-floored and the river meanders widely as far upstream as St. Thomas, about ten miles by river. Here the valley is still over 75 feet in depth. Several headwater streams converge at this point. Some of the drainage originates in the spillways between the morainic ridges. This may be considered as captured, since the whole development is recent and consequent upon the back slope of the moraine. A dam north of the city serves to impound the reservoir supplying St. Thomas. Port Stanley at the mouth of the river, is a summer resort, but also has important harbour installations. These make it a shipping port for St. Thomas and London.

Catfish Creek. The physical development of this system is quite similar to that of Kettle Creek. Its headwaters take rise in the long troughs between the low till ridges above the shoreline of glacial Lake Whittlesey, and after uniting near New Sarum the waters have cut a deep, steep-sided but flat-floored valley to enter Lake Erie at Port Bruce. Like Kettle Creek, Otter Creek, and Big Creek, it serves to drain the tobacco lands through which it takes its course and also provides water for irrigation.

Big Otter Creek. From its source near New Durham at an elevation of 850 feet, Big Otter Creek falls nearly 280 feet in a distance of 48 miles to Lake Erie at Port Burwell. It thus has an average grade of about six feet

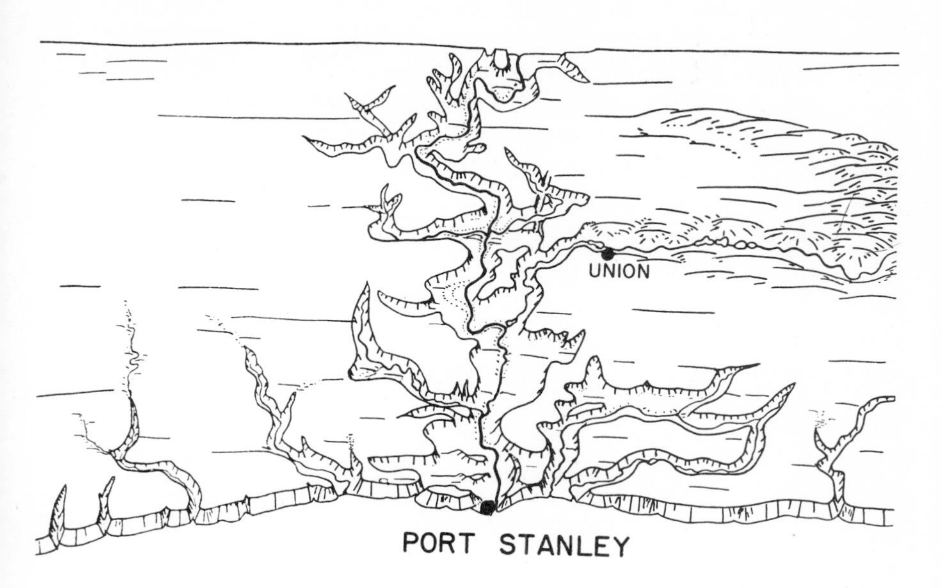

52. *Dissection near Port Stanley*

per mile. For the greater part of its course it flows in a steep-sided, rather narrow valley and nowhere has such a wide, flat valley floor as those found in the two previously described systems. In its deepest portion, between Tillsonburg and Bayham, the valley is over 125 feet deep and provides splendid sections of the local drift. Little Otter Creek, a tributary on the east, also flows in a narrow valley which for several miles is over 100 feet deep. Power sites along Big Otter Creek gave rise to a number of early settlements including Otterville, Tillsonburg, Bayham, and Vienna. The mouth of the stream provided the harbour of Port Burwell, which, having been improved and provided with rail facilities, serves as a port for the manufacturing towns of Tillsonburg and Ingersoll.

Big Creek. Big Creek rises in the Ingersoll moraine a few miles southeast of Woodstock at an elevation of 1,000 feet above sea level. In a total length of 55 miles it falls over 400 feet to Lake Erie. At first its course is easterly towards the wide plains of deltaic alluvium in Burford township where the Grand River at one time entered Lake Whittlesey. In a distance of 12 miles it drops 175 feet but has cut no significant valley because of its small volume. The stream wanders for a few miles southward over the old alluvium before it begins to entrench itself near Teeterville. Nine miles farther downstream at Delhi the valley is already 75 feet in depth, a depth which it maintains until within a few miles of its outlet into Inner Bay. The Lehman dam near Delhi utilizes a section of this valley for water storage, the main purpose of which is to increase the water available for irrigation of tobacco. For the most part the valley floor is narrow with only a few wide areas of flood plain as, for instance, near Carholme. There are also a number of remnants of alluvial terraces, and the many tributaries that join Big Creek have cut V-shaped gullies in the Norfolk sand plain. For a few miles near its mouth, however, Big Creek is without grade and its margins are marshy; in the last 25 miles of its course the average grade is about five feet per mile. In places, Big Creek and some of its tributaries have been utilized to run grist mills. The only settlement of any size on the banks of the creek is Delhi, now a processing centre for tobacco.

Grand River. The Grand River drains an area of 2,600 square miles on the dip-slope of the Niagara cuesta, the largest catchment basin in southwestern Ontario. The main stream rises northeast of Dundalk at about 1,725 feet above sea level and runs a course of 180 miles to Lake Erie at Port Maitland. The upper Grand and its branches, Willow Brook, Boyne Creek, and many unnamed streams serve to drain nearly all of the swampy upland around Dundalk and Grand Valley. Its tributaries, the Conestoga and Nith rivers drain the extension of that plain west of Arthur and around Milverton. The upper reaches of these streams intermingle with those of the Saugeen, Maitland, and Thames, the divides between them being indistinct, often consisting of a sprawling swamp from which drainage goes in two directions. Entering the main stream from the east at Preston, the Speed River drains the drumlins and gravel trains west of the Paris moraine. Between Paris and Brantford the Grand breaks through the Paris and Galt moraines. Below Brantford, only a narrow belt of the clay plain drains into the Grand and it is confined on the south by the Onondaga escarpment which forces the river eastward to Dunnville where a sag permits southward passage to Lake Erie.

The Grand is like the Thames in that it may be divided into an upper part where the river and its branches mostly flow in spillways, previously formed in the till plains, and a lower part where the river has made its own channel across a lake plain. The lower section is more nearly level than the upper; above Brantford the average grade of the main stream is 8.5 feet per mile while it is about two feet per mile below that point.

To measure the volume of flow in the Grand River, gauging stations have been maintained at Galt and Brantford for many years. The mean of

53. *Luther Dam (Hunting Survey Corporation photo)*

twenty-seven years' measurements at Galt is 1,210 c.f.s., the equivalent of 11 inches of rainfall. The mean flow for the flood months (March and April) is 3,400 c.f.s. For the driest month, August, it is 257 c.f.s., but in August 1936 it was only 47 c.f.s., with an absolute low of 26 c.f.s. The maximum flood was 46,280 c.f.s., on March 21, 1948.

In the extreme upper reaches there has been little valley cutting. Grades are low and deepening of the main valley is held up by bedrock. Limestone is exposed at Grand Valley about 75 feet below the level of the adjacent plain. Drainage is poorly established on the plain above Grand Valley and swamps are prevalent, the largest of which is the Luther bog that covers approximately 10,000 acres. It now serves as a reservoir behind a dam built in 1952.

From Grand Valley to Elora, 20 miles downstream, the bed of the river is controlled by the bedrock and the valley varies from 60 to 100 feet in depth. It is flanked by remnants of gravel terraces indicating its earlier history as a glacial spillway.

In this stretch of the valley, where it cuts through the Orangeville moraine, the Shand Dam was built in 1939. Constructed at a cost of over two million dollars, the dam impounds 12,483,000 gallons, or 46,000 acre feet of water, sufficient to provide a minimum of 200 c.f.s. summer flow. It creates a lake six miles long called Lake Belwood which in itself is a recreational asset.

In the vicinity of Elora there occurs a definite break in the slope of the bedrock and both the Grand and Irvine creeks have cut deep gorges in the limestone. Known as the Elora Rocks, their rugged beauty is a noted scenic attraction which has been made the site of a provincial park. The gorge in the rock ends a short distance below Elora, giving place to a wide winding valley in alluvial gravels or, occasionally, in till. Another conservation dam is to be built above West Montrose. From there to Paris the valley is 75 to 100 feet deep, and between Paris and Brantford the gap in the Paris moraine is over 150 feet in depth. The alluvial plains in which the river meanders are several miles wide at the maximum. It would be hard to find the equal of the volume of gravel and sand in this section of the Grand valley.

The Conestoga rises on the plain northwest of Arthur as small streams following shallow spillways. Several of these streams join near Rosebank to form the east branch in a valley already 75 feet in depth. Downstream it increases in size until at Hollen it is entrenched 150 feet deep in the till plain. The west branch of the river originates in a similar set of old spillways and has a valley nearly 100 feet deep above its confluence at Stirton.

Two miles south of Hollen the Conestoga valley is joined by that of Spring Creek from the west. The latter is a remarkably big valley for such a small creek. At the junction with Spring Creek the Conestoga swings sharply to the southeast in a rugged valley which is over 100 feet deep at Glen Allan. The capacity of both these valleys amounting to 45,000 acre feet is used by the Conestoga conservation dam above Glen Allan. Below

54. *Grand River below the Shand Dam*

55. *The Elora gorge on Irvine Creek*

Glen Allan the valley becomes shallower, except east of Hawksville where it crosses a moraine. The river meanders over a flood plain about 500 yards wide and it falls 200 feet in the 25 miles above Conestoga.

The Conestoga drains an area which receives more rain and snow than most of Southern Ontario. Its run-off, 334 c.f.s., is the equivalent of 15 inches of rainfall. However, its regime is highly unreliable, possibly because of the clay soil, a high proportion of which is cleared. In March and April the average flow is around 1,500 c.f.s., but it has fallen to 2 c.f.s., in dry summers.

The Nith River and its branch, Smith Creek, originate in small intermittent streams or ditches that drain the clay lands of Mornington township east of the Milverton moraine. Where they unite near Kingwood the valleys are 50 feet in depth, and at Nithburg the narrow valley reaches its greatest depth of over 100 feet. Below Lisbon the valley is shallower and more open but it is still cut in clay till. The river meanders widely and in places has two or three channels in operation. About two miles below Plattsville it meets one of the broad gravelly spillways by which glacial drainage escaped from the grand basin into the Thames. The Nith now flows up this valley as far as Ayr then turns down another spillway to the southwest. Finally near Wolverton the river swings to the southeast again towards the Grand at Paris. This lower stretch of the valley is 50 to 100 feet deep and is variable in width due to the wide meanders.

The Nith drains over 430 square miles and removes the equivalent of 12.3 inches of rainfall, which is 2.7 inches less than the run-off of the Conestoga basin. With an average volume of 388 c.f.s., a flow of about 1,400 c.f.s., during flood months, and minimum flows of slightly under 100 c.f.s., it is more reliable than the Conestoga River. The smaller amount of run-off is probably the result of less snowfall, and the better summer flow due to the more extensive area of porous soils. A dam to be built at Ayr will serve further to regulate the flow downstream.

The Speed River is remarkable in that throughout its whole length and that of its tributaries as well it flows in old spillways. It rises near Orton and drains the main part of the Guelph drumlin field. Many of its feeders come from the north, towards the Grand River; in fact if it were not for auxiliary earthworks Lake Belwood would overflow into the Speed drainage. South of Fergus, the Speed cuts across three fair-sized spillways and beheads Swan Creek, Cox Creek, and Hopewell Creek which flow down these valleys into the Grand. The fact that the Speed River has a fall of nearly 500 feet in 25 miles explains its ability to cut across drainage lines of more gentle gradient. It is surprising that such a rapid stream did not make a deeper channel.

The Eramosa River is the main branch of the Speed River, following the master spillway in front of the Paris moraine above Guelph. That swampy channel shows up clearly on the Guelph map sheet. It is seldom less than half a mile wide and is bordered by gravel terraces except where it encroaches upon drumlins. This tributary has a lower gradient than the Speed River, the fall from Brisbane to Guelph being 275 feet or an average of 12 feet per mile. Bedrock is exposed occasionally and the modern stream has accomplished very little cutting. The Eramosa receives very few tributaries, the most important being Blue Springs Creek which extends from Eden Mills towards Acton in another well-developed spillway through the Paris moraine.

The city of Guelph is situated at the confluence of the Eramosa and Speed rivers and that part which is built on lower terraces is subject to flooding. Below Hespeler the river has made a deep channel in the gravel terraces.

The total drainage basin of the Speed River covers over 250 square miles. Its mean flow is about 225 c.f.s. or the equivalent of 12.5 inches of rain, which is the same as for the Nith River. Flood months run up to 1,000 c.f.s., while minimum flows vary from 65 to 100 c.f.s. This river thus has the most reliable flow of all the larger tributaries of the Grand, no doubt because of the great inter-drumlin gravel beds and swampy valleys. Three dams are planned at Everton, Guelph, and Hespeler to help control the flow.

Below Brantford the valley of the Grand is cut in the silt and clay of Lake Warren and the underlying boulder clay. Around Brantford, soft silts have been highly dissected near the Grand and along Fairchild Creek. Prevention of further erosion is a problem. In the last 27 miles of its course the Grand has a fall of 27 feet, a gradient of exactly one foot per mile, and very little valley cutting has occurred. One-third of this fall is accounted for by the dam at Dunnville which was built to supply water to the feeder for the old Welland canal.

The area drained by the Grand River now has a population of about 500,000 persons and includes the cities of Brantford, Galt, Guelph, Kitchener, and Waterloo. The towns of Elmira, Fergus, Elora, Hespeler, Preston, Paris, and Dunnville and many villages are also situated on the banks of the river or its tributaries. In the early days of settlement the stream provided power, and power sites formed the nuclei of settlements. It was also considered navigable as far upstream as Brantford, while its upper reaches were used for log-driving. Now although this is one of the foremost manufacturing areas of Ontario, local waterpower is of little importance. Nevertheless, the river is very important from the standpoint of water supply and sewage disposal and for recreation. If the Grand is to continue to serve the rapidly increasing population of its valley, further measures to increase the summer flow will be necessary.

LAKE ONTARIO DRAINAGE

The drainage tributary to Lake Ontario includes the Niagara River and a number of smaller rivers. One or two such as the Trent and the Moira have fairly large drainage basins, but the others drain quite small areas.

Some of them, however, have an importance, both physiographically and in relation to the settlements on their banks, quite out of keeping with mere size.

Niagara River. The Niagara is a short river joining Lake Erie to Lake Ontario across the Niagara cuesta. In a distance of 32 miles its waters drop 326 feet from the level of Lake Erie (572 feet) to the level of Lake Ontario (246 feet). Approximately half of this drop occurs at the Niagara Falls, which, with the steep-sided gorge below is among the scenic wonders of the world.

The river starts at Fort Erie although it is worth noting that the difference in elevation between this point and Lowbanks is very small. Had the river taken the latter passage it would have gone down the Welland River to the present channel just above the falls.

The upper ten miles of the Niagara is a broad quiet river in a valley only about 25 feet in depth. It is split into two arms which enclose Grand Island. The western arm is also divided, enclosing the much smaller Navy Island. Below these islands the river is over one and a half miles in width but narrows to about one mile at the rapids above the falls. Here the river is again divided by Goat Island so that there are two falls, the American, which is about 1,000 feet in width and drops 165 feet, and the Canadian or Horseshoe Falls which drops 158 feet. The rim of the latter is about 2,600 feet in length and in the shape of a great curve bowed upstream because of the rapid erosion in the centre of the falls, amounting to about three and a half feet per annum. Over the falls, the river plunges into a gorge between 300 and 400 yards in width. This gorge, of varying width, extends northwards for six and one-half miles to the edge of the escarpment. In it the river flows tumultuously, especially in the narrower part above the whirlpool. Upon entering the Ontario lowland plain at Queenston, it flows quietly towards Lake Ontario in a channel half a mile in width.

It was thought, at one time, that the Niagara gorge was entirely the result of postglacial cutting and was thus a convenient measure of recent time. It is now known that part of it is the result of the re-excavation of a previously eroded gorge, a still-buried part of which extends from the whirlpool to St. David's. The rapids above the falls are thought also to be caused by the uncovering of the slope of this old valley. Moreover, it is now known that the volume of water in the Niagara River has varied greatly from time to time, hence its rate of cutting has been far from uniform. Regardless of its history, however, it is a splendid example of a river gorge produced by the retreat of a waterfall. The steep sides and

56. *Niagara Falls*

57. *Niagara gorge*

great depth are produced by the undermining of the soft shale beneath the hard dolomite of the cap-rock. The effects of early valley cutting, before the falls had migrated so far upstream, may be seen in the flat area of Niagara Falls Park and in similar remnants of rock terrace farther downstream. Records, kept for many years, show that the flow of the Niagara River has varied from 99,000 c.f.s., to 324,000 c.f.s. The yearly mean is about 195,500 c.f.s. representing an average depth of 10.5 inches of water from an area of 255,000 square miles, 94,000 square miles of which is open lake surface. Month-to-month variations are narrow, giving a reliability which makes the river especially valuable for power production. More than 2,700,000 h.p. of electrical energy is produced in the province of Ontario and an approximately equal amount in New York State.

Welland River. The Welland River drains part of the backslope of the Niagara cuesta east of Hamilton. It rises on the side of the sandy moraine near Ancaster at an elevation of about 800 feet above sea level, and takes a meandering course across the clay plain to the Niagara River at Chippawa. It falls 240 feet in 72 miles, an average gradient of a little over three feet per mile. About 200 feet of the total fall occurs in the first 15 miles, but there the stream is small and only a shallow channel is made in the southern flank of the Mount Hope moraine. Thus, east of Sinclairville the gradient is less than one foot to the mile, making the river little more than a sluggish ditch.

Actually the Welland River no longer runs into the Niagara River, but out of it, for the last four miles serves as the intake of the Chippawa-Queenston power canal, and as a final indignity has been forced to crawl beneath the Welland canal through a siphon in the city of Welland. Upstream from Welland the river still retains its original character. Welland is to be considered more as a child of the canal than of the river.

Scarp-face drainage. A number of small streams descend the Niagara escarpment through narrow notches and find their way directly to Lake Ontario. For the most part they bear names indicative of the distance from the mouth of the Niagara River. Among these are Four Mile Creek, Twelve Mile Creek, Twenty Mile Creek, Stoney Creek, and Red Hill Creek. Above the escarpment these streams all drain depressions between the clay moraines which parallel the Vinemount moraine. Mostly, they serve to drain off surface water and are apt to dry up in summer. They are sluggish streams and have cut only shallow valleys. In their descent of the escarpment they have their moment of glory, cutting deep V-shaped gorges and in some cases dropping over rock ledges as lovely little waterfalls. On reaching the base of the escarpment they proceed quietly and directly to Lake Ontario. Because of rising levels, the lake has submerged these river mouths. Ponded valleys almost closed off by gravel beaches at the shore are common to most of these streams.

Four Mile Creek is a rather insignificant stream which has its source in the mouth of the old St. David's gorge and crosses the Iroquois lake plain directly to Lake Ontario. Its valley is entirely recent and shallow.

Twelve Mile Creek rises on the north flank of the moraine at Fonthill. Here apparently there is a large drift-filled re-entrant in the Niagara escarpment. Twelve Mile Creek and its numerous branches have carved this into one of the most thoroughly dissected areas in Southern Ontario. At the edge of the escarpment is located Power Glen in which were built the old De Cew Falls power plant as well as the new development which makes use of the extra water added to the Great Lakes drainage by the Ogoki diversion in Northwestern Ontario. Turbine installations here total over 200,000 horsepower.

Above the escarpment in its original state, Twelve Mile Creek meandered in a valley about a quarter of a mile in width 75 feet in depth, south of the Vinemount moraine. About four miles from the Lake it was joined by a similar branch from the east. The lower part of the valley is drowned, forming a lagoon known as Martindale pond. Both branches as well as the trunk stream have been greatly altered by the works of man. The trunk and the east branch were used in the construction of the old Welland Canal. The west branch has now been straightened and protected by weirs in order to minimize erosion from the greatly increased flow of water from the power plants.

Twenty Mile Creek, draining an area of 113 square miles, descends the escarpment in a deep, narrow, V-shaped gorge in which actual waterfalls still exist. Its upper reaches resemble those of the Welland River since it, also, rises on the flank of the moraine at Ancaster and follows a parallel easterly course until captured by the steeper gradient of the outlet through the escarpment. The lower part of this valley is drowned but cut off from Lake Ontario by a sand bar, forming Jordan Harbour.

Forty Mile Creek issues from the escarpment at Grimsby but is of smaller volume, having captured a much smaller headwater area just in front of the Vinemount moraine at the top of the escarpment. It has no large harbour at its mouth.

Red Hill Creek flows in a rather large preglacial notch in the escarpment at the head of which is Albion Falls. In flood times this is a scenic feature worth visiting.

58. *Webster's Falls on Spencer Creek at Dundas, where it spills over the Niagara escarpment*

Though the streams of the Niagara peninsula are small they have influenced settlement considerably. St. Catharines grew up where the old Indian trail from Niagara crossed Twelve Mile Creek on a rude log bridge. Later, when mills were established and the various Welland canals built, the place became a thriving manufacturing and commercial centre. Now, of course, the early significance of the river has been lost. Port Dalhousie owes its site to the presence of the drowned valley indenting the lake shore. Grimsby, is another small town whose rise was occasioned by the crossing of a small stream (Forty Mile Creek) by an early colonization road. Smithville and Saint Ann above the escarpment and Jordan at its foot are related to the early power sites on Twenty Mile Creek as is the village of Stoney Creek to its own stream.

Dundas valley. The preglacial Dundas valley, now largely drift-filled, contains no important streams but does have a number of small ones which have deeply dissected the drift. Spencer's Creek finds its source in a number of small streams which drain the back slope of the Galt moraine into the Beverly swamp. From this issues a small permanent stream which eventually finds its way over the escarpment into the Dundas valley. Here, Webster's Falls, a miniature Niagara, forms the nucleus of a splended little park above the town of Dundas. A companion feature, Tew's Falls, is found about a mile to the east. Two small branches approach Dundas down the Dundas valley. Due to steep grades they and their short laterals have accomplished an amount of erosion out of keeping with their small size and the drift within this valley is badly gullied.

A large number of small creeks seam the steep face of the escarpment to the north of Hamilton Harbour, cutting deeply into the red shale which is but thinly covered by till. One of these, Waterdown Creek, flows in a preglacial notch deeply cut into the cap-rock. It serves to drain portions of the glacial spillways above the escarpment as well as becoming the master stream of the scarp face, finally emptying into the head of Hamilton Harbour. Waterdown arose at the point where this stream was crossed by the Dundas Road. Further importance attaches to this valley from the fact that it is climbed by the Guelph line of the Canadian Pacific Railway.

THE NORTH SHORE OF LAKE ONTARIO

Between Hamilton Harbour and the Bay of Quinte, the area which drains to Lake Ontario is bounded by the Niagara cuesta and the interlobate moraine, both of which are heights of land setting definite watershed boundaries. Consequently there are many short rivers in this area including Bronte Creek, Oakville Creek, Credit River, Etobicoke Creek, Humber River, Don River, Highland Creek, Rouge River, Duffin's Creek, Lynde Creek, Oshawa Creek, Bowmanville Creek, Wilmot Creek, Ganaraska River, and Cobourg Creek. West of the Humber, the source area is the Niagara cuesta, east of the Humber it is the interlobate moraine. The Humber, itself, derives water from both sources. Some streams, however, do not tap the source areas but arise from the drainage of the till plain. Such is the nature, for instance of the Etobicoke, the Don, and Highland Creek.

Coventry (29) has discussed the variations in flow, drought, and flood conditions in the streams between Toronto and Hamilton. Practically all of the streams having a permanent flow are fed from the area above the escarpment. Notable exceptions are the several spring-fed brooks in the sandy plain west of Port Credit—a benefit to real estate in that attractive area. The underlying rock in this lowland area consists of easily eroded shales and these rivers have cut steep-sided, narrow valleys to a depth of 75 to 100 feet. As in the area south of Lake Ontario, these rivers are drowned at their mouths forming harbours at Bronte, Oakville, and Port Credit. They are all utilized to the full by pleasure craft.

Credit River. The Credit River is the largest of the group, with a drainage area of 328 square miles. Its total length is 58 miles from northeast of Orangeville to Port Credit. It rises on the Niagara cuesta, giving drainage to borderlands between Caledon and Orangeville as does the Nottawasaga River farther north. Thus it rises in a hilly area of moraines, gravel terraces, and swamps in which forest is still fairly abundant. The several branches mostly follow long swampy valleys in the Hillsburg and Orangeville areas where they have failed to cut deep channels.

The Credit leaves the plateau through a deep notch in the escarpment at Cataract while the west branch descends through a similar gulch at Belfountain. Both contain waterfalls and rapids. Cataract for a number of years was the site of a small electric power plant, while Belfountain was developed as a pretty little private park. The streams unite at the Forks of the Credit, a picturesque spot at which the escarpment wall towers 400 feet above the valley floor. Downstream the river has built a wide alluvial plain traversed by two distributaries which combine again before reaching Inglewood. Here the river is joined by a branch from the east that occupies a wide gravelly valley at the foot of the Caledon mountain.

59. *Small river valley, Oakville*

For about ten miles to the vicinity of Georgetown, the river flows in a narrow valley between the escarpment and the till plain. Terrace remnants mark this as the route of former ice-border drainage, but the modern valley has been cut deeply into the till and underlying shale. Numerous, small, rapid feeders enter this portion of the Credit from the face of the escarpment. They are actively eroding the shaley slope and in a few places gullying has proceeded so far as to produce veritable badlands.

At Glen Williams the river swings southeastward away from the scarpfoot and follows the slope of the till plain towards Lake Ontario with a gradient of more than 20 feet per mile. At Norval it is joined by another deeply entrenched branch which drains a section of the escarpment behind Georgetown and Stewarton. From this point to Port Credit the valley is cut in the boulder clay and often into the underlying shale. Near Huttonville and Erindale it is over 75 feet deep, while a wide alluvial terrace occurs near Churchville and Meadowvville. At Erindale the level of glacial Lake Iroquois is reached. The river cuts through a well-developed terrace into the shale and then is deflected northward by a great barrier beach.

The last mile of the river before entering Lake Ontario is drowned and marshy. The quiet section between the Queen Elizabeth Way and the Canadian National Railway serves for canoeing in summer and provides a huge open air skating rink when frozen over in the winter. Between the railway and Highway No. 2, the marsh has been filled in and serves as a parking lot. The waves of the lake continue to attempt the building of a bar across the river mouth but have been defeated by continued dredging. Port Credit is now maintained as one of the most efficient and thoroughly used small-boat harbours on the shores of Lake Ontario. Large lake vessels do not use the mouth of the river. Oil tankers anchor offshore and serve the refineries in the western part of the town through underwater pipe-lines; while just to the east of the old harbour mouth, a large pier has been built out into the lake for the handling of general cargo. Port Credit's annual record of freight handled in recent years is in excess of 1,000,000 registered tons.

Despite its small size the geographic significance has been considerable throughout the history of settlement. Erindale at one time had a mill and, indeed for a short time, a small hydroelectric generating station. Streetsville still has flour mills operated by water-power, and there are other mill sites upstream. The river banks from Port Credit to Streetsville have become favoured residential suburbs of Toronto. Erindale has been chosen by the University of Toronto as the location of one of its off-campus colleges. At Streetsville, Eldorado, and Huttonville, parks have been developed, providing recreational facilities attracting people from Toronto and from the surrounding areas. Especially because of its recreational value to this suburban area, the Credit River has received careful study with a view to maintaining a good flow of clean water. To this end control dams have been planned for sites near Orangeville, Cataract, Belfountain, Georgetown, and Silver Creek.

Etobicoke Creek. It would unduly prolong this section to discuss each small river in the same detail as the Credit; nevertheless certain points and problems concerning other streams must be mentioned.

Etobicoke Creek, which has a drainage area of only 74 square miles, lies entirely in the lowland. Yet, because portions of the town of Brampton and the village of Long Branch were built upon land which belongs, physiographically, to the river, a single flood has been known to result in damage to the amount of more than half a million dollars. Two major controls have been built by the river authority. In Brampton the stream has been provided with a new concrete-lined channel, diverting the water which formerly menaced the downtown area, part of the old channel being filled and added to the park. In Long Branch the channel has been improved and given a clear exit to Lake Ontario, while the adjoining land has been taken into the Metropolitan park system as Marie Curtis Park.

The Humber and the Don. Two small rivers of great interest to Metropolitan Toronto are the Humber and the Don. Both of them have helped to modify its physical landscape and to influence its growth and development throughout the history of settlement.

The Humber River drains an area of 220 square miles. Its main course is 63 miles in length from the top of the Niagara escarpment near Mono Mills to the shore of Lake Ontario at the western outskirts of Toronto. In general it parallels the Credit, but with a somewhat more tortuous course. It has several tributaries; chief of which are the west branch which drains a part of the Peel Plain, and the east branch which rises in the interlobate moraine. In the vicinities of Bolton, Kleinburg, and Woodbridge, the Humber valley is more than 100 feet in depth. The headwaters, also have rather intricately dissected the morainic area. Below Thistletown the river has cut through the unconsolidated Pleistocene overburden into the underlying Palaeozoic shales. The vertical valley walls in these stratified rocks are well displayed near Weston.

The Humber was in the main path of Hurricane Hazel on October 14–15, 1954, when up to 8 inches of rain fell, most of it within 16 hours. The resulting flood caused extensive property damage and great loss of life. At Weston a row of houses built on a low terrace near the river was swept downstream, carrying some of their occupants to their deaths, an awful price to pay for having built upon land belonging to the river.

The Don River system is smaller, draining an area of 143 square miles in the Peel Plain and the south slopes but without reaching any part of the interlobate moraine. It consists of two main branches, the east Don and the west Don which effect a junction just upstream from the point where the waters have cut through the old Iroquois beach to seek the level of Lake Ontario. The valleys of the Don, in some places a hundred feet in depth, have considerably hindered the expansion of Toronto to the northeast. Like that of the Humber, the lower Don has been drowned by the rising waters of Lake Ontario.

Both the Don and the Humber systems are included in the Metropolitan Toronto Conservation authority which has planned a series of recreation parks water control dams throughout the area. The first one, now completed, is located at Clairville, on the west branch of the Humber. Other locations are at Ebenezer, Pine Grove (Boyd), Nashville, and Bolton.

Between Toronto and Trenton, many small streams descend the southern slope of the Oak Ridges moraine, most of them originating in springs at an elevation of about 900 feet. There are practically no streams on the sandy crest of the moraine, the rain percolating into the porous beds to reappear again as springs on the shoulder of the ridge. Until they reach the Iroquois shoreline these are rapid streams with high gradients, often 30 feet or more per mile. Consequently they have cut sharp gullies in the till plain. Towards Lake Ontario these gradients lessen and in time of freshet real floods accumulate in the lower valleys.

Ganaraska River. The Ganaraska is the best known of these streams because of the flood damage which it has caused in Port Hope at its mouth, and because of the pioneer study undertaken by A. H. Richardson (115). It drains an area of over 100 square miles. Although its main branch is only 22 miles in length, its average gradient is 30 feet per mile and it has a large number of tributaries. Floods occur every year, and during the past hundred years no less than twenty disastrous floods have inundated Port Hope. That this is in part the result of the unwisdom of the townspeople who have occupied land belonging to the river, is certainly true. Structures encroach even upon the channel as well as the whole width of the flood plain. Nevertheless, proper reforestation and rehabilitation of the watershed would help greatly. It is recommended in the above-mentioned report that an area of 20,000 acres be set aside as the Ganaraska Forest. Much reforestation has been done and a dam for flood control has been built at Garden Hill.

Trent River. Several fair-sized rivers drain into the Bay of Quinte, including the Trent which is the largest of all the rivers in Southern Ontario. All of them cross the boundaries of the Canadian Shield, having their headwaters in the region of Precambrian rocks and their lower courses on Palaeozoic limestones which, in this region, have very little overburden.

The Trent drainage area comprises 4,790 square miles. From the rocky forested hills of the Shield a number of rivers run in a general southwesterly region towards the Kawartha Lakes which serve as great reservoirs for the Trent system.

The Kawartha Lakes lie along the juncture of the Palaeozoic limestones and the crystalline rocks of the Shield. They occupy portions of preglacial valleys which formerly drained towards the southwest and are probably somewhat overdeepened by ice action (151) but which have been blocked in the south by morainal debris. These lakes drain from one to the other over the lowest saddles in the rock between them, thus accounting for their zig-zag pattern. This was also for a long time the outlet of various stages of Lake Algonquin, in consequence of which the connecting streams flow in wide rock-floored valleys swept clean by the much greater flow of water. The outlet was not occupied long enough to permit complete grading of the course and a number of waterfalls still remain, as at Fenelon Falls and Burleigh Falls. From Balsam Lake to Kaichiwano Lake the total drop is about 78 feet.

From the south the system also receives a number of tributary streams draining that portion of the drift-covered plain north of the Oak Ridges moraine. These rivers flow in swampy valleys of low gradient. Scugog River, for instance, is twelve miles in length between Lake Scugog and Sturgeon Lake and falls only six feet, practically all of which is accounted for by the dam at Lindsay. Pidgeon River is another sluggish, north-flowing river in a wide swampy valley.

The Kawartha Lakes drain to Rice Lake by way of the Otonabee and Indian Rivers. The Otonabee was undoubtedly the main route of the Algonquin drainage, although the Indian River valley, a few miles to the east, also shows evidence of a former much greater stream. In the ten miles from Lake Kaichiwano to Little Lake in the city of Peterborough the Otonabee River falls 140 feet, but only nine feet in the 20 miles between Peterborough and Rice Lake.

Rice Lake, which extends for about 23 miles in a southwest-northeast depression, seems also to occupy a portion of a preglacial valley although the rock is completely obscured by drift except at the outlet in the northeast. This body of water is notable for the number of drumlin islands which it contains.

From the outlet of Rice Lake, now controlled by the dam at Hastings, the Trent River flows northeastward in the old valley for a distance of seven miles, then turns eastward in a rather aimless fashion across a limestone plain for six miles to Healey Falls. Here it tumbles into another old limestone valley approximately 75 feet in depth, making a complete, right-angled turn to flow southward. The northward continuation of this valley contains the stream known as Crow River which drains three fair-sized lakes situated, like the Kawartha Lakes, on the border of the Shield. The southward-flowing portion of the Trent is ten miles in length and falls approximately 155 feet to Percy Reach. The latter name is given to a broad eastward-trending portion of the river, about 13 miles in length and having almost no gradient whatsoever. Bordering the western portion of Percy Reach is a swampy area over four miles in width, while eastward the river divides to enclose the large flat area of Wilson Island. Again the river encounters a southward trending valley, the upper continuation of which is occupied by Rawdon Creek. From this point to the Bay of Quinte at Trenton, the river has only a shallow channel in limestone. This lower section has a number of tributaries including Mill Creek, Salt Creek, Cold Creek, Squire Creek, and Rawdon Creek, all of which flow in swampy hollows among drumlinized uplands.

The whole Trent River system is an example of the youthful, deranged, non-integrated type of drainage resulting from Pleistocene accidents. The lakes, broad reaches, and marshy areas contrast sharply with the rapids and waterfalls of the intervening areas. Since a large portion of the drainage area is strongly drumlinized there are a great many linear and oval inter-drumlin swamps as well.

The chain of lakes and the good-sized river led to the idea of a canal between Lake Ontario and Lake Huron. Long before it was finished, however, the need for such a water-route was past and the only real benefit conferred has been the by-product of power which has helped the growth of local industry at Peterborough, Hastings, Campbellford, and other places. The only craft on the river and the lakes are those of the sportsmen, since the Kawartha region is one of Ontario's most accessible and best recreational areas.

Moira River. Lying just east of the Trent and paralleling it in many ways, the Moira drainage basin has an area of 1,090 square miles. The headwaters of the Moira rise in the rocky highlands of Hastings and Lennox and Addington counties about 60 miles north of the Bay of Quinte. The upper streams are very crooked and pass through a succession of lakes, swamps, and rapids. The Skootamatta for instance, travels over 50 miles in an airline distance of 27 and falls more than 700 feet.

Stoco Lake lies approximately on the border between the Precambrian and Palaeozoic rocks. From this lake to Belleville the Moira runs on limestone. In the first 13 miles it follows a well-marked valley in the bedrock and falls 110 feet, an average gradient of over eight feet per mile. At Plainfield the river enters a broader valley with less gradient and turns westward. The tributary in the upper portion of this valley, is known as Parks Creek, issues from two marly lakes near Marlbank. Through the sandy area between Plainfield and Foxboro, the Moira is broad and sluggish. From Foxboro it turns south through a shallow channel to Belleville, falling almost 80 feet in the last five miles.

The mean discharge measured at Foxboro is 1,080 c.f.s., equivalent to 13.5 inches of rainfall annually. However, the discharge during individual years has varied from 418 c.f.s. (5.2 inches of rainfall) to 1,980 c.f.s. (about 25 inches). The lowest flow recorded is 15 c.f.s., and the highest 12,460. For the most part April is the flood month and September the time of weakest flow.

Despite the large forested area and the numerous lakes on the Shield, the flow of the Moira River is notoriously unreliable. Perhaps the limestone plains of the lower part of the basin are responsible for the flash floods which occur periodically. The problem of flood control on this river is important because the central portion of Belleville is on the narrow flood plain. Because of this, control dams have been built at Lingham Lake and Deloro.

Salmon River. Seven miles east of Belleville the Salmon River enters the Bay of Quinte. It has its sources far within the Shield in several winding streams, the most important of which is Story Creek, draining a series of small lakes near Cloyne. From this source to its mouth the stream course extends approximately 80 miles with a total fall of 700 feet. The system has many lakes lying between the rock ridges of the Shield, among which may be mentioned Kennebec Lake, Big Clear Lake, Buck Lake, Bull Lake, Horseshoe Lake, and Sheffield Lake. The latter is apparently connected with Clare Creek of the Moira System. Lying at the edge of the limestone plain are Beaver Lake and White Lake occupying part of the valley which is drained by Parks Creek to the Moira. The Salmon River basin upstream from these lakes might be called territory diverted from the Moira drainage area by the accident of glaciation. Beaver Lake is also worthy of note from the fact that it is almost divided in two by a large esker which traverses it from end to end.

Leaving Beaver Lake in a northeasterly direction, the river wanders rather indefinitely in a swampy area before entering the shallow valley leading south past Tamworth to Croydon. At the latter point it enters the larger valley occupied by Black Creek and, swinging southwesterly, follows this valley to the Bay of Quinte. This valley is well entrenched in the limestone plain, being over 100 feet deep in mid-course. The upper section is broad and swampy, the mid-section well drained, while the lower few miles are evidently drowned by the rising water level of Lake Ontario. The gradient of the river in the last 23 miles is about 200 feet or 8.7 feet per mile.

Napanee River. The Napanee River enters at the northeastern angle of the Bay of Quinte. It flows in a well-marked valley of preglacial origin, the direction of which is roughly parallel to that of the Salmon River. This valley is from a mile to a mile and a half in width, nearly 150 feet deep in places, and about 25 miles from the Bay of Quinte to the large swampy region on the edge of the Shield which contains Napanee Lake. The preglacial origin of this valley was proven many years ago by Mather who discovered old drift in the floor of the valley near Newburgh. The modern river has removed a large amount of drift-filling from this valley. The chief headwaters of this system are Cameron Creek, Depot River, and Hardwood Creek, each of which drains a group of rocky lakes. A flood control dam has been built at Second Depot Lake. Two fairly large lakes should be mentioned, namely Mud Lake and Varty Lake, which occupy depressions in the border of the limestone platform.

Rather in contrast with the Salmon River valley the valley of the Napanee is lined with settlements including Napanee, Strathcona, Newburgh, Thompson's Mills, Camden East, Yarker, Colebrook, and Petworth. We should also include Bellrock on Depot River and Verona on Hardwood Creek. Early power sites on the river gave rise to most if not all of these settlements. They have persisted, however, largely because the Napanee valley is paralleled by both railway and highway.

OTTAWA VALLEY DRAINAGE

East of the Frontenac axis the drainage of Ontario is almost entirely towards the Ottawa. From the Thousand Islands to the Quebec boundary the north bank of the St. Lawrence receives but a few short streams, the water divide being within a few miles of the river bank. The Ottawa valley is deeper than that of the St. Lawrence and apparently, in preglacial time, it contained the master stream of the drainage system and, even under the greatly changed conditions of the present, it controls the local drainage.

The rivers of this region which will be discussed include the South Nation, Rideau, Mississippi, Madawaska, Bonnechère, and Petawawa. Only the first of these is entirely within the lowland plain; most of them rise and have parts of their courses in the Shield, while the last mentioned flows entirely within the area of the Shield. All of them, however, are of geographical significance for the region under discussion.

South Nation River. The South Nation River drains an area of about 1,430 square miles in Eastern Ontario. From its source, a few miles north of Brockville, to its confluence with the Ottawa, the stream traces a course of 110 miles and descends 275 feet. The average gradient therefore, is approximately 2.5 feet per mile. Except for its slight over-all gradient towards the Ottawa, the South Nation owes little to preglacial influences and its geomorphic features all post-date the withdrawal of the Champlain Sea.

In spite of the fact that it drains an almost flat plain, there are regional differences which should be noted. The gathering ground of the head-

60. *South Nation River near Cass Bridge*

waters is on a limestone plain which has little overburden apart from peat bogs, and the small streams wander from one swampy depression to another without entrenching themselves. The stream drops 100 feet, however, in the first seven miles. In the next 23 miles it drops another 75 feet at an average and almost uniform gradient of three feet per mile. In this stretch the river passes through a region of slightly deeper drift, partly till, but more dominantly composed of water-laid sands overlying clay. The stream has cut a shallow valley. Dams have been installed at Spencerville and Ventnor, but with a very low head. Near the village of South Mountain, however, the river enters the "Nation River valley" or Winchester clay plain, and in the next 17 miles to beyond Chesterville the drop is 25 feet or 1.45 feet (17½ inches) per mile. A low dam has also been built at Chesterville. Between Chesterville and Crysler there is more of an actual valley, deepening below Crysler until the solid rock is exposed at Casselman.

Below this point the river swings west and north again to cut a canyon, 75 feet deep, through a broad sand plain. Within this narrow valley, terraces were left showing the stages of valley cutting. Emerging near Lemieux, the Nation enters upon the broad clay plain which forms the floor of an old estuary of the Ottawa River and for the next 15 miles has hardly any gradient and only a very shallow valley, held up by the rock sill at Plantagenet.

The Nation River has a number of fairly important branches, the largest being Bear Brook and Castor River. Both of these have shallow channels and drain very flat areas. The Nation River has not yet been able to establish drainage in its whole territory, and there are numerous peat bogs including the Alfred, Mer Bleue, Winchester, Moose Creek and other bogs not so well known.

There are only five incorporated villages in the whole watershed and of these only two, Chesterville and Casselman, are actually situated on the river. Finch is situated on the Payne River where it is crossed by the Canadian Pacific Railway. The other incorporated places also have rail connections. The river and its branches, however, have caused the location of many unincorporated villages and hamlets including Plantagenet, Lemieux, St. Albert, South Mountain, Ventnor, and Spencerville on the main river, Russell and Embrun on the Castor River and Berwick on the Payne River. The river provides the domestic water supply of some of these settlements.

The river is remembered most for its recurring annual floods. Spring floods are to be expected in a region of heavy snowfall and low winter temperatures. Occasionally, however, as in 1945, there are midsummer floods as well. There are four chief flood areas: Oak valley, which lies in the flat region near the junction of the Smith Branch; the main river about ten miles upstream from Chesterville; along Bear Brook; and in the Cobb Lake and Riceville area in South Plantagenet township along the main river. This last is by far the larger area and has experienced floods to a depth of 18 feet. Smaller areas around Crysler, Ventnor, and Spencerville sometimes also experience floods. At one time a scheme was proposed for canalizing the river and joining it to the St. Lawrence, for purposes of navigation. Recently another plan has been brought forward for connecting the upper part of the river to the St. Lawrence in order to divert drainage in times of flood.

Physiographically, the South Nation is an inadequate river, a boy trying to do a man's job. It is a stream which is prevented from reaching maturity by its lack of gradient, which even now is less than that of many a senile river. It therefore presents a difficult problem of drainage and flood control.

Rideau River. The Rideau River enters the Ottawa River at the city of Ottawa. At its mouth it falls over a ledge of limestone in a beautiful curtain of water, from which it was given its name by the French explorer Champlain. The river rises on the Shield in the same general area from which the Salmon, Napanee, Cataraqui, Gananoque, and Mississippi rivers also emanate. Along the boundary between the Precambrian rocks and the Palaeozoic lowland of eastern Ontario in large rock basins are the Rideau Lakes which serve as reservoirs for this system. These lakes and the others which abound in the neighbourhood constitute one of Ontario's finest recreational areas. They lie at an elevation of about 400 feet above sea level and the Rideau in its 62-mile course to Ottawa falls about 225 feet or less than four feet per mile. While part of this course is over the limestone plains, a long section is in a shallow valley cut in drift.

The Tay is a river falling into the lower Rideau Lake a few miles above Smiths Falls. It is 55 miles long and also drains a number of lakes to the northwest of the Rideau Lakes.

The Rideau is remarkable for the fact that it has been canalized throughout almost its entire length. The Rideau canal is 126 miles in length between Ottawa and Kingston. It was built by Colonel John By in the years 1827–32 at a cost of about $4,000,000. Its construction was fairly simple since there was a minimum of excavation and most of the waterway was secured by throwing dams across the Rideau and the Cataraqui. There are 24 dams and 47 locks, each of the latter being 134 feet long, 33 feet wide, and having five feet of water on the sills. In the early days it was a busy waterway and contributed a great deal towards the settlement of the region. The building of railways and the canalization of the St. Lawrence rendered it obsolete within fifty years, but commercial navigation lingered until 1930 when the last "cheese boat" ceased running. Since that time it has been used only by pleasure boats and maintenance craft.

The Rideau is in many ways a sister to the Trent, rising on the Shield then flowing across limestone plains and through areas of deeper drift. It has large headwater reservoirs, complete canalization, and controlled flow.

Mississippi River. The Mississippi River resembles the Rideau in having numerous headwater lakes in the Shield and in traversing both limestone and clay plains. The Mississippi has not been turned into a canal, however, although it has at times suffered improvement to facilitate the driving of logs. From its source at the head of Mazinaw Lake to its confluence with the Ottawa near Galetta its course is over 120 miles in length. Mazinaw Lake lies at 875 feet above sea level and the confluence at about 250, giving a total fall of 625 feet, or about five feet per mile. However, with rivers in such an ungraded condition, average gradient means very little. Only in the lower 28 miles of its course, below Carleton Place, has the Mississippi had a chance to cut a normal valley, and even here it has uncovered ledges of hard rock. The Clyde River from the north and the Fall River from the south are important tributaries. The latter drains from Sharbot Lake.

A gauging station is located at Appleton, about 24 miles from the mouth. Above this the drainage basin has an area of 1,150 square miles.

The mean annual flow is 1,110 c.f.s. which is equivalent to a run-off varying from 6.8 inches to 22.8 inches. The minimum recorded flow is 100 c.f.s., the maximum flood stage, 9,190 c.f.s. September is the month of drought and April the usual month for floods, although about one year in five the greatest freshets come in May.

Power sites on the Mississippi have been important since the early days and have given rise to a number of towns and villages. Those on the lower river include Galetta, Pakenham, Blakeney, Almonte, Appleton, and Carleton Place. Lanark is on the Clyde, a few miles above its mouth. Hydro-electric power plants are located at Galetta and at High Falls more than 60 miles up the river. Together they deliver about 4,300 h.p. Other potential power sites exist.

Madawaska River. The Madawaska is the largest of the Ontario drainage systems tributary to the Ottawa, the area being 3,300 square miles. The Madawaska is a long river, the distance from Source Lake in Algonquin Park to the confluence with the Ottawa at Arnprior being over 200 miles. Some important branches are: the Opeongo which drains Opeongo Lake in the central part of the Park; York River, which drains a large area in Haliburton, northern Hastings, and northern Lennox and Addington; and Waba Creek, entering near Arnprior and draining White Lake, a large body of water on the boundary between Lanark and Renfrew counties. For a short distance above Arnprior only, the river and its tributary Waba Creek cross the clay plain. Here the river has cut a valley up to 75 feet in depth. Within the Shield the river flows through many scenic gorges. At Burnstown, for instance, the valley is over 200 feet in depth.

Flow records are kept at several places along the river. At a point six miles above Arnprior, the mean annual flow is 3,260 c.f.s. or the equivalent of 14 inches of run-off. Annual flows have varied from 1,270 c.f.s. (5.4 in.) to 5,630 c.f.s. (24 inches of run-off). The lowest flow ever measured was 409 c.f.s. while the highest flood stage carried 20,280 c.f.s. April and May are almost equal as flood months while September is usually the month of weakest flow.

Because of its great volume and steep gradients the river is a valuable source of power. The Ontario Hydro-Electric Commission is carrying out a programme of development. There are eight large regulating dams and reservoirs on the system and three power plants in operation which together generate 146,000 h.p.

Bonnechère River. The Bonnechère rises in Algonquin Park at an elevation of well over 1,000 feet above sea level. It takes a general southeasterly course largely controlled by the great bedrock faults of the region, finally falling into the Ottawa at Castleford. Its total length is about 100 miles and the area of its drainage basin 935 square miles. Included are a number of lakes, the largest being Golden Lake, Round Lake, and Lake Clear. The course of the stream is almost entirely controlled by the block-faulted relief of the area. The upper part of the valley lies in a graben whose bounding escarpments are more than 500 feet in height. The valley of the river is hardly entrenched at all and drift barriers cause expansion into broad lakes. In the vicinities of Golden Lake and Round Lake, there is a great deal of sand and gravel which was undoubtedly deposited by glacial meltwater. Below Eganville a rather deep valley, interrupted by waterfalls, is cut in the Palaeozoic limestones preserved in the floor of the graben. The Bonnechère caves are found in this area. In the clay plain between Douglas and Renfrew the river has cut a narrow valley up to 50 feet in depth. At Renfrew again there are waterfalls below which the river has cut a trench over 100 feet deep in Pleistocene deposits, and here there is a succession of deep lateral gullies.

The gauging station at Castleford records the mean flow of the river at 633 c.f.s. which is the equivalent to 9.3 inches of run-off. The annual flow, however, varies from 254 c.f.s. (3.7 inches) to 1,210 c.f.s. (17.5 inches). August through October is definitely a season of low water, while April and May are months of freshet. The lowest record, 58 c.f.s. was made on September 19, 1948, while the greatest April flood was 10,200 c.f.s.

Muskrat River. The Muskrat and Indian rivers, which join to make one entrance to the Ottawa at Pembroke, are small but interesting rivers. The Muskrat is closely aligned with the Muskrat fault and drains Muskrat Lake, some ten miles in length and extraordinarily deep. Other smaller lakes are found in this chain also. The Snake River, a tortuous tributary falling into Muskrat Lake, drains Mink Lake and Lake Dore. All of these lakes are in the down-dropped block and associated with limestone outliers. The Indian River, on the other hand, issues from the block mountain uplands by way of the deep valley of the Gardez-Pieds cross-fault. Along this valley there are very interesting gravel terraces associated with the Fossmill outlet of the upper Great Lakes.

Petawawa River. The Petawawa River is a large stream falling into Allumette Lake, an expansion of the Ottawa River, at a point about ten miles upstream from Pembroke. In the lower four miles of its course it falls about 100 feet, cutting a deep valley in the Petawawa sand plain. The upper reaches of the river and its tributaries are associated with the

61. *Ottawa River valley, Bissett Creek*

deep valleys in the uplands of Algonquin Park. These are aligned with the faults of the Bonnechère graben and, although they have never been adequately studied, may be assumed to be part of the same system. Petawawa and Barron rivers both were greatly increased during the period of the Fossmill outlet of glacial Lake Algonquin and the large Petawawa sand plain is a delta built in the Champlain Sea.

The Petawawa drains an area of 1,572 square miles. Its mean annual flow is 1,630 c.f.s., the equivalent to a run-off of 14 inches. It has, however, been known to vary between 660 c.f.s. (5.7 inches) and 2,640 c.f.s. (23 inches of run-off). The minimum recorded flow of 204 c.f.s. took place in October while the maximum flood, 12,700 c.f.s. occurred on April 16, 1951. On the average, both September and October are dry months while May is usually the flood month.

Petawawa River and its tributaries flow almost entirely within the uplands areas of the Shield and, although they have been of considerable use in lumbering in the past, they have had no influence upon permanent settlement.

Ottawa River. The Ottawa River forms the eastern boundary of the province of Ontario for a distance of approximately 360 miles from the head of Lake Timiskaming to the Carillon Rapids. Approximately half of that distance is along the border of the region under discussion. Hence it is appropriate that a brief section be devoted to the Ottawa River itself. Not only does it provide a boundary (of convenience only be it noted) but it is a most striking physiographic feature. It is, like most of the other rivers which issue from the Canadian Shield, still in a youthful or ungraded state. Despite the fact that for long distances it flows in a deep and rather narrow valley cut into the old peneplain of the Shield, it inherited most of this from a preglacial ancestor.

Like many other rivers in glaciated territory, the Ottawa seems to consist of a chain of stillwaters or lakes connected by sections of steeper gradient often with rapids and waterfalls. Among the lake-like expansions of the river are Allumette Lake (el. 366 feet), Lower Allumette Lake (352 feet), Coulonge Lake (239 feet), Lac Rocher Fendu (257 feet), Lac des Chats (239 feet), and Lac Deschenes (192 feet). All of these lie in the sections of the river above Ottawa. Between the lakes are various famous rapids including Allumette Rapids, Paquette Rapids, Chute du Rocher Fendu, Chats Falls and, at the foot of Lac Deschenes, the great Chaudière Falls. Below the city of Ottawa, lakes and rapids are fewer but the Long Sault Rapids and Carillon Rapids combine to drop the river from L'Orignal Bay (131 feet) to Lake of Two Mountains at 73 feet above sea level.

In recent decades the Ottawa has provided still further wealth in the development of hydro-electric power. There are four Ontario Hydro generating stations, the largest operating at Des Joachims to produce nearly 500,000 h.p. The Otto Holden and Chenaux plants generate 430,000 h.p., while 164,000 h.p. is produced at Chats Falls. The province of Quebec has built its own plant at Carillon Rapids to be shared equally by Ontario and Quebec.

The river drains an area of about 56,000 square miles in Ontario and Quebec. The average flow at Grenville is 68,700 c.f.s., which is equivalent to 16.6 inches of water from the whole area. Yearly records have varied from 44,700 c.f.s., (10.9 in.) to as much as 87,300 (21.3 in.). This degree of regularity is, of course, influenced by the presence of the many lakes and dams along the river and will be further affected when other dams are constructed. In recent years the minimum flow has been 23,270 c.f.s., while the highest flood stage has reached 287,000 c.f.s., which is, after all, only about four times the normal.

For a large river the Ottawa does not have an impressive valley. In some of the more rapid parts, banks up to 100 feet in height are developed. Bordering the Petawawa sand plains there are bluffs 75 feet in height, but they stand half a mile to a mile back of the present very gently sloping river bank. The shoreline of Lac des Chats, especially towards the lower end, is low and swampy. Elsewhere the river has low banks, 25 to 50 feet in height and out of proportion to the size of the river. However, where the river has uncovered ancient rock escarpment, as at Parliament Hill in Ottawa, the scenery takes on a more impressive appearance.

The Ottawa River has been an important factor in the historical development of Canada. It served as an important route for the early explorers and fur traders. Later it was the route by which squared timber was shipped to the markets of the world. Even today it carries logs and pulpwood to mills located along its banks. It is navigable by small ships of shallow draught as far as Ottawa, which is also the terminus of the Rideau Canal. At one time it was proposed that the Ottawa be fully canalized to provide a navigational outlet to the upper Great Lakes, but this ambitious project was found to be too costly. The great value of the river today is in its full utilization for the production of hydro-electric power.

THE SIGNIFICANCE OF SOUTHERN ONTARIO STREAMS

The chief rivers of Southern Ontario have been described in this section as physical features, as units in the physiographic pattern. They form boundaries, barriers, and corridors of geographical importance. The rivers, also, have physiographic functions. They are the natural agents which, during the postglacial age, have carried on most of the work of geological erosion and transportation; they are gradually, and in some areas rapidly, cutting into the drift and carrying down the materials to be spread out on the floors of the Great Lakes as well as smaller lacustrine basins. Gradually, also, some shallow retaining basins are being drained as rivers cut down their outlets.

No river, however, has everything its own way, even though it is always in alliance with the force of gravity. It would have, if the glacial deposits remained bare as they were at the disappearance of the last ice sheet. However, in place of ice, the surface soon acquired a cover of soil and vegetation, a matted binder to hold the surface in place. It must be admitted that it has been effective, for most of Ontario's surface is still a glaciated plain, and only a small area is given over to recently cut valleys. In unsettled country, forested country, such as most of Ontario was until a century or so ago, nature preserves a balance and the work of rivers is remarkably slow. Floods, there have always been, and always will be, but under primeval conditions their effects were not excessively severe.

In settled country such as Southern Ontario now is, a new environment has been created in which the river must become the agent of man as well as the agent of nature. Depending upon capacity and location, we look upon the river for drainage and for water supply; if large enough, it is used for navigation, or it is canalized; large rivers, also, have become our most important sources of power.

The river has become one of the most important cases where the principle of the multiple use of resources may be put into play, and it must be kept in play if the greatest benefits are to be obtained. It is manifestly impossible to restore primeval conditions since this would involve complete reforestation and population evacuation of the whole area. However, properly managed, contour-farmed crop lands and pastures are almost as retentive of moisture as are forests. Add to this the scientific management of the river itself, and it is not inconceivable that the managed regimen might be an improvement upon the original. River valley authorities, the Ontario Hydro-Electric Commission, and the Ontario Water Commission are all part of the machinery by which the people of Ontario are trying to apply scientific principles of river management so as to obtain the maximum benefit from all of Ontario's streams.

4 PHYSIOGRAPHIC REGIONS

Within Southern Ontario as we have seen, there are many different land forms that may be recognized and mapped. In terms of its gross structure, which is based mainly on the bedrock, the area consists of four natural divisions. These are: (1) the broad half-dome that slopes from the Niagara escarpment to Lakes Huron and Erie; (2) the Niagara escarpment itself; (3) South-Central Ontario between the edge of the Canadian Shield and Lake Ontario; and (4) the lowlands between the St. Lawrence and Ottawa rivers. The edge of the Shield might be taken as another division even though it is outside the borders of the area under discussion. The local land forms, usually glacial drift forms, occur within the major divisions. While their distribution pattern is not simple it is possible to place them satisfactorily in the fifty-two minor physiographic regions that appear in Figure 62.

Southwestern Ontario is a large unit in which the most striking contrast is between the rolling till plains of the upland and the ancient lake bottoms bordering the Great Lakes. The topographic contrast is accentuated by variations in climate and natural vegetation, and in the specialized agriculture on the sand or clay plains and the more generalized farming of the uplands. The body of this elephant-shaped area is divided into twenty-three regions while Manitoulin, Cockburn, and St. Joseph islands are also included.

The Niagara escarpment is dealt with as a single unit except that the two large re-entrant valleys, the Beaver and Bighead, and the Cape Rich area are discussed separately. It should be mentioned too, that the section between Caledon and Creemore which is buried by moraines and sandy terraces is excluded.

In south-central Ontario the major relief is provided by the massive interlobate moraine rather than by the bedrock. The crest of this moraine is mapped under the name of Oak Ridges. The north and the south slopes together with the adjacent till plains form two regions, while a third till plain or group of separate till plains comprise the uplands of northern Simcoe county. The Peel and Schomberg are both uniform clay plains, but both the Algonquin and Iroquois plains on the other hand are highly variable. In discussing the areas of shallow drift on limestone, Prince Edward county or peninsula seems to stand out by itself and is described separately from the Napanee plain.

Eastern Ontario is mainly dealt with in eight regions with three more small regions taking care of as many odd areas of distinction. The large tract of limestone plains lying adjacent to the Shield is similar to the Napanee plain. Only two of the regions are dominated by glacial forms, the bulk of this division being taken up by sand or clay plains.

NIAGARA ESCARPMENT

The Niagara escarpment extending from the Niagara River to the northern tip of the Bruce peninsula, and continuing through the Manitoulin Islands, displays an association of landforms not found elsewhere in Ontario. Vertical cliffs along the brow often mostly outline the edge of the Silurian dolomite formations while the slopes below are carved in red shale. For some distance back from the brow the dip-slope of the cuesta in many places has been stripped of soil and overburden. Flanked by landscapes of glacial origin, this rock-hewn topography stands in striking contrast, and its steep-sided V-shaped valleys are strongly suggestive of non-glaciated regions. While the escarpment stands out boldly in the Niagara Peninsula and along the shore of Georgian Bay, there is an intervening area in which the slopes are mantled by morainic deposits and, particularly in the townships of Caledon, Albion, Mono, and Mulmur, long stretches are almost completely hidden.

From Queenston, on the Niagara River, westward to Ancaster, the escarpment is a simple topographic break separating the two levels of the Niagara Peninsula. In general, the base is followed by the 350-foot contour while the top of the cliff reaches the 625-foot level. In some places, particularly just east and west of St. Catharines, and again near Beamsville, there is a broad sloping bench between the escarpment and the old Iroquois shoreline. It is covered by several feet of boulder clay. In places also, as near Grimsby, there is a narrow rock shelf near the top of the escarpment which has a similar but shallower covering. There are a few notable breaks in this stretch of the escarpment. Among them is the valley of Four Mile Creek near St. David's, a buried former valley of the Niagara River, which affords an easy highway over the cuesta to Niagara Falls. Power Glen, the valley of Twelve Mile Creek, is a preglacial notch of considerable size just west of St. Catharines; its slopes are mantled by very much dissected beds of sand, silt, and glacial till. Another large

62. *Fifty-two minor physiographic regions of Southern Ontario*

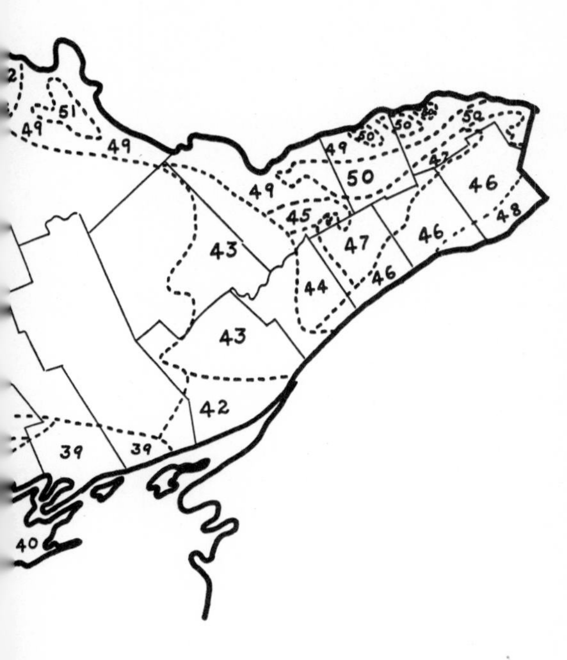

1. Niagara Escarpment
2. Beaver Valley
3. Bighead Valley
4. Cape Rich Steps
5. Horseshoe Moraines
6. Flamborough Plain
7. Dundalk Till Plain
8. Stratford Till Plain
9. Hillsburg Sand Hills
10. Waterloo Hills
11. Guelph Drumlin Field
12. Teeswater Drumlin Field
13. Arran Drumlin Field
14. Oxford Till Plain
15. Mount Elgin Ridges
16. Caradoc Sand Plains
17. Ekfrid Clay Plain
18. Bothwell Sand Plain
19. St. Clair Clay Plains
20. Pelee Island
21. Erie Spits
22. Norfolk Sand Plain
23. Haldimand Clay Plain
24. Saugeen Clay Plain
25. Huron Slope
26. Huron Fringe
27. Bruce Peninsula
28. Manitoulin Island
29. St. Joseph and Cockburn Islands
30. Oak Ridges
31. Peterborough Drumlin Field
32. South Slope
33. Peel Plain
34. Schomberg Clay Plains
35. Simcoe Lowlands
36. Simcoe Uplands
37. Carden Plain
38. Dummer Moraines
39. Napanee Plain
40. Prince Edward Peninsula
41. Iroquois Plain
42. Leeds Knobs and Flats
43. Smiths Falls Limestone Plain
44. Edwardsburg Sand Plain
45. North Gower Drumlin Field
46. Glengarry Till Plain
47. Winchester Clay Plain
48. Lancaster Flats
49. Ottawa Valley Clay Plains
50. Prescott & Russell Sand Plains
51. Muskrat Lake Ridge
52. Petawawa Sand Plain

63. *Badly gullied hillside north of Georgetown on Queenston shale. The light horizontal streaks are greenish grey shale interbedded with soft red shale*

64. *Terraces of outwash sand on the Niagara escarpment at Dunn Hill*

preglacial valley is that now occupied by Red Hill Creek which tumbles over the escarpment in a curtain fall which in flood times is quite imposing. Fifteen Mile, Sixteen Mile, Twenty Mile, and Forty Mile creeks, all descend the escarpment also, but by narrow and precipitous valleys which do not afford easy passage.

The Dundas valley is the most notable break in the southern part of the escarpment, extending inland eight miles from the west end of Lake Ontario. The rim is sharply outlined by rock bluffs, but within the valley there is deep drift the surface of which is deeply cut by many gullies. A peculiar occurrence is beds of sand and silty clay in alternate layers. Agricultural considerations are no longer of much concern, since the valley is nearly all urbanized. It provides some highly attractive building sites. From the days of early settlement this valley has provided the route for roads and railways from Lake Ontario to southwestern Ontario.

From the Dundas valley northward to Credit Forks the brow of the escarpment increases in elevation from 800 feet to about 1,450 feet above sea level in a distance of 50 miles. In this section, also, the escarpment is cut by numerous creeks. Several fairly large valleys are found near Waterdown, Lowville, Campbellville, and Limehouse, as well as the scenic gorges at Credit Forks. With the exception of the valley of Bronte Creek at Lowville, all of these passes are used by both road and railway. In this region a rather broad belt of red shale is exposed and the long lower slopes of the escarpment are highly eroded. North of Aldershot and again north of Cheltenham they have the appearance of veritable badlands. There are several mesa-like outliers of the escarpment. The largest one, located near Milton, has an area of about four square miles and is separated from the main body of the upland by a deep valley partially filled with glacial stream deposits. The promontory of the southern end of this valley is known as Rattlesnake Point. For a mile or more back from the edge of the escarpment there was, apparently, very little drift deposited and the meltwater from the ice carried away most of the finer particles, leaving a very shallow and rocky soil. Much of this area has been left in forest so far and should remain so for the protection of the headwaters of the Credit River and Bronte and Oakville (Twelve Mile and Sixteen Mile) creeks.

From the Forks of the Credit, northward for about 30 miles to the valley of Pine River the elevation of the cuesta increases to over 1,600 feet above sea level. Most of the rock slopes, however, are obscured by hummocky, bouldery, morainic ridges and deposits of sand and gravel. In some places, as at Mono Centre, vertical cliffs of dolomite are exposed where they served as river banks for the meltwater floods. Here, too, are found small mesa-like outliers of the Lockport dolomite. The escarpment slopes are dissected by a number of deep valleys, including those which carry the headwaters of the Humber, the Nottawasaga, Sheldon Creek, the Boyne, and the Pine. These postglacially deepened valleys have captured the drainage of the great glacial spillways running parallel to the escarpment; short lateral branches now serve the adjacent sections, while intervening sections are left intact and now may even contain swampy stretches. Abandoned alluvial terraces, often as much as 100 feet above the present valley bottoms, are prominent features of this area. The steep slopes resulting from the extreme dissection are conducive to soil erosion where unwise clearing has taken place.

The highest and most picturesque part of the escarpment is the Blue Mountain section near Collingwood, which stands over 1,000 feet above the waters of Georgian Bay. Here the dolomitic cap rock is exposed in cliffs 150 feet high, while huge blocks breaking away from the wall have left the deep crevasses known as "the caves."

The scenic value of the sector is greatly enhanced by the Pretty River valley, a deep preglacial notch in the rim. The chief members of the Horseshoe Moraine system lie above the escarpment at this point but, during the retreat, the ice pressed against the escarpment for some time and built a strong moraine across the valley. During this period a small lake, several hundred feet deep existed, finding an outlet southwestward

65. *Dissected terraces and moraines, Mulmur township*

through the upper course of the Beaver River. Since the disappearance of the ice, the Pretty River has cut a deep notch in the moraine and has also greatly dissected the floor of the small enclosed basin. In strong contrast to this rugged landscape, the lowland to the east is floored by the flat deposits of Lake Algonquin, the old shoreline of which skirts the base of the escarpment.

Ten miles south of Collingwood, Glenhuron lies in the valley of the Mad River which traverses the escarpment in the same manner as the Pretty River. Here a similar, but not quite so striking, transverse valley-moraine was built. The little village of Singhampton lies at the head of the valley, on the plateau. Between the two places runs the narrow, steep-sided, scenic canyon known as the Devil's Glen, where a provincial park and a ski run have been established. A mile and one-half north of Singhampton, overlooking Edward Lake, stands the highest point of the Niagara cuesta, a stony moraine on a rocky butte with a summit approximately 1,790 feet above sea level. Several other hills in the vicinity also rise above the 1,750-foot contour while fairly large areas lie above 1,700 feet.

Just north of Creemore, there is a curious outlier of the Niagara cuesta consisting chiefly of Ordovician shales without any protecting cap of Lockport dolomite, with an area of about five square miles, this upland is separated from the plateau to the west by a deep through-valley. The northern end of the valley, however, is blocked by a moraine which forces the waters of the Mad River to adopt a very circuitous route in order to reach Georgian Bay. This valley is traversed by an abandoned line of the Canadian National Railway, which at one time served the villages along the foot of the escarpment.

In this vicinity a narrow shelf or ledge is often manifest about one hundred feet below the edge of the escarpment. This feature is caused by the presence of a second dolomitic formation, the Manitoulin, which becomes more important farther north. It is quite noticeable just behind Glenhuron.

West of Collingwood and immediately between Craigleith and Camperdown, occurs the steepest and most mountainous part of the escarpment. This is the highest hill in southern Ontario and is well known now as the site of the Georgian Peaks and Blue Mountain ski runs. In the curious flat-topped mesa near what was known locally as the "Deer Park," the brow of the escarpment approaches within a mile of the Bay, and the wooded slope of this promontory is a well-known landmark.

Immediately west of Georgian Peaks, the edge of the cap rock retreats southward again, in the great re-entrant of the Beaver valley, which extends for 25 miles into the heart of the upland. A few miles farther west lies the valley of the Bighead River. Between the two valleys stands another promontory, but the dolomite cap at the "Griersville Rock" is over four miles from the shore of the bay. A broad and dissected shale slope intervenes between the "Rock" and the ancient shorecliff of Lake Algonquin. The Bighead valley is somewhat broader than the Beaver valley and contains a large number of well-developed drumlins. Because of the interest which attaches to them, these two valleys will be treated as separate regions in this report.

Between the Bighead valley and Owen Sound, the escarpment advances to Cape Rich where it ends abruptly in a striking shale cliff, the "Clay Banks," which rises sheer from the edge of the water to a height of almost 400 feet. Here the Algonquin shoreline has been undercut by the wave work of the present water level. The Cape Rich promontory comprises a great deal of eroded terrain underlain by the red Queenston shale as well as areas of flat mesa on the Manitoulin dolomite. The highest point of the peninsula, over 1,375 feet, at the St. Vincent triangulation station is on the Lockport dolomite which hereabout is over four miles from the bay.

The eastern border of this upland is marked by the precipitous Algonquin shorecliff, over 200 feet in height, while the slopes below, also nearly 200 feet high, are mantled by various wave-worked features. The

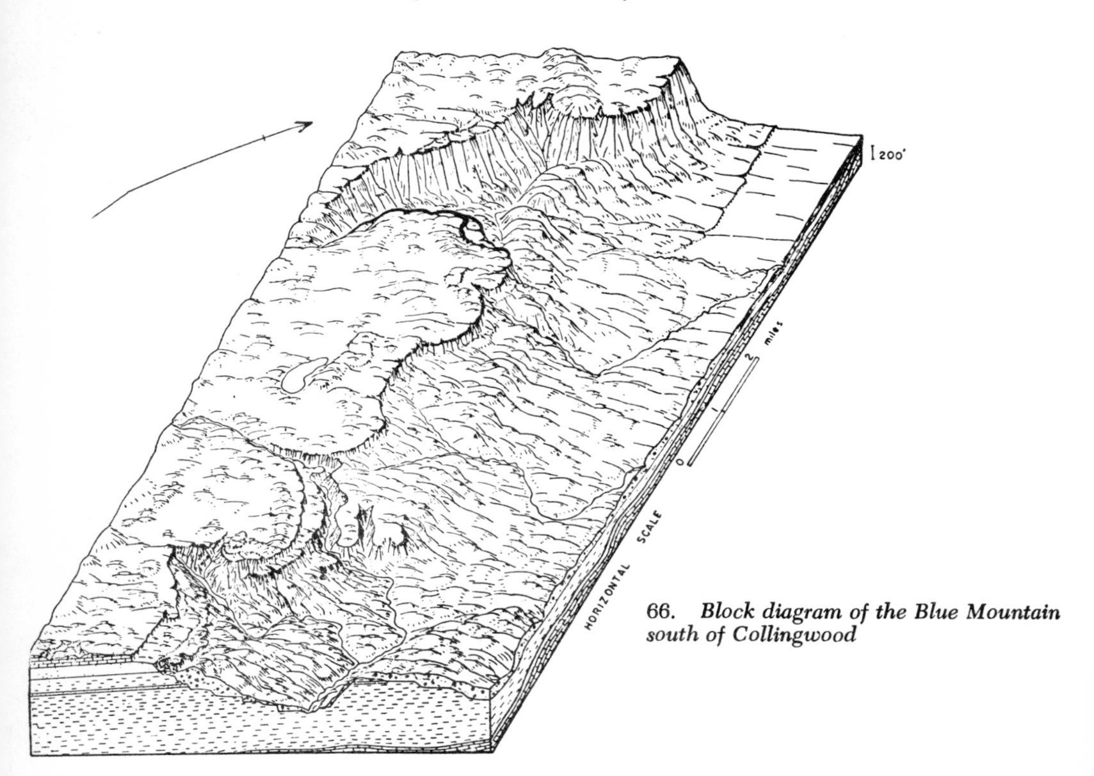

66. *Block diagram of the Blue Mountain south of Collingwood*

67. *Blue Mountain west of Collingwood. Note also the beach ridges in the midground.*

Cape Rich foreland, itself, is composed of Algonquin beach ridges. Mountain Lake (also known as Carson Lake) lies in a deep valley cut in the shale and is impounded by the Algonquin beach. The narrow lowland at Cape Rich was being developed as an orchard area, but in 1942 the whole northern end of St. Vincent township was acquired by the Government of Canada as a tank range.

The western slope, to the shore of Owen Sound, is more gradual and exhibits a wealth of glacial formations and abandoned beach ridges. The edge of the escarpment, however, can be traced as a cliff, 50 to 150 feet high, which takes a sinuous southwesterly course from Morley to Inglis Falls.

Between Owen Sound and Colpoys Bay occurs "Stony Keppel," a moraine-strewn, though thinly mantled, limestone plateau with an average elevation of 900 feet above sea level. The higher level on the Lockport dolomite is bordered on the north and east by cliffs averaging 150 feet in height, although at Skinner Bluff, Esther Cliff, Halliday Hill, and Dodd's Hill, they are over 200 feet. Pyette Hill is a curious little outlier separated by more than a mile from the main plateau. Below the cliff is another rock plain developed upon the Manitoulin dolomite, ranging from half a mile to two and a half miles in width. The surface of this plain is quite smooth and the soils are extremely shallow except where developed upon the Algonquin beach ridge. This Manitoulin shelf is also bordered by a steep cliff which in places is nearly 100 feet in height. Below it is a narrow, sloping remnant of the Algonquin lake plain. In part it is covered with thin water-laid sediments and in part its surface is composed of badly eroded red shales.

Both Owen Sound and Colpoys Bay are deep valleys; soundings taken off Cape Commodore show more than 50 fathoms of water. The total relief of the Niagara escarpment in this vicinity is thus over 1,200 feet. Both Owen Sound and Wiarton have well-protected harbours.

On the north side of Colpoys Bay stands the jutting promontory of Cape Croker which is a detached mesa of Manitoulin dolomite almost surrounded by water. Other lower remnants of the plateau without the protecting dolomite form three islands (Griffith, White Cloud, and Hay) lying at the mouth of Colpoys Bay. Behind Cape Croker the edge of the dolomite formation is marked by Malcolm Bluff, Kings Point Bluffs, Jones Bluff and Sidney Bay Bluff, all of which reach an elevation of about 950 feet above sea level. Towards the north Cape Dundas and Lion's Head are slightly lower, but none the less impressive. Farther north the escarpment becomes lower; at Dyer Bay the cliffs are only about 125 feet high. The surface of the plateau rises again, however, the west bluff at Cabot Head standing more than 300 feet above the level of Georgian Bay. The cliffs diminish rapidly westward along the north end of the Bruce peninsula. Finally at Tobermory, the cuesta is entirely broken through, allowing the union of Georgian Bay and Lake Huron. Here on one of the small islands off the coast are found those curious "stacks" or remnants of the dolomite formation which are known as the "Flowerpots."

All along the Bruce peninsula section of the escarpment there are to be seen many good examples of wave work, construction as well as erosion. At Esther Cliff, for instance, stands a great Algonquin beach ridge behind which there was a bay of about two square miles. A small remnant of this still remains, known as the Slough of Despond. Lower

68. *The Niagara escarpment at Lion's Head. In this area only the upper dolomite bluffs are exposed, the lower slopes being under water.*

69. *Stacks formed in dolomite by wave erosion on Flowerpot Island*

Algonquin beach ridges are well displayed at the heads of both Sydney Bay and Hope Bay. At Barrow Bay a lagoon exists, cut off from the present lake by a wide barrier beach. Beach formations nearly enclose Wingfield Basin at Cabot Head. At the heads of some small bays, notably at Dunks Bay, fine sandy beaches have been built.

LAND USE AND SETTLEMENT

Land use and settlement in southern Ontario have been greatly influenced by the presence of the Niagara escarpment. Its steep slopes and shallow, rocky soils have inhibited agricultural use and tended to preserve the forest. As a topographic break it stretches across the whole of peninsular Ontario like a great wall through which there are only a few good "gates," thus tending to concentrate the building of roads, railways, and urban settlements at certain definite points. The largest cities, however, have grown up where the influence of the escarpment was combined with that of a port on the Great Lakes. Water falling over the escarpment is a source of power; the smaller streams, amenable to the manipulations of a pioneer technology, became the foci of early industry and settlement; the modern development of Niagara power by the Ontario Hydro-Electric Commission furnishes energy to a large part of Southern Ontario. Worthy of note, also, is the fact that throughout the history of settlement, the escarpment has served as a source of building material including cut stone, lime, shale for the making of bricks, and more recently for enormous quantities of crushed limestone in all sizes.

The direct effect of the escarpment upon agriculture has been almost entirely negative. In many places the slopes approach the type found in non-glaciated regions, giving rise to similar problems of soil management and erosion control. During the pioneer stages of settlement steep slopes and shallow soils that should have remained under forest were cleared and cropped. Perhaps the growth of mills and villages in the valleys encouraged cultivation of nearby areas. The hard clay soils developed on the red shales, however, were easily eroded, and much land has been badly gullied. Most of it, of course, has long since been retired to sod, but even under permanent pasture the more steeply sloping fields still suffer from erosion. The change from horses to tractors has been the cause of withholding the plough on many farms, but the operation of most types of machinery has always been difficult in the fields along the escarpment. The present tendency, especially toward the north, is for land holdings to increase in size while the intensity of use is lessened. Fortunately, cattle thrive on the well-drained pasture.

Nevertheless, the escarpment also affords gentle slopes, shelves and terraces, where topography is favourable for the plough. In the Niagara peninsula such shelves near both the base and the brow of the escarpment are used for vineyards, and even flat or gently sloping areas along the brow itself are so used.

The relationship of the Niagara escarpment to transportation patterns in southern Ontario is worth a little study. The earliest routes were, naturally, water routes. To these the Niagara escarpment presented a formidable barrier to be overcome only by taking to the land and establishing a portage route. Most notable of these was undoubtedly the Niagara Portage Road from Queenston to Chippawa established at the request of Lord Dorchester shortly after 1790. It climbed the steep face of the escarpment by a switchback route which required constant improvement. It is reported that during the season of navigation as many as 60 wagons per day would be loaded at Queenston to make the trip. The road was too near the American boundary, however, and traffic suffered during the war of 1812. This, in addition to the natural difficulties of the route, increased the demand for the building of a canal across the escarpment.

In 1824 a canal across the escarpment was begun under the private initiative of William Hamilton Merritt, a miller on Twelve Mile Creek near St. Catharines. Because of water shortage in the summer months, he planned to divert the Welland River over the escarpment to increase his power supply. Finding support among his friends, the idea was expanded to include navigation facilities. Eventually, a barge canal with wooden locks was opened in 1829, providing a navigable connection between Lake Ontario at the mouth of Twelve Mile Creek and Port Robinson on the Welland River. Difficulties on both the Welland and Niagara led to the building of an extension to Port Colborne which was completed in 1833. Since then the canal has been rebuilt or modernized four times and is finally to have twin locks throughout its whole length from Port Colborne to Port Weller. Besides providing an essential link in one of the greatest waterways in the world, the canal has helped to focus industry and urban development in the escarpment area. Greater St. Catharines in 1961 had a population of more than 95,000 people, while the Niagara Frontier (Lincoln and Welland counties) ranks among the most important manufacturing regions in Canada.

It has seemed logical to many people looking at a map of Ontario that Hamilton, rather than Toronto, should have become the entrepôt and primary city of the region. There were several inhibiting geographical factors in the situation and in fact the urban nucleus of

Hamilton was not laid out until ten years after Simcoe's city had been started. Not till 1826 was a navigable channel actually cut through the Burlington bar; earlier cargoes had all to be transhipped in small boats. The Hamilton waterfront itself was not easily accessible to the roads leading to the newly settled areas to the north and west. In 1837 the Desjardins Canal was cut to the turning basin in Dundas, thus bringing goods several miles farther inland, and giving direct connection to newly developed roads to Guelph, Galt, and Brantford. The opening of the Great Western Railway in 1853 dealt a mortal blow to the Dundas entrepôt, and Hamilton became the centre of a rail-net rather than a leading port.

Later, and mainly during the present century, Hamilton has become Ontario's centre of heavy industry, in part because of its central location, and in part because of the escarpment which has furnished great quantities of limestone for use as flux in the smelters. Because of the necessity to bring in the iron ore and coking coal the port now handles a greater cargo tonnage than its rival, Toronto, but is much less diversified. Industrialization enables Greater Hamilton to support a population of 400,000, but this is only a small part of the Mississauga urban region that encircles the western end of Lake Ontario, from Niagara to Oshawa.

The Niagara escarpment is an enormous and continuous outcrop of useful rock. Its importance to the steel industry of Hamilton has already been noted, but this is only a small part of the total production of limestone. In addition there is the use of the underlying shale for the making of brick, tile, and other ceramic products. Also there must be mentioned the great deposits of sand and gravel which have collected in various areas along the escarpment and which are in large part derived from it. As might be expected the exploitation of all of these materials is carried on much more actively in the south within easy reach of the large centres of population.

More than a dozen large quarries dot the escarpment from Queenston to Georgetown; two of them, near Dundas and Burlington, each have a production rating of more than 1,000,000 tons of crushed limestone per year. Another notable centre of activity is found in the vicinity of Milton. The most easterly quarry of all is at Queenston; it has for more than a century been the source of most widely used building stone in Canada. Another building stone quarry is worked near Thorold. In the northern part of the escarpment small quarries are worked near Collingwood, Owen Sound, and Wiarton.

Beneath the dolomite there is a sandstone formation (Medina-Cataract) which has long been quarried for building stone although its use is not so extensive as it once was. Quarries producing the Credit Valley sandstone are still in operation near Georgetown.

The Queenston shale which underlies the lower slopes of the escarpment provides one of the most important sources of brick-making material in Ontario (66). Between Niagara Falls and Cheltenham there are more than a dozen plants located on the Queenston shale and, besides, considerable quantities of shale are transported to plants located elsewhere. Formerly there were also brick plants located in the northern part of the escarpment. Hamilton, Burlington, and Milton are important centres of production.

For vitreous tile, the weathered soil horizons without free carbonates are used; the unweathered shale is unsuited to this product. For sintered, lightweight aggregate this shale will give an expanded product, but the sintering range is too narrow for bloating of the separate pieces to produce coated aggregate. While not outstanding, Pleistocene deposits of gravel and sand along the escarpment have supplied some of the material used in road construction and concrete. The continuously increasing demand for coarse and fine aggregate have resulted in pits being opened in nearly every outwash deposit between Queenston and Caledon. Two of the larger ones are in the buried bedrock valley near St. Davids and the kame-terrace west of Fonthill. These are shaley and somewhat indurated sand and fine gravel deposits and are used because granular materials are scarce in the Niagara peninsula. North of Hamilton, the deposits cannot be mentioned individually, but the largest ones are near Lowville, Campbellville, Georgetown, and Inglewood. Indurated strata and shale are still problems in some of these deposits but in others the quality is better. North of Caledon there are extensive gravelly terraces along the escarpment which supply local needs. Algonquin beaches at the base of the escarpment have been excavated in the Georgian Bay area.

The prominence, continuity and special qualities of the Niagara escarpment have given it a unique place in the discussions on resource conservation and the provision of recreation facilities in southern Ontario. Coventry (29) called attention to the fact that the source areas of many permanent streams in south-central Ontario were to be found on the escarpment. Many years ago we suggested that the escarpment area might be designated as a special conservancy area. The Conservation Branch in recent years under the direction of A. H. Richardson has made extensive studies of the river systems which take rise along the escarpment including among others, the Sixteen Mile, the Credit, the Humber, and the Nottawasaga. It is necessary only to mention the famous Niagara

Parks which were set up by an Act of the Ontario Legislature in 1885 to protect the beauty of Niagara Falls, and later enlarged to include 3,000 acres along the river. Since then the province, various municipalities, and conservation authorities have provided public parks in many sections of the escarpment. They are particularly numerous in the Hamilton area. Others such as Rattlesnake Point, Belfountain, Glen Haffy, and Devil's Glen also capitalize the unique scenic qualities of their particular sections of the escarpment, providing facilities for picnics, swimming, and other outdoor recreation. A number of private parks have also been established as well as summer recreation; the escarpment provides the sites for many ski runs such as those of the Hockley valley, the Blue Mountain, the Beaver valley, and other spots. A recent development also is the Bruce Trail which seeks to maintain a foot path along the brow of the escarpment from the Niagara River to Tobermory at the northern tip of the Bruce peninsula. From the vantage points along this trail many scenic parts of the escarpment may be appreciated which are neither visible nor accessible from the roads.

In the settlement geography of Southern Ontario, the Niagara escarpment is no longer so largely the negative factor that it once was. In the great urban developments of St. Catharines and Hamilton and in smaller places as well, residential locations with a view, located on the brow or on the slopes of the escarpment, are highly prized. The rural areas of the esarpment, also have been dotted with estates, and with less pretensions country houses. The Ontario highway system now provides for rapid all-year commuting from any part of the southern portion of the escarpment. Even in the northern and more isolated stretches of the escarpment many rural residences are beginning to appear. For many reasons, therefore, the Niagara escarpment must be regarded not only as the outstanding physical feature of Southern Ontario but also as a highly significant factor of its human geography.

BEAVER VALLEY

One of the most scenic features of Southern Ontario, the Beaver valley, is a small but well-defined region of 77 square miles, occupying a sharply cut indentation in the Niagara cuesta, opening upon Georgian Bay. From Thornbury, on the shore to Flesherton at the upper end of the valley, is a distance of 22 air miles. At its widest point, between the "Griersville Rock" on the west and Georgian Peaks on the east, the valley is six and a half miles across. This width is largely maintained to a point halfway between Heathcote and Kimberley, about nine miles from the Bay, where the valley suddenly contracts to a width of a little over two miles. Its southward continuation is a narrow, tapering, steepsided notch which comes to an end just below Flesherton. From Heathcote to Kimberley, the flat floor of the valley (about 760 feet above sea level) is about 650 feet below the rim.

The upper rim of the valley is the edge of the Niagara dolomite formation which appears as an almost vertical cliff, 50 to 100 feet in height. Below this is the Manitoulin dolomite formation in the form of a flat shelf which, towards the mouth of the valley, may be half a mile in width. This horizontal shelf is particularly well developed at Griersville, west of Heathcote, and along the eastern shoulder of the valley near Duncan, Ravenna, and Loree. The mesa behind Georgian Peaks, already mentioned, is a flat-topped mesa of Manitoulin dolomite, almost completely severed by stream erosion from the rest of the upland. Standing at an elevation of over 1,400 feet above sea level, its rim is less than a mile from the shore of Georgian Bay (*ca.* 580 feet) more than 800 feet below. This high shoulder, the apex of the Blue Mountain, effectively shuts off the valley from the east. The edge of the Manitoulin shelf is also marked by an abrupt descent, though usually less sharp than that

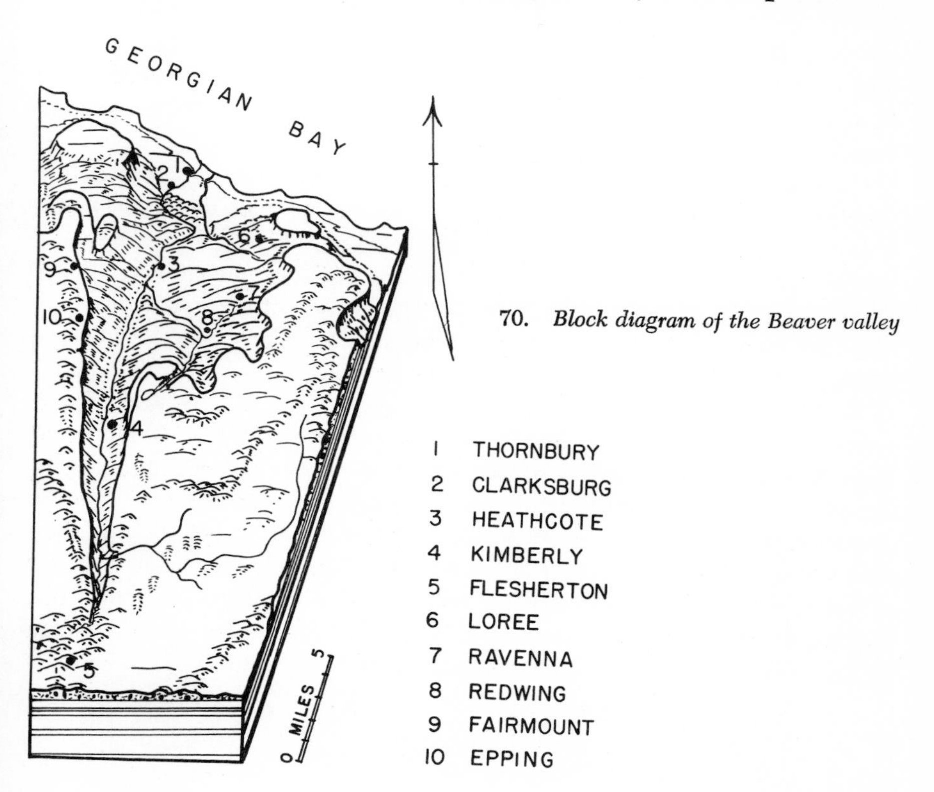

70. *Block diagram of the Beaver valley*

of the upper rim. Below this the more gently sloping, but nevertheless steep, sides of the valley are carved in softer shales.

The wider, northern part of the valley contains several tributary streams and one fair-sized stream, Indian Brook, which enters Georgian Bay independently, about a mile east of the mouth of the Beaver River. The gradient of the slopes on the shale in this open part of the valley is roughly 125 feet per mile and the streams have cut very sharp juvenile valleys, those of Indian Brook and Duncan Brook being over 100 feet in depth. Grier Creek on the western flank of the valley is also actively dissecting the shale formation.

The physiographic development of this valley is quite interesting. The greater part of its erosional history occurred in preglacial times when the forerunner of the Beaver River was a tributary to the stream which carved the deep valley of Georgian Bay. The advance of the glacier up the valley, possibly several times, served to smooth off all the protruding spurs which must have resulted from river erosion, thus leaving it an open, steep-sided, broad-bottomed feature almost comparable to the U-shaped valleys resulting from Alpine glaciation.

Across the valley, just north of Heathcote, is a belt of very undulating till which is a moraine marking a temporary halt in the recession of the ice. At this time the valley must have contained a long lake, several hundred feet in depth, in which much of the material of the flat floor above the moraine was deposited. The steep slopes of the valley are mantled by a thin covering of till, red in spots because of the incorporation of material from the Queenston shale.

In comparison with the Bighead valley, drumlins are so scarce that they may be pointed out individually. A mile and a half north of Heathcote the 12th line of Collingwood runs over the back of a drumlin; a second one may be seen just south of Heathcote in concession X.

Glacial Lake Algonquin covered all the land in this area below the 775-foot contour. Its well-marked shoreline lies along the back slope of the trans-valley moraine at a distance of two and a half miles inland from the present shore of Georgian Bay. Here there is a well-developed barrier beach with its highest crest slightly above 775 feet and with at least ten subsidiary crest lines marking successive, lower water levels. At the highest level a long lagoon occupied the flat-bottomed Beaver valley, but with the lowering of the lake level, the beach was broken through initiating the present Beaver River. Through the beach and the moraine the inner valley of the Beaver is now over 50 feet in depth, but the flat-floored region between Heathcote and Kimberley is still poorly drained.

71. *Ski run in the Beaver valley near Kimberley*

The crescent-shaped lake plain between the big Algonquin beach and the present shoreline is far from uniform. In fact, it includes just about all the possible components of a lake plain in good form. The barrier beach extends unbroken between the Beaver River and Indian Brook, while beyond the rivers in either direction the shoreline takes the form of a bold bluff with a boulder-strewn terrace at its base. On the eastern flank of the valley this strip of boulder pavement is quite broad. The body of the plain is occupied by a delta in which the texture of the surface materials varies from sand to fine silt.

The plain is terminated towards the north by a steep bluff marking the shoreline of the Nipissing Great Lakes. The base of this bluff is about 630 feet above sea level or 50 feet above Georgian Bay. Below it is a terrace swept out upon limestone bedrock covered with numerous gravel beach strands and strips of boulder pavement derived from the former covering of till. Fortunately the strip of almost useless land is only a quarter to half a mile in width. Its main use is for cottages along the shore.

Thus, even though small and obviously a unit, the Beaver valley exhibits considerable complexity of land forms including lake plains, beaches, moraines, steep valley sides, and vertical cliffs. The coherent ensemble of these elements resembles in many ways the landscapes of the hilly country of northern England and gives it rank among the most scenic areas of the province.

In the Beaver valley the soils of the lake plains around Thornbury and Clarksburg resemble those in the broader area of the Algonquin plain in southern Simcoe county. The boulders on the terraces near the shore of Georgian Bay have prevented cultivation of those soils, which are pastured or are covered with cedar thickets. Rose briars and dogwood abound on the stony pastures. A little of the dry gravelly soil associated has been planted to apples, but usually it is not so distinguished. The most valuable soils in the area are the deep silt or fine sandy loams above the Nipissing shorecliff, these being preferred for apples except where drainage is poor. These soils allow deep root range and are drought-resistent. A broad area with imperfect drainage lies between the Beaver River and Indian Brook. The driest sand on the west of Thornbury produces heavy crops of apples when well fertilized. The gravelly soil on the big barrier beach is poor in dry seasons, and recently some of it in orchards has been irrigated.

The clay loam soils in the main body of the valley resemble those on the drumlins in the Bighead valley although with fewer steep erosive slopes. They are adaptable and durable. On the upper slopes the soils are those of the Niagara escarpment. The reddish shaly soil is hard to till and inclined to droughtiness. On the shelves of Manitoulin dolomite the soil is usually shallow and stony.

While most of the valley carries on a generalized type of agriculture very similar to the other hilly areas in that part of the province, the lake plain below the old Algonquin beach has become the scene of intensive specialization. The advantage, already noted, of well-drained loamy soils is coincident with a climatic advantage derived from the sheltering hills and the nearness of Georgian Bay. Consequently this area has a longer and more reliable frost-free season than many areas much farther south. The old general farms have been replaced by apple orchards until very little unplanted land is left. The long open autumn season permits fruit to ripen properly and the apples, Northern Spies in particular, enjoy a reputation for colour and flavour which is second to none. In 1961 Collingwood township was reported to have over 3,000 acres of orchard, practically all of which is located in this section of the Beaver valley. Mixed farming is still carried on, on the slopes, with emphasis upon pasture for beef cattle. Alfalfa has been found to thrive on the shaly soils and is grown not only for forage but for the manufacture of alfalfa meal. The greater part of the flat floor of the mid-portion of the valley remains in forest because of the aforementioned imperfect drainage.

This part of the province has been settled for barely 100 years but it shows considerable adjustment to topographic influence. Here, again is seen the absurd effect of running a pre-determined set of survey lines across a new territory, regardless of local land forms. These survey lines diverge from the axial direction of the greater part of the valley by about 30 degrees. The resulting zig-zag of the highway adds several miles to its length. The angular pattern of farm lines upon the slopes also creates some odd effects.

The small nucleated settlements are, in general, located on the river. Thornbury (population 1,100) has boat-landing facilities, but its site was determined because there was power enough available for a mill. Its railway and highway advantages became important later. Less than two miles upstream were other power possibilities since the gradient of the lower part of the Beaver river is rather steep, and these gave rise to the development of the village of Clarksburg (400). Kimberley and Redwing are examples of hamlets growing up around mills located on tributary streams while Heathcote seems to have been located simply by the river crossing. Ravenna is a small cross-roads hamlet. Eugenia is located where the largest branch of the river falls into the valley. The power possibilities of this site have been developed by the construction of a hydro-electric power plant in the bottom of the valley and a large reservoir covering 1,400 acres located on the plateau.

The Beaver valley is a picturesque place. The greenish-blue waters of Georgian Bay, the orchards near the shore, the mosaic of fields upon the gentler slope and forest belts upon the valley floor, and the steeper slopes just below the limestone crags, all combine to form a geographic personality and a natural beauty which once seen is not soon forgotten. It is at its best, perhaps, in mid-October, Thanksgiving time, when the ripe apples are on the trees and the hillsides are transfigured by the autumn colours of the maple leaves.

BIGHEAD VALLEY

The Bighead valley is an indentation in the Niagara escarpment back of Meaford. It is about eight miles in width between "Griersville Rock" and Bayview, and ten or twelve miles in depth. The bed of the Bighead River which drains this box-like valley lies about 600 feet below the shoulders of hard dolomitic cap-rock. Not so well known or so scenic perhaps as its neighbour, the Beaver valley, it is nevertheless of more interest from a physiographic point of view.

Like the Beaver valley it was largely eroded in preglacial time, and, in so far as its rocky outlines are concerned, only slightly modified by ice erosion. The depositional effects of the ice, however, are very striking. The shoulders, sides, and floor of this valley are completely covered with drumlins, over 300 being recognizable within an area of about 80 square miles. Oddest of all, perhaps, is the fact that the ice which moulded these oval hills was not a local lobe advancing up this deep valley, as seems to have been the case in the neighbouring valley of Owen Sound, but an ice sheet which crossed the valley in a north-south direction, apparently in complete disregard of the underlying topography. Thus the axes of the Bighead drumlins are oriented almost directly north-south in agreement with those around Durham and the Teeswater drumlins many miles to the southwest.

Excellent views of this valley are obtainable from a number of points along the southern rim such as the "Griersville Rock," or better still at Blantyre or on the ridge above Minniehill. From either of these lookout points the valley appears almost as a basket of eggs and it seems particularly appropriate to term it "a nest of drumlins." Good views are obtainable from the northern rim, also, but they are not nearly so accessible as those just mentioned.

Other features of this valley should also be noted. Across the mouth of the valley, about two and one-half miles inland from the present shoreline of Georgian Bay and approximately 200 feet higher, is the old shoreline of Lake Algonquin which in this sector is marked by a huge barrier beach. This great gravel bar, over four miles in length, ties together several drumlins and forces the Bighead River to swing close to the southern side of the valley before finding an exit behind Meaford.

At the highest Algonquin level a lagoon of several square miles in area was impounded in the valley. Into this, a great amount of alluvial silt and sand was dumped by the tributary streams, giving a broad flat floor from which the crests of the drumlins emerge as "islands" much as they did when surrounded by the waters of the lagoon. No drumlins stand outside this beach today, although it seems probable that originally there were some which were completely levelled by Algonquin waves. West of Meaford a sandy ridge standing above the sand plain is probably a fragmentary moraine. Within the town of Meaford one finds another bouldery terrace backed by a 25-foot bluff marking the highest water level of the Nipissing Great Lakes. The steep-sided valley by which the Bighead makes its way across the old shorelines is more than 50 feet in depth.

The drumlins are formed of a greyish-brown calcareous till which shows a considerable admixture of red shale. In this respect it is not unlike that found in the drumlins of New York State south of Lake Ontario.

While the unique locality is undoubtedly confined to the valley below the escarpment rim, the Bighead drains some drumlinized territory on the plateau above. In addition, drumlins belonging to this same group are found beyond the confines of the drainage system in the vicinity of Hoath Head, Rockford, and Chatsworth. Around Hoath Head they are low, rather widely spaced, and rest on an almost bare dolomitic rock plain. Near Chatsworth they are from 100 to 150 feet in height, making up a very rough landscape hardly to be distinguished from that of the nearby moraines.

The well-drained stony clay loam on the drumlins is a productive and adaptable soil. Containing more shale and igneous rocks than most of the drumlins in Southern Ontario, it is a slightly acid soil like the Milliken loam of Scarborough township. Only where the subsoil has been exposed on the steepest slopes is there free lime carbonate in the surface soil. The leached A_e horizon has the pale buff colour which is generally associated with shale. The profiles are 18 to 30 inches deep with a deep B horizon of slightly blocky clay loam. On the drumlins above the escarpment in the Chatsworth area the soils are more stony and calcareous like those of the Arran drumlin field.

In the lake plain near Meaford the sandy soils are closely related to those in the Algonguin bed around Alliston, but have a slight reddish cast due to Queenston shale. Near the Algonquin beach the water table is high and there is some swamp. Towards the present shore, above the bluff of Lake Nipissing, drainage is better, and the sandy soil is well adapted to apples and truck crops which can pay the cost of fertilization.

The Bighead valley appears as one of the better sections of Grey county. Here again, by a fortunate circumstance, the survey pattern conforms with the grain of the country and permits cultivation along the contour rather than diagonally up and down the slopes. Where the drumlins are big, the slopes are too steep for cultivation and a great deal of land is left in grass. From the fact that this area lies in four different townships it is not possible to quote representative statistics at present. The pattern of land utilization, however, appears to be much the same as that of the Guelph and Teeswater areas, the economic mainstay of agriculture being the production of beef and pork.

On the loamy terraces near Meaford, which also have some climatic advantages, a number of farms are specializing in apple growing with a minor emphasis upon small fruits. It would seem that here, no less than in the Beaver valley, considerable opportunity for specialization exists.

Meaford (3,800), although offside, is undoubtedly the trading centre of the valley, but some trade from the upper end is drawn towards the city of Owen Sound. Meaford, however, owes as much, if not more, to its position on Georgian Bay, as it does to the wealth of its hinterland. Within the valley itself Bognor is the only hamlet of any size. Walters Falls, at the point where Walters Creek breaks over the escarpment, is perhaps a little larger. Both of these developed as mill sites, as did Massie in a similar position on the Bighead itself.

CAPE RICH STEPS

The northern half of St. Vincent and Sydenham township stands between the Owen Sound and Nottawasaga Bay. In preglacial times it was the upland between two river valleys leading to the master stream that flowed down the Georgian Bay depression. From the water's edge at 580 feet the

72. *Block diagram of the Cape Rich foreland*

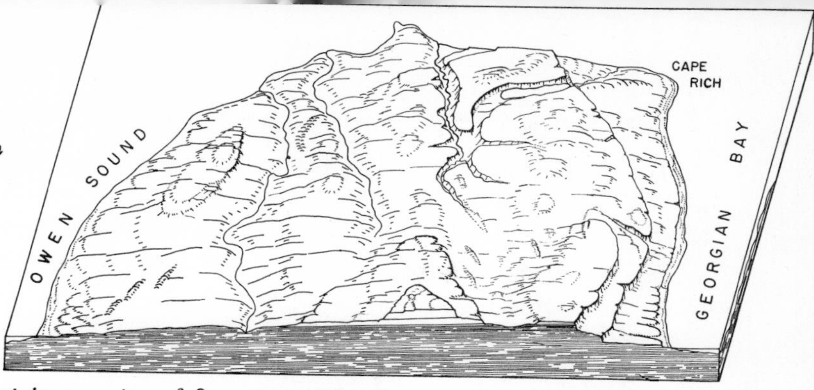

land rises 500 feet in a series of five steps. The first two are the work of Lake Nipissing and Lake Algonquin and are narrow terraces near the shore of Georgian Bay. Above the Algonquin level the next "tread" is a broad gentle slope leading up to the edge of the Manitoulin dolomite. It is based on red shale and there is very little glacial till on it apart from the few drumlins around Annan. Manitoulin dolomites overlies the red shale on a mesa-like remnant standing 1,100 feet above sea level, situated halfway between Cape Rich and Meaford in concessions VIII and IX of St. Vincent. A narrow saddle of red shale separates this tableland from a shelf of similar dolomite, around the upper cap of Lockport dolomite. A small, low mesa of Manitoulin dolomite lies north of Johnson, on the Owen Sound side. The upper step may be recognized as the brow of the Niagara escarpment in this section.

The two lower terraces are cut in shale and are strewn with boulders or gravel beaches. Around Cape Rich and Balaclava, stratified clay occurs on the second terrace. The beds are dissected by gullies in the Balaclava area, while just above the two bluffs gullies cut by rapid streams appear frequently.

The only lake in this little region is Mountain Lake or Carson Lake. It is long and narrow, and lies in a preglacial valley dammed by a great gravel beach. Sucker Creek runs for nearly two miles in this same valley farther west.

The ancient lake benches south of Cape Rich must be considered as part of the Georgian Bay apple belt, although since 1943 this land has been part of a military reserve. The red clay soil which comprises the greater part of the area may be appraised by the fact that much of it was in grass before 1943 and used only for hay and pasture. The shallow soil on the dolomite is best used for pasture or forest. There is some good land on the drumlins. The southern border of Georgian Bay is somewhat droughty in summer and crops, such as alfalfa, which resist drought or winter wheat that ripens before the hottest part of the summer are best suited to these conditions.

Well water is scarce on many of the farms, which is another reason for not carrying on intensive farming on most of these 50 square miles.

HORSESHOE MORAINES

The Port Huron morainic system forms the core of a horseshoe-shaped region flanking the upland that lies to the west of the highest part of the Niagara cuesta. The associated meltwater stream deposits are also included giving the region two chief land-form components: (*a*) the irregular, stony knobs and ridges which are composed partly of till and partly of kamey deposits; and (*b*) the more or less horizontally bedded sand and gravel terraces and swampy valley floors. The northern section, in Grey county, includes several tracts of shallow, stony drift on the Niagara cuesta and, also, a few scattered groups of drumlins. The toe of the horseshoe region lies on the highest part of the upland south of Georgian Bay at about 1,700 feet above sea level, while the two heels are about 900 feet lower. The total area is approximately 2,158 square miles.

The southwestern limb of the region, typically as seen in the southern part of Huron county, has a fairly simple landscape. Structurally it consists of two, and in some places three morainic ridges composed of pale brown, hard, calcareous fine-textured till, with a moderate degree of stoniness. The western flank is marked by the low gravel beaches of glacial Lake Warren beyond which the slightly sloping lacustrine plain extends westward to the shorecliff of Lake Huron. The eastern flank is marked by the old spillway containing flat sand and gravel terraces and some linear, undrained swampy areas. The wide sand plain southwest of Hensall represents a delta where the spillway debouched into Lake Whittlesey. South and southwestward the clay ridges become slightly lower and smoother and considerably broader in the northern part of Lambton county. There are small sand plains near Ailsa Craig and Warwick.

HURON CLAY LOAM

A_h 4–6 inches, friable, granular, dark grey-brown, clay loam; pH 6.5–6.8.
A_e 4–8 inches, slightly platy, friable, pale brown clay loam; pH 6.5–6.8.
B 10–12 inches, dark brown to dark grey-brown, moderately and irregularly blacky clay; few stones; pH 7.0.
C pale brown, moderately stony clay till, hard and fragmented when dry, somewhat plastic when wet; pH 7.5+.

Huron clay loam is the most representative soil type on the morainic ridges and it occurs quite widely in other well drained areas as well. The average depth of the Huron profile is from 18 to 20 inches and even though the slopes are generally moderate, the top soil is susceptible to erosion. More gently sloping or almost flat areas also occur on the moraines. In these the subsoil does not develop a true B horizon and tends to remain wet and plastic. A variety of crops including wheat, beans, and corn may be grown but extensive areas on these clay ridges are devoted to pasture. North of Clinton this region is more complex. The clay till ridges and the intervening spillways are in good form as far north as the great depression containing the Greenock Swamp. East of the clay ridges, there is a belt of gravelly hills, about five miles in width which has been called the Wawanosh moraine. It contains large areas of coarse and poorly sorted materials as well as areas of well-sorted sandy materials. The soils developed on the former are included in the Donnybrook series, so named from the little hamlet near which the Maitland river breaks through the moraine; the soils of the latter are included in the Waterloo series because of their resemblance to similar soils in Waterloo county.

In the Walkerton-Hanover area, the morainic belt is rather narrow, being restricted by the Saugeen clays to the north and the Teeswater drumlin field to the south. It is also deeply cut through by the valleys of the Teeswater and Saugeen rivers.

Instead of the Huron clay loams of the till ridges to the southwest, the till soils of this area are closely related to the Harriston of the Teeswater drumlin field; the soils of the intervening terraces are similar to the Teeswater loams.

The toe of the "Horseshoe" lies on the high country or plateau in the central part of Grey county. The townships of Bentinck, Normaby, Holland, Glenelg, Euphrasia, Artemesia, Osprey, and Collingwood are covered by a complex of till ridges, kame-moraines, outwash plains, and spillways, interspersed with more smoothly moulded till plains and drumlinized areas. Local relief among the moraines and drumlins is often more than one hundred feet. The area also contains many small lakes and streams, and numerous swampy areas. The tills of this area tend to be loamy and contain numerous stones and boulders, in large measure derived from the Lockport dolomites. On these loose, stony tills the well-drained soils are classified in the Osprey series. They are stony, have usually a very shallow profile and are alkaline in reaction, being referred to the Brown Forest soil group. From Singhampton south to Caledon the moraines lie along the brow and slopes of the escarpment. There are three moraines with trains of outwash between them, the latter sand and gravels terraces often being deeply cut by gullies.

From the edge of the escarpment in Caledon township the moraines trend somewhat west of the Niagara escarpment forming a belt with moderately hilly relief passing east of Acton and Guelph to Galt and Paris. South of Paris the moraines tend to flatten out and finally disappear under the sands of Norfolk county. Associated with the moraines is a system of old spillways with broad gravel and sand terraces and swampy floors. These are larger than those already noted in the southwestern limb of the "Horseshoe." Good cross-sections of this landscape may be seen along Highway 7 from Rockwood to Georgetown, along Highway 6 south from Guelph to Puslinch, or along Highway 401 across Puslinch township. Some of it is very hilly with a local relief of more than 100 feet, often with steep irregular slopes and small "kettles" or enclosed basins which contain water in the spring and early summer. The soil material is a coarse, open, stony till composed largely of dolomite with traces of red shale. The soils of these rough and stony surfaces are referred to the Dumfries series, so named from the townships south of Galt. Dumfries profiles are usually from 18 to 24 inches in depth. The soil profile which is often seen in ditches, road-cuts, and borrow pits may be represented as follows:

DUMFRIES SERIES

A_h	2–4 inches, dark greyish brown, friable stony loam; pH 7.0.
A_e	7–9 inches, pale brown, friable, stony loam; pH 6.6.
B	6–10 inches, brown to dark brown, blocky, stony, clay loam; pH 7.2.
C	pale brown, sandy, stony loam till; highly calcareous.

The soils of the associated outwash gravels are classified mainly in two series, Caledon and Burford. The former usually has a greyish brown sandy loam surface soil underlain by a deep, light yellowish brown A_e horizon, and a dark yellowish brown B horizon, the whole profile often being three feet in depth. The Burford, on the other hand is usually a gravelly loam with a depth of about 18 inches and slightly darker colours. The surface soils of both series are only slightly acid while the underlying materials are highly calcareous.

LAND USE AND SETTLEMENT

The patterns of land use and settlement have been variously affected by the physical patterns of the Horseshoe moraines. There is much hilly and droughty land and, whether on the hard Huron clay or the various knobby gravelly areas, this leads to extensive pasturage. Farms therefore tend to be larger than the average for southern Ontario but since, in many cases, only parts of townships are involved, it is difficult to get a statistical measure of the full effect. Throughout the whole area one sees many abandoned farmsteads and vacant building sites but the land is still in use, more often as pasture than crop land. In several townships less than 50 per cent of the total acreage is recorded as improved land while a relatively high percentage is in woodland. In Glenelg and Holland townships which represent the extreme condition in the "toe of the Horseshoe" in Grey county only 20 per cent of the total land area is devoted to crops.

Throughout this extended region agricultural land is used chiefly for livestock production. In most parts of the area farms average 170 acres in size, with about two-thirds classed as improved land. The census of six representative townships in 1961 showed populations of about 18 cattle, 11 hogs, 4 sheep, and 50 poultry, a total, perhaps of 21 animal units per hundred acres of farm land or 34 animal units per average farm. (An animal or livestock unit is reckoned as 1 horse, 1 cow, 5 hogs, 7 sheep, or 100 hens.)

There is little doubt that the Horseshoe moraines now have less than half the rural population which they held at the beginning of the century. In the central part of Grey county there are only ten farm people per square mile. There is also a non-farm population of about five per square mile.

Quite a different story, however, can be told of Puslinch township. Although most of its area is made up of rough moraines, gravel spillways, and swamps, and only 50 per cent of its area is reported as crop land, it now has a considerably larger total population than it had at the turn of the century. Its farm population is considerably more dense than the average for southern Ontario (30 per square mile); farm families are somewhat larger and farms are somewhat smaller than the average. In addition, quite a considerable majority (60 per cent of the total) is non-farm population. This is because the township is a close neighbour to two growing cities, Galt (30,000) and Guelph (40,000), and is crossed by two major provincial highways and several county roads. Obviously, agriculture is not the major source of employment, the suitability of the land for agriculture is not a matter of major concern, and even many of the farm families obtain a large part of their living from non-agricultural sources.

Whereas the Niagara escarpment has been cited as the most important location of rock quarries in southern Ontario, the moraine and spillway systems which almost encircle the upland of peninsular Ontario provide the largest workable sand and gravel deposits. Particularly important are the pits in the spillway terraces, in the Grand valley near Paris, and Preston, and along the Speed River near Guelph (69). The Caledon pits have already been mentioned. Another large source of gravel of excellent quality has been worked for many years near Durham in Grey County, and there are, of course, a great number of small pits. The main deleterious constituent in these gravel deposits is siltstone derived from the shale formations below the Niagara escarpment and it is most abundant near the escarpment. Generally the quality is very good.

Despite the fact that land use in the region, as a whole, tends to support a somewhat sparse rural population, there are a number of urban settlements. The largest and most important of these are the Grand River towns including Brantford (55,000), Paris (6,000), Galt (30,000), Preston (12,000), Hespeler (5,000), Kitchener (80,000), Waterloo (25,000), and the Speed River city of Guelph (40,000). Such centres were not built by and for the moraine-spillway area alone but to serve much wider regions. Indeed they have become part of the great manufacturing complex of southern Ontario. However, as has already been indicated, they may have an overpowering effect upon land use in nearby moraine areas. Two smaller towns on the eastern limb of the complex are Acton (4,500) and Orangeville (5,000).

Even more closely associated with these hilly lands is the northern group of little towns including Walkerton (4,000), Hanover (4,500), Durham (2,250), Flesherton (525), Markdale (1,100), and Chatsworth (400). After driving through the area one is almost pursuaded that the size and prosperity of these settlements is a very direct reflection of the intensity of the agricultural activity in the surrounding countryside. Of course, Walkerton is a county seat, and Hanover is well known for its furniture and other manufacturing activity, but the moraine is less bulky and large areas of clay plain and drumlinized till plain provide the basis for better soils. Gravel terraces are more widespread than kamey hills, and the gravels are more deeply covered with loam than in some other places. The course of this section of the Saugeen River is definitely related to the trend of the moraine, and the settlements were originally located near power sites. Their growth and importance, in turn, was the cause of the development of the patterns of railways, provincial highways, and country roads, which serve to connect them with their umlands and with more distant parts of Southern Ontario. Since the turn of the century the townships of their trade areas or umlands have lost 40 per cent of their population but these towns have doubled in size. Durham, another power site on the Saugeen, was at one time larger than Hanover and, in pre-railway times, had closer connections with urban Ontario. Durham is centrally located to serve four townships but its umland has a lower agricultural potential and has lost 50 per cent of its population. The town has gained 50 per cent. It shows less commercial and industrial activity than its downstream neighbours but has long had an emphasis on wood products.

Flesherton, Markdale, and Chatsworth are obviously related to the old colonization road leading to Owen Sound. Railway services were later also provided along the same route. Each village has, essentially, a cross-road location with respect to its surrounding area. The region has a low agricultural potential, a fact which is emphasized by the reforestation of a fair-sized area. The surrounding townships have lost more than 50 per cent of their population, while the villages, despite improved transportation, have gained less than 25 per cent.

One cannot leave this area without mention of some of the smaller hamlets. Most of them with a population of a hundred and fifty persons or less. Many of them were associated with small water-power sites which sustained the sawmills and grist mills. Among these might be mentioned Feversham, Walters Falls, Priceville, and Holstein. Hidden away from the main routes of travel, they perform the ancient role of service centre and help to preserve some of the tradition of rural Ontario.

FLAMBOROUGH PLAIN

An isolated tract of shallow drift on the Niagara cuesta north of Hamilton has been named the Flamborough plain since it spans Flamborough township and takes in parts of Nassagaweya and Beverley townships. It is an area of about 150 square miles, bounded on the northwest by the Galt moraine, and on the south by the silts and sands of glacial Lake Warren. A few drumlins are found scattered over this limestone plain and swamps are plentiful. The limestone has been washed bare in places, particularly near the edge of the escarpment on its eastern border. The plain slopes to the south from 1,200 to about 900 feet above the sea. What little overburden there is on the bedrock, apart from the drumlins, is either bouldery glacial till or sand and gravel.

Some of the drumlins are well moulded and typical, and those north of Freelton have irregular contours and dimensions. In fact several of them appear to consist of two or three roughly formed drumlins joined together. Gravel bars on their crests are quite common.

The plain is drained mainly by two small streams. Spencer Creek serves the Beverley swamp and the area north of it and flows into the Dundas valley. The town of Dundas taps it for its water supply. East of Beverley swamp four small streams tributary to Bronte Creek must be crossed on the way to Campbellville, while the upper reaches of Oakville Creek take over north of that village. The swamps and gravels of the upland provide these two streams with permanent flows.

Good soil is not plentiful in this little region: the soil is either wet or stony and shallow. Thus most of the area is still in woods or pasture. The cultivated soil is found mainly on the drumlins and deeper gravel terraces. The swamps serve as water reservoirs and produce cedar posts and other wood. In one instance, black muck and marly soil are dried and sold as soil conditioner, and in another, marl is utilized.

DUNDALK TILL PLAIN

The "roof" of peninsular Ontario in the counties of Dufferin, Grey, and Wellington comprises an area of 925 square miles of gently undulating till plain. North and west of Dundalk, low drumlinoidal swells appear with their long axes oriented northwestward. A few low drumlins are also included in the west, adjacent to the Teeswater drumlin field. The main part of the area is a fluted till plain, the flutings running southeastward; in other words the surface appears planed but scored by shallow troughs which are barely perceptible to the eye. The region is bounded on the east by moraines and some morainic ridges lie inside the boundary near Shelburne and Orangeville.

With an elevation of 1,400 to 1,750 feet this region forms the watershed from which issue the headwaters of the Saugeen, Maitland, and Grand rivers, as well as those of the Nottawasaga which take a shorter exit eastward over the Niagara escarpment. Numerous small flat-floored valleys

73. *Oblique aerial view of the Dundalk plain west of Shelburne, showing numerous depressions and swamps*

form a network over the plain and connect with either the Grand or the Maitland spillway systems. The valleys are frequently swampy, containing small underfit streams or no streams at all.

The plain is characterized by swamps or bogs and by poorly drained depressions. Indeed, it is related that the original surveyor, having run his lines around the areas now constituting the townships of Luther and Melancthon, simply entered them on his map as "all swamp." Receiving orders to subdivide them into farm lots he did so, but commented that it was the meanest land he had ever surveyed. He determined to name them after the meanest men he had ever heard of, so being a devout Roman Catholic he could think of none more appropriate than the leaders of the Protestant reformation. Attempts have been made to alleviate the meanness by township drainage schemes but without much economic benefit.

One such large scheme involved the drainage of the Luther bog, one of the important source areas of the Grand River. While nothing was gained in the sphere of agricultural land use, the results had a deleterious effect upon the water supply of the river. After acquiring the land the Grand River Conservation Commission placed dams or dykes across the outlets of the swamp, creating a reservoir of some 10,000 acre-feet with which to supplement the summer flow of the stream.

Most of the area carries a superficial deposit of silt, probably windblown and comparable to the much deeper deposits of loess in the Mississippi valley. The material is usually less than two feet in depth and often only a few inches so that it is all included in the weathered soil profile. It is deepest north of Shelburne around Hornings Mills and Honeywood where fairly extensive beds four or five feet thick occur.

In the northeast, a small part of this plain is the work of the Lake Simcoe ice lobe. The drumlins in this area point southwestward and thus were built by an ice sheet advancing over the Niagara escarpment from the east. A small sandy moraine appears four miles west of Redickville to mark the limit of this advance. The till left in this area is more pervious than the Dundalk till. In front of the Singhampton moraine there are shallow deposits of outwash which vary from gravel to fine sand. On the surface this area has unusually deep deposits of the uniform aeolian silt that blankets this whole plain. The end result of these deposits is well-drained soil, especially important in this high, cool region where poorly drained soil is general. This is probably aided by vertical drainage into cracks in the bedrock, a feature of the area just back of the Niagara escarpment. The effect on the quality of the farms shows up unmistakably between Hornings Mills and Maple Valley. This is an area specialized to potatoes. With its red barns the prosperous appearance of this small area is not surpassed in any other part of this region.

The original vegetation of the better-drained areas was a hardwoods association of maple, beech, and some birch; but swamp forest containing elm, ash, cedar, and tamarack probably occupied a larger area. Although usually classed as a mixed hardwood forest section, the altitude of this region had sufficient effect to bring into both upland and lowland forest a sprinkling of northern species. White and black spruce, white pine, tamarack, balsam fir, alder, white birch, aspen, and willows are fairly common. Many diverse species of sedges and rushes inhabit old sods and roadsides where silverweed also abounds.

In spite of the elevation, drainage is slow on this high plain and gleisolic soils are more extensive than zonal soils. Furthermore, although the zonal profiles must be classed in the Grey-Brown Podzolic group they are more or less modified. The area is one of transition, as might be inferred from the spruce intermingled with the hardwoods. It is moister and cooler than most of southern Ontario and shallower brown podzolic or weak podzol profiles are found superimposed on the Grey-Brown Podzolics. The most noticeable effect of this is the darkening of the upper part of the A_e horizon of the Grey-Brown Podzolic profile.

The soil map of Northern Wellington includes large areas classified in the Huron, Perth, Brookston, Harriston Listowel and Parkhill series. In all cases the surface soils are loams or silt loams, regardless of the nature of the underlying till. This is due, actually, to the presence of a separate geological deposit, quite probably loess or windborne material which overlies the pebbly till to a depth of 12 to 24 inches. The physical effect of this material, which is more pervious than the boulder clay beneath, is to form a watersoaked layer that is slow to dry out in the spring, thus preventing early cultivation of the land. In many cases, also, this material controls the depth of the weathered profile. Soil descriptions show the A and B horizons to be stone-free in comparison with the underlying till. Eastward, in northern Dufferin county the surface deposit may be from four to six feet in depth; soil weathering here is considerably deeper than in the rest of the plain. Named after a small village in the area, the soil series is known as Honeywood. The profile may be developed to a depth of two and one half to more than three feet without reaching the base of the deposit.

HONEYWOOD LOAM

A_h 3–5 inches, dark greyish-brown, friable crumb-structured loam; pH 6.6.
A_e 10–16 inches, brown to pale brown, friable loam; pH 6.2.
B 10–14 inches, yellowish brown, slightly blocky, friable loam; pH 6.8.
C calcareous yellowish brown loam.

On the lower slopes and in the hollows except where there is muck or peat, Parkhill silt loam and Brookston silt loam are mapped. These have very dark humified surface soils which may vary from six to twelve inches in depth, underlain by glei horizons which are light brownish grey with olive mottling and of a sticky consistency. These areas are usually too wet for cultivation and if cleared are devoted to pasture. Usually where they have been neglected a weedy growth of dwarf willows has been permitted. Indeed, the presence of "willow flats" is one of the characteristics of this landscape.

LAND USE AND SETTLEMENT

On much of this undulating till plain, forest forms a very small part of the cover while grain crops, especially if the observation be made at harvest time, seem to form a very large part. Hay and pasture are dominant but at this time of year many sod fields have been ploughed in preparation for next year's seeding. The upturned soil on the slopes is brownish grey in colour, but in the flats it is definitely grey with occasional black patches. The grey colour indicates the dry glei of a poorly drained or gleisolic soil, which in its moist state is a medium olive grey, while the black colour indicates a shallow layer of muck. Only where ditches are maintained is it possible to cultivate this land.

Census figures for six townships within the region throw some light on the agricultural development. Despite the poor drainage most of the land was occupied by pioneer settlers. At the turn of the century the population averaged about 30 per square mile, most of them being resident on farms. By 1931 population had dropped to about 19 per square mile and in 1961 it was less than 16. At no time has non-farm rural population been more than 2 or 3 per square mile. Coincident with the decline in population farm size increased to about 140 acres in 1931 and 170 acres in 1961. Only 10 per cent of the total area is unfarmed. About 77 per cent of the farm land is improved while 50 per cent is used for the growing of crops. About 21 per cent is in hay, 10 per cent in oats, and 11 per cent in mixed grains (oats and barley sown together). There are small areas of barley, wheat, corn, flax, potatoes, and buckwheat. Less than one per cent of the farmland was planted to corn in 1961, but in recent years some large new silos in the area and large cornfields are in evidence, testimony to the success of hybrid corn in Ontario. Fall wheat is seen occasionally in the southern and better drained fields of the area, as are a few fields of flax, while in the northern part of the area some buckwheat is grown. In 1961, there were more potatoes than corn, most of them planted in Melancthon and Mulmur townships on Honeywood soils. Sod is represented by 21 acres of hay and 24 acres of pasture in each hundred acres of farm land. Only 9 per cent remains in woods but other land, chiefly swamp and rough pasture, makes up about 15 per cent.

The importance of livestock is shown by the fact that, on the average, there are 24 animal units per hundred acres or about 40 per farm. Using census averages again, each hundred-acre lot has 19 cattle (only 4 milk cows) 13 pigs, 4 sheep, and 115 poultry. In comparison with pre-war years the region has almost 50 per cent more cattle, a few more hogs, only one-third as many sheep, while horses have nearly disappeared, about one remaining for each two farms. Tractor power is universal with an average of 1.2 tractors per farm and an array of associated machinery.

The Dundalk plain has the coolest and shortest growing season of any farming area in southern Ontario; consequently many crops which are commonly found in other areas cannot be grown. On the other hand it is preferred over the warmer, drier climates for potatoes. For the most part the soils are a little too heavy and high in lime for this crop; also they are often either poorly drained or stony. However, the crop is an important specialty on the Honeywood soil around Redickville and Honeywood where the well-drained, silty soil supports a group of excellent farms. Aside from this and a slight concentration of dairying around the small towns, the agricultural income of the plain is derived from rather extensive mixed farming with an emphasis on beef cattle, supplemented by the sale of hogs, cream, poultry, and eggs.

This high tableland was somewhat isolated from the early waves of settlement in southern Ontario, most of the land being taken up during the latter part of the nineteenth century. Early settlement was more or less tied to the colonization roads Hurontario Street and the Garafraxa Road which have now become highways 10 and 6 respectively. The rectangular township surveys were made later and had to be fitted around these early surveys. A slight modification of existing nucleated settlement took place after the building of the Toronto Grey and Bruce Railway in the 1870's, but in large measure it sought to service existing settlements along the routes to Owen Sound. The railway tended to parallel the main roads, which in turn have become provincial highways; thus the more important towns have maintained their original positions, while railway junctions such as Fraxa and Saugeen have not developed at all.

The chief centres on the plain are Dundalk (850), Shelburne (1,250), Arthur (1,200), Mount Forest (2,700), and Listowel (4,000), while Orangeville (5,000), Grand Valley (650), Harriston (1,650), Palmerston (1,550), and Drayton (650) are on the margins. A few small settlements developed at railway stations, but for the most part the names on the map are those of former post offices, discontinued when rural mail delivery was instituted.

In comparison with other sections of Ontario the Dundalk plain has rather a severe landscape with some resemblance to the plains of Saskatchewan. The forest was almost completely cleared and few farmers planted many trees about their homesteads. Perhaps an old ragged orchard may persist, but the rows of elms and maples, so characteristic of other areas, are conspicuous by their absence.

74. *Dundalk plain, East Garafraxa township. The smooth gently undulating terrain and the scarcity of trees and shrubs is characteristic.*

With a few notable exceptions, farm buildings follow a simple plan. The houses are small and square, with either a pyramidal or gable roof, and are of brick, cement block, or frame construction. Usually the barns are grey and square with a gable roof and a field-stone basement with insufficient window openings. Usually there is a low implement shed and a garage, but few other outbuildings. A windmill is often part of the farm equipment.

The austerity of the plain is emphasized by the many square barns standing alone against the sky, by deserted houses, and stone foundations from which the buildings have long since vanished. These vacant farms should not, however, be considered as abandoned lands, for each hundred acres provides pasture for twenty or twenty-five beef cattle. The farms should have been larger from the beginning.

These impressions were derived from field notes recorded many years ago, yet they need only a little retouching to bring them up to date. Farms are larger and more homesteads have been vacated. Some barns have been painted and the basements whitewashed. New silos have appeared, houses have been repaired, and electric services now appear to be universal. But it is still a lonely region with only about half the people it once had. There has, as yet, been very little of the building of rural non-farm residences that dot the concessions and side-roads of so many Ontario countrysides. The old rural Ontario still survives on the Dundalk plain.

STRATFORD TILL PLAIN

The city of Stratford is situated on a broad clay plain of 1,370 square miles, extending from London in the South to Blyth and Listowel in the north with a projection towards Arthur and Grand Valley. It is an area of ground moraine interrupted by several terminal moraines. The moraines are more closely spaced in the southwestern portion of the region; consequently that part resembles the Mount Elgin ridges. The northern half of the region is mostly level, modified by one or two moraines. The over-all slope is towards the southwest, from approximately 1,500 feet to 900 feet above sea level. The highest section is drained by the Conestoga and Nith rivers, tributaries of the Grand. The Maitland River serves another small part, but most of the central and southern portion is within the Thames watershed. The divides between these watersheds are vague, however, at times amounting to belts in which drainage has not been established. This plain differs from the Dundalk till plain in having a faint knoll and sag relief rather than a fluted surface.

Throughout this area the till is fairly uniform, being a brown calcareous silty clay whether on the ridges or the more level ground moraine. It is a product of the Huron ice lobe. Some of the silt and clay is limestone flour, probably a good deal of it coming from previously deposited varved clays of the Lake Huron basin. The silt and clay contents vary within certain limits and so does the stoniness, but it is seldom a stony till. Shallow surface deposits of silt are less prevalent than on the Dundalk till plain, but they cover sizable tracts in northern Middlesex county and elsewhere. Sand or gravel is often present in the inter-morainal valleys south of St. Mary's.

Gravel for road building and other construction is not plentiful in this region. The area north of Seaforth is the best supplied from eskers and kames. One occasionally sees kames farther south, but there the most common sources of gravel are the terraces along the Thames River.

The Stratford till plain, being part of the slope east of Lake Huron receives more rain and snow than is usual in Southern Ontario. It is a muddy country and late spring seedings are an unfortunate probability that must be balanced against the advantage of few summer droughts. In winter the difficulties of transport in this snow-belt are well known.

The soils of the Stratford plain may almost all be classified in the Huron catena based on the heavy-textured calcareous till which is so widespread in the region once covered by the Huron ice lobe of the Wisconsin ice sheet. The moderate ridges of the Lucan, Mitchell, and Milverton moraines, as well as some other well-drained areas, have a Huron profile which is a fully developed Grey-Brown Podzolic; the gently sloping flanks of the ridges and the slightly undulating parts of the plain also have a Grey-Brown Podzolic profile, but with a tendency to glei formation at the base of the profile, known as the Perth series; the more even parts of the plain, often so flat as to resemble the floor of an ancient lake, are poorly drained and have developed the rusty mottled, olive grey, glei horizon which is characteristic of the Brookston series. The surface soils may be either silt loam or clay loam depending upon the presence or absence of the surficial silt deposit. Both the Brookston and the Perth soils require artificial drains to permit a fully developed cropping system. Municipal

---

HURON

A_h 4–5 inches, very dark brown, medium granular, friable, stone-free, silt loam or clay loam; pH 6.9.

A_e 6–7 inches, light yellowish brown, weakly platy, friable, almost stone-free, clay loam; pH 6.8.

B_1 4–5 inches, light greyish brown, medium blocky, slightly stony, clay loam; pH 7.0.

B_2 6–8 inches, dark-brown, medium blocky, hard, slightly stony, clay; pH 7.0.

C brown, blocky, stony, hard calcareous till.

The total depth of the solum may be from 20 to 25 inches.

PERTH

A_h 5–6 inches, very dark greyish brown, granular, friable silt loam or clay loam; pH 6.8.

A_e 2–4 inches, yellowish brown, weakly platy friable clay loam; pH 6.9.

B 7–12 inches, dark brown, blocky, hard, slightly stony, clay.

C light greyish brown, slightly stony, calcareous till.

BROOKSTON

A_h 6–7 inches, black, finely granular, stone-free silt loam or clay loam; pH 7.0.

G_1 5–7 inches, greyish brown, mottled with olive brown, blocky, stone-free, clay loam; pH 7.0.

G_2 12–15 inches, light brownish grey strongly mottled with yellowish brown, coarsely blocky, plastic clay; pH 7.2.

C greyish brown, blocky, slightly stony, calcareous till.

The solum may vary from 18 to 27 inches in depth.

ditches are integral features of the landscape, while tile drains have been installed in thousands of acres and more are being installed every year. The more sloping Huron soils are susceptible to erosion. In many places the surface soil has been washed from the hillsides, leaving sticky brown clay or even the greyish subsoil exposed, with the result that crops on these fields are very uneven.

The situation in the Stratford plain provides a convenient opportunity to demonstrate the soil group known to soil scientists as the catena. The Huron, Perth, and Brookston series form such a natural group, their differences being exhibited in these profile descriptions.

The soils of the Stratford plain have good natural fertility, a good supply of lime in the subsoil, and few obstacles to cultivation once the necessary drainage facilities have been supplied. This is one of the more productive agricultural areas of the province.

LAND USE AND SETTLEMENT

Census figures show that there is no vacant land. Every rural lot, and some of the areas within urban municipal boundaries, is part of a farm operation. This was the case in the 1920's also, when farm size averaged 96 acres, while today the farms are roughly 124 acres in area and there is 23 per cent fewer farms. The total improved land has actually increased from 87.5 to 90 per cent, leaving only 10 per cent to be accounted for by woodlots, marsh, and rough pasture. Crops are grown on 61.5 per cent of the farm area leaving 28.5 per cent for pasture, orchards, feed lot, homesteads, and farm lanes. There is only about 1.2 acres of pasture for each grazing animal, not counting the calves. On the average 124-acre farm, 76 acres are in crops, including wheat, 2 acres; oats, 14 acres; barley, 1 acre; mixed grains, 24 acres; hay, 27 acres; corn, 5 acres; and other crops, 3 acres. In Perth county, at the time of writing, about five-sixths of the corn acreage was used for silage, but the production of shelled corn was on the increase. Other crops included flax, beans, soy beans, sugar beets, oats cut green as a soiling crop or dried for hay, and a variety of other forage crops.

On this clay plain, as on the one to the north, livestock raising is important. Each hundred acres of farmland supports about 37.5 animal units, making about 47 per farm. Each farm has about 30 cattle, including 10 milking cows, 34 hogs, 1 sheep, and 230 poultry. There is only one horse for each four farms on the plain, while each farm averages 1.4 tractors. Automobiles and hydro-electric service are found on almost every farm. Census figures show that the average acre of farm land on this plain yields a 70 per cent greater gross return than the average for farm land in Ontario. This stands in such strong contrast with already discussed Dundalk plain, lying almost continuously with it, that we need a close look at the land use practices and their results. Comparisons might, also, usefully be made with the Guelph, Oxford, and Teeswater till plains which have fluted for drumlinized surfaces and better drainage.

Over 95 per cent of the gross agricultural income is from livestock and livestock products; most important being cattle (28.5 per cent), hogs (26.5 per cent), milk (23.0 per cent), poultry and eggs (16.3 per cent). Perth county is usually shown as part of the southern Ontario dairy belt and it has an important concentration of dairy cattle with almost as many cows as Oxford county; yet in 1961, 40 per cent of the cattle population was kept for beef and the percentage seems to be increasing. Thousands of steers are brought from western Canada each year to be finished in barns and feed lots or on pasture. The total out-turn of hogs is greater than that of any other county, although the concentration is not so great as in Waterloo; the same can be said for poultry production. In recent decades this plain has lost little of its farm land (less than 2 per cent in 30 years), and has lost only one-fifth of its farmers. In 1941, the average farm had an area of 124 acres instead of 98 which was the average in 1931; this is an increase of 25 per cent. This degree of stability is shown by few other areas in Ontario. The province as a whole has lost 37 per cent of its farmers, while farms have grown from 118 to 154 acres in size, an increase of over 30 per cent, while nearly 20 per cent of the occupied land has been abandoned or passed to other uses.

There are no large cities on the plain; Stratford (21,000) has gained an average of only 100 persons per year and shows little tendency to sterilize adjoining fertile land. The small towns and villages, also, show only modest growth. The population of Perth county (57,000) has gained only 11 per cent in 30 years (15 per cent in 60 years) and it remains one of the more rural counties in Ontario. Neighbouring counties, containing the growing cities of London (180,000), Brantford (57,000), and the Galt–Kitchener–Waterloo complex (155,000) help to provide markets for agricultural produce, as do the more distant Metropolitan Toronto (1,825,000) and Hamilton (400,000) urban areas.

The relationship of Stratford to its plain is of some interest, for it is not centrally located. Its inception about 1830 during the days of the Canada Company was as a stopping point and stream-crossing on the colonization road, which just happened to coincide with the junction of five survey lines which also became travelled routes. Having thus come into existence it later became the focus of five railway lines to other centres of population in Ontario. One cannot but conclude that the Avon, insignificant as it may be among the rivers of Ontario, is the prime focus of Stratford, and Stratford has made the most of it. Dammed to provide a shallow lake in a scenic park it has given Stratford an attraction which many other towns might have copied but have not. No other town but Stratford could have a Stratford Festival, of course, but the existence of the old spillway trough, incised across the clay plain, provides the basis of the scenic qualities of the site. Thus inspired, man has done the rest.

Because of the off-centre, or threshold, situation of Stratford, the relationship of the clay plain is that of hinterland, rather than umland. Smaller local centres, therefore have arisen, including the towns of Exeter (3,000), Mitchell (2,200), Listowel (4,000), St. Mary's (4,800), and Seaforth (2,300); the villages of Hensall (950), Milverton (1,100); and numerous small unincorporated places of 100 to 500 population. St. Mary's is a special case, existing not so much to serve the countryside, as by reason of its own resources, chiefly quarrying and cement manufacture. In any case, it lies pretty well within the London sphere, to which the southern end of the clay plain belongs, having no independent villages of its own.

HILLSBURGH SANDHILLS

The Hillsburgh sandhills form a natural boundary on the southeastern flank of the Dundalk till plain. Extending from Orangeville to Hillsburgh and Belwood, they cover approximately 64 square miles. Although this was the first land to appear as the glacier melted, it must not be concluded that it is the highest part of the upland. However, it is all over 1,400 feet above the sea and the highest ridges reach 1,600 feet, which is only 150 feet less than the highest part of the Dundalk plain 35 miles to the north. The rough topography, sandy materials, and the flat-bottomed swampy valley running through the moraine from Orangeville to Hillsburgh are the outstanding characteristics.

The steepest slopes are those along the sides of the big spillway north of Hillsburgh, but knobby hills are general. The point to be stressed about the soil materials is the prevalence of fine sands, which is not typical of kames. In this the area is like the Waterloo hills and the sandy ridges of Albion township to the east. Three miles south of Orangeville, Caledon Lake stands in the bottom of the glacial spillway. It is a shallow, marly lake that appears to owe its existence to a mass of sandy drift dropped in the spillway just south of it. The Grand River cuts through this moraine at Belwood; that section of the valley is now the stretch occupied by the artificial lake created by the Shand dam. West of Belwood the moraine is smaller and the sand gradually gives way to boulder clay.

The Hillsburgh sandhills are noted for potatoes, due in no small measure to the finer sandy loam which is better than the soil ordinarily associated with kame-moraines. Except on the steeper slopes which are droughty and erosive, the soils are quite well suited to this cash crop. Beef cattle and hogs constitute the standard farm produce of the area.

Hillsburgh (460), an unincorporated village lying on the floor of the old spillway is a railway stop, and the largest rural supply point in the area. Orton and Belwood are smaller centres.

WATERLOO HILLS

The Waterloo Hills region occupies about 300 square miles or 192,000 acres, lying chiefly in Waterloo county but extending into Blenheim township in Brant county and North Easthope in Perth county. The surface is composed of sandy hills, some of them being ridges of sandy till while others are kames or kame moraines, with outwash sands occupying the intervening hollows. The general elevation is from 1,000 to 1,400 feet above sea level and in some places the hills are quite rough. A peculiar characteristic is the preponderance of fine sand, particularly on the surface, and in this the region resembles the Hillsburgh sandhills. Adjoining the hilly region is an extensive area of alluvial terraces of the Grand River spillway system which, although more nearly horizontal, contains similar but more uniform sandy and gravelly materials.

The Conestoga River and the till plain lying north of it separate the northern end from the main body of this area. Similarly, the sandhills in Easthope township are set apart by the Nith River valley. Both of these outlying moraines have broad swampy valleys associated with them. In the main part of the region a number of kettle lakes appear, for example, Spongy Lake, Hofstetter Lake, and Sunfish Lake. Small swamps are even more numerous.

The Baden hills, so conspicuous from No. 7 Highway, are splendid, though rather extreme, examples of kames. They stand up as domes about 200 feet high and consist entirely of gravel and sand. Around the base is an apron of outwash sand, and a kettle lake north of the highway is a typical complement.

The soils of the hilly areas are well drained and have developed as mature Grey-Brown Podzolic soils. For the most part they may be classified in four series: Guelph, Harriston, Dumfries, and Waterloo. The Guelph and Harriston are found on gentle slopes in the areas where loamy tills occur, while there are some nearly level areas which may have the imperfectly drained London and Listowel soils. Dumfries soils are found in rougher parts where the parent material is a loose gravelly till. Waterloo sandy loam is found on most of the rounded, sandy hills of the area but there are associated small areas of many other types.

WATERLOO SANDY LOAM

A_h	3–5 inches, dark brown, friable, slightly stony, sandy loam; pH 6.4.
A_e	10–14 inches, light yellowish brown, very friable, stone-free sandy loam; pH 6.4.
B	8–10 inches, brown, finely blocky, dense, almost stone free, loam; pH 7.0.
C	light grey, loose, calcareous sand.

The original forest consisted of splendid pines and hardwoods such as sugar maple, beech, wild cherry, and red oak. Only about 10 per cent of the forest cover remains, and that in a depleted condition. However, the sandhills have more remaining forest than the nearby till plains.

LAND USE AND SETTLEMENT

Waterloo county was settled in the early 1800's by people of German descent who came to Upper Canada from Pennsylvania in their Conestoga wagons. The tradition they brought with them was one of mixed field crop and livestock agriculture. Crops normally occupy about 65 per cent of the farm land but in some cases there is even more. A sample area studied by Lee (89) was found to have 75 per cent in crop. Usually about 60 per cent of the crop land is in cereals, mainly oats and mixed grains.

Ontario Department of Agriculture statistics for 1963 show that each hundred acres of farm land had 4.5 acres in wheat, 8 acres in barley, 17.3 acres in oats, 11.0 acres in mixed grains, 8.8 acres in corn, and 21.3 in hay. The most noticeable changes in the cropping system in recent years have been the increase in both grain and silage corn, and decreases in wheat and mixed grains. Seeded pasture has also increased, probably through the improvement of much which was formerly "natural pasture." Pasture occupies about 16 per cent of the farm land.

Agriculture is geared to intensive livestock production. Each hundred acres of farm land carries about 42 cattle, one horse, one sheep, 38 hogs, and 520 poultry. This is a total of roughly 42 livestock units. Most intensive of all, Woolwich township carries somewhat over 50 livestock units per hundred acres. Needless to say, no farm is an average farm: some specialize in beef production, some in hogs, and a growing number in poultry. The number of dairy farms has decreased although the population of dairy cattle has increased.

Like the rest of agricultural Ontario, Waterloo has lost some of its farm population, but less than the average. Its farm families are about the same size as they were 30 years ago and its farms are only 15 per cent larger. More than 40 square miles of farm land, however, have been absorbed by urban expansion.

The whole of the Waterloo Sandhills is included in the umland of the Greater Kitchener urban complex, which embraces Waterloo, Kitchener, Preston, Hespeler, and Galt, containing a total of 155,000 people in 1961. There can be no doubt that the market represented by this concentration of consumers is one of the reasons for the intensification of agriculture in the region.

The region is undoubtedly prosperous. Houses and barns are in good repair and usually painted. Much modern equipment is to be seen. Agriculture has made little concession to landform in this area, and the effects of a grain, hog, and poultry economy may be seen on some eroding slopes. Nevertheless the recently increased interest in cattle has resulted in more effective use of sod for both hay and pasture, which is a necessity in maintaining light textured soils on sloping land.

GUELPH DRUMLIN FIELD

Centring upon the city of Guelph and Guelph township, the Guelph drumlin field occupies an area of 320 square miles lying to the northwest, or in front, of the Paris moraine. Within this area, including parts of Wentworth, Waterloo, Wellington, and Halton counties, there are approximately 300 drumlins of all sizes. For the most part these hills are of the broad oval type with slopes less steep than those of Peterborough drumlins.

75. *Baden hills: good examples of kames*

In this they more closely resemble those of the Teeswater field. They show some variation in the trends of their axes, but all of them are referable to the ice thrust which radiated from the western end of the basin of Lake Ontario. The axes of the main group, near Guelph, bear about north 70 degrees west, while in Puslinch and Beverley townships, to the south, the direction is almost due west.

The drumlins of this field are not so closely grouped as those of some other areas and there is more intervening low ground, which is largely occupied by alluvial materials. The till in these drumlins is loamy and calcareous, and was derived mostly from the Lockport dolomites so strategically exposed along the Niagara cuesta. In addition it contains fragments of the underlying red shale which is exposed below the escarpment. In consequence it is pale brown in colour, or not so grey as the Peterborough till. Mechanical analysis of Guelph till samples have shown an average of 50 per cent sand, 35 per cent silt, and 15 per cent clay. While this represents the composition in the central part of the area, it should be noted that towards the southwest the till is considerably heavier. On the other hand it is progressively more sandy as one proceeds towards the northeast. Moreover, towards the northeast the surfaces of the drumlins are covered by a few inches of stoneless fine sand and silt which is probably wind-deposited. The till throughout is rather stony, with large surface boulders being more numerous in some localities than others.

Over-all, this is a sloping plain between 1,000 and 1,400 feet above sea level, with an average gradient of about 20 feet per mile from north to south. It is underlain by dolomites of the Lockport and Guelph formations which dip gently towards the southwest. These rocks are exposed only in the valleys since the drift cover is seldom less than 50 feet in depth.

The ice which moulded this drumlin field advanced from the southeast and the front of the melting glacier retreated in the same general direction, that is, down the slope of the plain. The drainage of the ice front was consequently able to find progressively lower and lower outlets, so that the drumlin field is furrowed by more or less parallel valleys running almost at right angles to the trend of the drumlins themselves. There are also numerous interconnecting cross-valleys which occupy deeper depressions between drumlins. Along the sides of these valleys there are broad sand and gravel terraces, while the bottoms are often swampy. The general landform pattern, then, consists of drumlins or groups of drumlins fringed by gravel terraces and separated by swampy valleys in which flow sluggish tributaries of the Grand River. Incidental to this pattern are the several gravel ridges or eskers which cross the plain in the same general direction as the drumlins, with gaps cut through in places by the meltwater spillways. The best example is the one which can be traced from Guelph to West Montrose, while two more are found within five miles to the north. In the presence of so many large stream terraces, these eskers are not so highly regarded as a source of gravel as those of the Dundalk plain.

Within this area the dominant soil materials are the stony tills of the drumlins and deep gravel terraces of the old meltwater spillways. Both types of material usually have a shallow deposit of loam which reduces the stoniness of the surface soil.

GUELPH

A_h 4–6 inches, very dark brown, finely granular, friable, slightly stony loam; pH 6.8.

A_e 8–12 inches, brown to pale brown, medium granular, slightly stony loam; pH 6.6.

B 6–8 inches, dark brown, medium blocky, friable moderately stony clay loam; pH 7.0.

C greyish brown, coarsely blocky, hard, somewhat stony, loamy calcareous till.

The depth of the profile usually ranges from 20 inches to slightly over two feet.

LONDON

A_h 5–7 inches, very dark greyish brown, granular, friable, slightly stony loam; pH 6.9.

A_e 4–6 inches, yellowish brown, mottled, somewhat blocky, friable, slightly stony loam; pH 7.3.

B 6–8 inches, yellowish brown, mottled, medium blocky, friable slightly stony, loam or clay loam; pH 7.5.

C greyish brown, blocky loam, calcareous moderately stony till.

The profile is usually somewhat shallower than the Guelph ranging from 14 to 18 inches in depth.

PARKHILL

A_h 6–8 inches, very dark brown, medium granular, friable, stone-free loam; pH 6.9.

G_1 6–8 inches, light olive brown mottled with olive yellow, medium blocky, friable, slightly stony loam; pH 7.3.

G_2 8–10 inches, light brownish grey mottled with yellowish brown, coarsely blocky, friable, slightly stony loam; pH 7.5.

C light brownish grey with brown mottling, weakly blocky, hard slightly stony, calcareous till.

The soils of the drumlins are classed in the Guelph catena which contains three members on the basis of drainage: Guelph, London, and Parkhill. Toward the north some of the soils of the drumlinized plain have been classified in the Harriston catena because of the yellowish brown colour and somewhat more silty texture of the parent material. Otherwise the soils are fairly similar.

The loamy soils of the extensive gravel terraces are referred to the Burford catena and are widely distributed in the Grand River basin. Only the well-drained member, the Burford series, is described here. Some of the spillways have poorly drained and even swampy stretches, which for the most part have not been cleared for cultivation.

BURFORD

A_h 3–4 inches, dark greyish brown, finely granular, friable, gravelly loam; pH 6.8.

A_e 8–10 inches, brown to light yellowish brown, granular, friable, slightly cobbly loam; pH 6.6.

B 4–6 inches, dark brown, medium blocky, friable, cobbly clayloam; pH 7.2.

C light yellowish brown, loose, calcareous, gravel.

The profile varies from 16 to 20 inches in depth.

Both the Guelph and Burford are good general purpose soils. The open subsoil of the Burford is somewhat of a disadvantage in dry seasons, but the Guelph on the other hand suffers a comparative disadvantage when the spring is late or the autumn wet. On the whole it is fertile, easily worked, and adaptable to many crops. It will be a durable soil if proper measures, such as strip cropping and grassed waterways, are used to control surface erosion.

The rather large areas of London loam are also quite fertile, but often need systematic tile drainage in order to be of full use. This is also true of the poorly drained Parkhill areas.

LAND USE AND SETTLEMENT

Agriculture on the Guelph till plain was traditionally of a rather generalized type but at the present time it shows signs of specialization toward the production of beef cattle, hogs, and poultry. According to the 1961 census, the average farm is about 140 acres in area with 80 per cent improved and less than 10 per cent left in woods. About 55 per cent of the land is in crop and 25 per cent in improved or seeded pasture. Considerable areas of unimproved land are also grazed. Much of the former "natural pasture," often poorly drained land has been improved by drainage and seeding, and now has a high carrying capacity. There are approximately 30 livestock units per hundred acres of farm land, or 42 per average farm, showing a great improvement over previous decades.

Each farm carries about 32 cattle, 28 pigs, five sheep, and 300 poultry. In total livestock carrying capacity this area is not far behind the Oxford till plain, long regarded as one of the more productive livestock regions of Ontario. The chief difference is that whereas about two-thirds of all farms in Oxford are dairy farms, only about one-third of those in the Guelph till plain could be so regarded, dairying being relatively well developed only near the city of Guelph.

The city of Guelph (42,000 including the suburbs) is the social, cultural, and commercial centre of this region. Founded in 1827 by John Galt of the Canada Company, it was located on a gravel terrace at the confluence of the Speed and Eramosa rivers or, what is really more significant, the confluence of the two big glacial spillways which were the forerunners of these small modern streams. As the city has grown it has spread over the surrounding hills. The Roman Catholic cathedral surmounts a drumlin at the end of Macdonald Street, while the Guelph Collegiate Institute is located on a sister summit to the west. The big drumlin to the east of the Speed River is crowned by hospital buildings, while to the northwest, Woodlawn Cemetery and Mary Mount Cemetery lie on drumlin slopes. South of the city, the University of Guelph, including the Ontario Agricultural College and Experimental Farm, occupies another group of drumlins. Guelph rivals Peterborough as a "city of drumlins" but since the slopes are less steep the pattern is less noticeable. Manufacturing firms, for the most part, seek level sites. In an earlier period many of these plants were located in the southeastern part of the city on the Pleistocene terraces adjacent to the Eramosa River. The postwar industrial expansion of Guelph, however, has largely been toward the northwest where extensive flattish areas also exist. The city limits have been expanded to include more than 5,600 acres, but urban building has extended into the township in several directions.

Although Guelph dominates its umland almost completely, there are a few other, much smaller centres. Fergus (4,000) and Elora (1,500) toward the northern edge of the area were founded at mill sites on the Grand River and became service centres for small areas. Because of the development of special facilities for the manufacture of farm equipment Fergus has become the larger settlement. Eden Mills (300), Rockwood (900), and Everton are mill sites on the Eramosa River. Erin (1,100), at the far northeastern corner of the area, is located in a continuation of the same spillway although on a branch of the Credit River.

The Guelph countryside has a pleasing appearance. Through some unknown accident the early surveyors laid out the road in accordance with the grain of the country, that is, the roads are in line with the drumlin shapes. The fields are thus less inconvenient than those of the Peterborough area while the gentler slopes are easier to till and tend to be less erosive. In any case it is easier to arrange and carry out systems of contour cultivation. Roadsides are commonly planted with rows of maple trees. Farmsteads are well built with large barns set on stone foundations. The houses, also, are often built of field stones. Woodland is more plentiful here than on the flat plains to the north and, because of the rolling terrain, it appears to be even more abundant than the 10 per cent recorded in statistics. There are fewer abandoned farms than in the adjacent morainic area. Although one sees some evidence of soil erosion on the steeper slopes, it is kept in check fairly well by the cover of vegetation. The streams in the great spillway valleys, though small, have some water in them even in the driest summers, indicating the great reservoir capacity of the Pleistocene gravel beds. Although they have been extensively worked in the vicinity of Guelph, these terraces also still retain a great reserve of useful building material.

TEESWATER DRUMLIN FIELD

Taking its name from the village of Teeswater and from the river which drains a large part of the area, the Teeswater drumlin field occupies approximately 575 square miles in the counties of Bruce, Grey, Huron, Perth, and Wellington. Lying in front of northwestern limb of the Horseshoe moraine system, its position is somewhat analogous to that of the Guelph drumlin field with respect to the eastern limb of the Horseshoe. As in the case of the Guelph group also, toward the outer margin of the field the drumlins become weaker and gradually fade into an undulating till plain. They are not confined by any well-developed terminal moraine. The orientation of the drumlin axes varies from almost north-south in the Wingham and Teeswater areas to almost northwest-southeast in the country between Palmerston and Harriston.

As in the Guelph field, the characteristic drumlin is a low, broad, oval hill with gentle slopes. The till is loamy in texture, moderately compact, highly calcareous, and pale brown or yellowish brown in colour, since the material is largely derived from the soft, pale brown limestone of the region. The Teeswater till has fewer boulders than the Guelph till because these rocks are softer and were less resistant to glacial grinding than the Lockport dolomites which predominate in the Guelph till.

The Teeswater field was traversed by large meltwater rivers draining the ice-fronts to the north and west of "Ontario Island." They cut the

'6. *Glacial till, Wroxeter. Note the* *ınsorted material and mixture of boulders* *nd pebbles of various sizes.*

broad valleys which now carry branches of the Saugeen and Maitland rivers. For the most part these valleys are cut in till but occasionally, as in the case of Formosa Creek, the down-cutting was continued deep into the underlying limestone. Associated with most of these valleys are broad terraces of sand and gravel, filling much of the low ground between the drumlins and creating the same sort of "drumlin and gravel flat" landform pattern that is seen in the vicinity of Guelph. Well-formed drumlins are found in parts of Turnberry, Howick, and Minto townships but they are rather rare in Normanby, Carrick, Culross, and Wallace, though the rolling till plain everywhere has a drumlinized or fluted aspect.

The continuity of the drumlin field is broken in several places by the presence of kames and associated outwash. One large group of sandhills lies in Carrick township directly south of Mildmay and Neustadt, while a third is found near Pike Lake, between Clifford and Mount Forest. Sandhills and drumlins are much intermingled along the western border of the plain. On the surface of all of these formations quite generally there are shallow deposits of silty, probably wind-deposited material which tends to give the surface soils a silty texture as well as more than ordinary depth.

On the drumlins, gravel terraces, kames and moraines, drainage is usually good; however, this drumlin field, like most others, contains many areas of swampy land. Toward the northwest it borders on the Greenock swamp, while in the southwest near Wingham there are other wide swamps related to the old spillway system. The effect of drainage differences is seen in the distribution of forest vegetation. On both till and outwash the well-drained areas attract sugar maple and beech while the wet lands are clothed largely with elm, soft maple, ash, and white cedar.

The soils of the Teeswater till are similar to those of the Guelph till but have a more definite buff or yellowish brown colour. They are also more silty in texture and somewhat less stony. The soils have been classified by the Ontario Soil Survey in the Harriston catena which contains the well-drained Harriston and imperfectly drained Listowel series. The surface of the latter is smooth and with less relief than in the Harriston, and many areas require the installation of tile drains to provide the best conditions for crop growth. Depressions and flat areas have been classified as Parkhill loam, indistinguishable from the flat areas associated with the Guelph catena.

The soils of the gravel terraces are similar to those of the Burford catena found in the Grand River basin, but because the parent material is composed largely of limestone rather than dolomite these soils have been placed in a different catena, the Teeswater. Only the well-drained member will be described. The soil profile may be from two to three feet in depth depending upon the depth of the silt deposit, but the weathered profile nearly always extends into the underlying gravel. Good drainage makes this a desirable soil for wheat, corn, and alfalfa.

HARRISTON

A_h 3–5 inches, very dark greyish brown, medium granular, silty, slightly stony loam; pH 6.8.

A_e 12–18 inches, brown to light yellowish brown, friable, soft, stone-free silty loam; pH 6.6.

B 6–8 inches, medium brown, nuciform, friable, usually stone-free loam or clay loam; pH 7.0.

C light yellowish brown, medium blocky, compact, slightly stony, calcareous till.

The total depth of the profile may range from 20 to 30 inches depending upon slope and the depth of the original deposit of silt.

LISTOWEL

A_h 4–6 inches, very dark brown, medium granular, friable loam, few stones; pH 7.0.

A_e 6–12 inches, yellowish brown, slightly mottled, soft, silty loam; pH 6.6.

B 8–12 inches, mottled yellowish and dark brown, somewhat blocky, firm, clay loam, variable stoniness; pH 7.2.

C light yellowish brown, medium blocky, hard, moderately stony, calcareous till.

The total depth of the profile may range from 18 to 24 inches.

TEESWATER

A_h 3–4 inches, dark greyish brown, granular, friable, stone-free, silt loam; pH. 6.6.

A_e 15–20 inches, yellowish brown to light yellowish brown, very finely blocky, friable, stone-free silt loam; pH 6.2.

B 7–10 inches, dark yellowish brown, medium blocky, firm, somewhat gravelly, silty clay loam; pH 7.0.

C light yellowish brown or pale brown calcareous, sand and gravel.

LAND USE AND SETTLEMENT

The Teeswater drumlin field is considered to be one of the good general farming and livestock districts of Southern Ontario, carrying on a system of agriculture very similar to that of the Guelph drumlin field. Its climate is similar to that of the Guelph area, but economically the region suffers due to the greater distance from the markets of the large metropolitan centres.

Despite the fact that the area contains some poorly drained land as well as some dry, gravelly hills, it was fully occupied by agricultural settlers during the third quarter of the nineteenth century, reaching its population peak in the early 1880's. Population steadily declined thereafter, largely through the loss of the many members of large farm families who went elsewhere to live. There was little actual land abandonment, the census of 1931 showing an average of 634 acres per square mile still under agricultural occupation. At that time farms averaged 113 acres in area, while 73.5 per cent of the farm land was improved, and 51.3 per cent was recorded as under crops. The chief crops (and per cent of farm land) were: hay (18.5 per cent), oats (18.3 per cent), mixed grains (6.4 per cent), wheat (2.5 per cent), barley (2.3 per cent), and improved pasture (17.5 per cent); about 13 per cent remaining in woods and 14 per cent in other unimproved land.

In 1961 occupied farm land totalled 617 acres per square mile; therefore some land has been abandoned since 1931. Farms now average about 145 acres in area while the number of farm units has decreased by about 24 per cent. Crops are grown on about 45.5 per cent of the total farm land, the chief ones being: hay (18 per cent), mixed grains (11 per cent), oats (9 per cent), silage corn (2 per cent), barley (1 per cent), while turnips, husking corn, flax, potatoes, and other crops share the remaining 4.5 per cent. Seeded pasture occupies about 28.5 per cent, woodland about 14 per cent, and other unimproved land about 10 per cent. It would seem that seeded pasture has been increased both by taking some of the former crop land and by improving some of the former natural pasture.

The livestock population of this region has been greatly increased. In 1931 each hundred acres of farm land supported about 20 animal units, about 16 per cent of the land being required to support the horses which provided the power for cultivation. In 1961 each hundred acres supported 33 animal units, making about 48 per farm. Only a few places in Ontario have a better record for animal production, the average farm carrying about 40 head of cattle, 28 pigs, 3 sheep, and 300 poultry. Less than one quarter of the operating units could be classed as dairy farms, although milk and cream are also sold from some beef herds.

In recent years, beef cattle, hogs, and poultry have been on the increase. Only a few flocks of sheep are to be seen, perhaps 100 in the whole area. The horse is a rare animal; one would probably have to visit half a dozen farms before finding a working team; thirty years ago horses were as numerous as the human population on these farms. It might also be remarked that farms are fewer and larger than they used to be and, rather surprisingly, that the average farm family has also increased in size since 1931.

The Teeswater drumlin area is still one of the more strictly rural parts of Ontario. Its total population is now only about one half what it was at the beginning of the century. There are no nearby cities to attract population to adjoining rural areas. The largest towns, Walkerton (4,000) and Wingham (3,000) have both grown considerably in the post-war period. The other small market villages, Harriston (1,650), Palmerston (1,550), Teeswater (900), Mildmay (850), Brussels (850), Clifford (550) and Neustadt (500), have changed very little since the beginning of the century. There are a number of small unincorporated places also, including Ayton, Formosa, Fordwich, Gorrie, Wroxeter, Bluevale, and Belgrave. Most of these points were mill sites, some still operating, others simply enlarged cross-roads hamlets. Few of these show any growth and some are actually declining.

The early surveyors divided this land into blocks of one thousand acres, each a mile and a quarter square. The roads are usually well graded and gravelled and many of them are oiled. Not nearly as many of the roadside maple trees have been destroyed as in some other areas. The old farmhouses and barns are kept in good repair and farmsteads are usually surrounded by trees. There is, of course, some change: new silos, poultry houses, hog pens and cattle-feeding facilities gradually make their appearance, yet probably the native, returning for a visit after many years of absence, will find the charming, rolling landscape but very little changed.

ARRAN DRUMLIN FIELD

Between Owen Sound and Southampton, lying mainly in the townships of Arran, Amabel, Keppel, and Derby, is an area of 200 square miles containing several hundred long narrow drumlins. Since the township of Arran is central to the area and is almost entirely drumlinized its name has been applied to the whole field. In part of the area lacustrine clay occurs between the drumlins.

This group lies to the north of the Horseshoe moraines and is younger than the Teeswater drumlin field farther south. In addition to the difference in shape already noted, these drumlins are oriented almost northeast-southwest having been formed by the advance of an ice lobe from the basin now occupied by Georgian Bay. They are thus almost parallel to Owen Sound and Colpoys Bay, two deep indentations in the Niagara escarpment which undoubtedly influenced the direction of ice flow.

While these drumlins were being uncovered the glacier apparently underwent a series of fluctuations, building eight or ten thin moraines

running east and west across the field in the vicinity of Tara and Derby Mills. They are formed of till of the same nature as the drumlins and often form connecting ridges between several drumlins.

The two morainal strands to the south provide a definite boundary to the Arran drumlin field. On the most southerly of these strands, the Gibraltar moraine near Chesley, there is a good bar of beach gravel at 925 feet above sea level, high enough so that the greater part of the drumlin field must have been under water as the ice front receded. This may account in some measure for the rather stony surfaces of some of the Arran drumlins and the shallow stratified clay in the hollows.

The Algonquin shoreline in Amabel township provides a clear-cut western boundary to the drumlin field. The old shorecliff is very steep and many drumlins stand half cut away by wave action. Beyond this line not a single drumlin remains, although the presence of some very bouldery patches on the old lake floor undoubtedly indicates that there were other drumlins which the waves completely destroyed. This is in contrast to other parts of the Algonquin shoreline where drumlins remained as islands and indicates the strength of wave action in this vicinity.

The till of this drumlin field is a highly calcareous, heavy loam containing a great many stones and boulders of both Lockport dolomite and Precambrian rocks. The mantle of ground moraine is not very deep and many of the drumlins, especially towards the northeast, stand isolated upon the dolomite bedrock. As in other drumlinized areas, the interdrumlin depressions are often occupied by swamps, but a few of them contain small lakes including Gould Lake, Chesley Lake, and Arran Lake. The latter is over three miles in length. The area is traversed by the Sauble River which follows a very crooked course as it winds around the various drumlins. It has, however, hardly begun to cut a valley and is rather ineffective as an agent of drainage.

The Arran drumlins are higher and steeper than those of the Teeswater field to the south and the surface is more stony. Harkaway silt loam, the dominant soil of the Arran drumlins, is akin to the Otonabee loam of the Peterborough drumlins, not only because of topographic similarity, but also because of stoniness. The Arran till is highly calcareous and the surface deposit of silt, thin or absent. Brown Forest soils, rather than Grey-Brown Podzolic profiles, have developed from the highly dolomitic till, the profile depth ranging from ten to eighteen inches. Free carbonates

77. *Vertical aerial photograph of Gould Lake area in Amabel township. The oval peninsula is a drumlin, and the striped pattern is also due to drumlins. The small whitish dots in some fields, particularly those south of the right-hand arm of the lake, are stone piles. The Algonquin beach confines the lake on the north (top).*

are often present in the surface soil of cultivated fields. It is a good soil for grazing and general farming although somewhat limited in its moisture-holding capacity. The control of erosion is necessary while the rebuilding of soil on the spots already badly eroded is a major problem.

LAND USE AND SETTLEMENT

Travelling through the Arran area one does not get the same impression of agricultural well-being that is gained from observations in the Teeswater area. Here, again, the surveyors obeyed their compasses in defiance of the lay of the land, placing the roads diagonally across the drumlins. Farm layouts are apt to be very inconvenient with many odd corners and difficult slopes. One gets the impression that there is a reasonable amount of uncleared land, despite the census record of only 12 per cent in wood and waste.

In Arran township 97 per cent of the land is reported to be occupied by farms which average about 164 acres. Most of them are 100 or 150 acres in size, but three are reported over 600 acres. About 77 per cent of the farmland is improved with 36 per cent in crop and 41 per cent in improved pasture. Hay, mixed grains, and oats are the chief crops. Wheat and barley, formerly of some importance, are now little grown, but corn for ensilage seems on the increase. Animal production is important, each hundred acres carrying about 29 animal units or an average of about 48 per farm. Less than one-fifth of the cattle are dairy cows, beef animals dominating the scene as they do in most of Bruce county. On numerous farms regular cropping has ceased and the land has been given over to range. On the whole the type of agriculture is conservational as befits sloping land; sod dominates and there is only a small area of intertilled crop. It is true that many farmers still plough up and down the slopes, but contour cultivation is hard to arrange because of the placement of the line fences.

Being an isolated part of Ontario and dominated commercially by the port city of Owen Sound (17,500), it is not surprising that the Arran drumlin field contains no towns. One incorporated village, Tara (500) grew up at one of the few favourable mill sites on the Sauble River and, because of its central position, has maintained itself as the farm supply point of the southern part of the region. It is also a cultural centre and the home of one of the oldest agricultural societies in the province. Another small village, Allenford, grew up at the point where the Owen Sound–Southampton road (now Highway 21) crossed the Sauble River.

Here, also, there was opportunity for a mill site. Allenford later suffered from the fact that it was bypassed by the railway which established its station about one mile to the east, thus effectively preventing the consolidation of the village. The village of Shallow Lake is situated farther north where the dolomite rock appears on the surface between the drumlins. This village achieved some prominence when the small lake for which it is named was drained in order to excavate the marl on its floor for making cement, but these workings have long been abandoned. The whole area is one of gradually declining population as farms are gradually being increased to more economic size.

OXFORD TILL PLAIN

The Oxford till plain occupies a central position adjacent to the Stratford till plain in the peninsula of southwestern Ontario. It covers about 600 square miles, or 385,000 acres, mostly in Oxford county. An upland surface ranging from 1,000 to 1,200 feet above sea level, it is crossed by three well-marked valleys cut by glacial meltwater streams. The surface is drumlinized, with good drumlins appearing south of Woodstock, where the glacier apparently overrode an older moraine, and faint drumlins and flutings farther north. Both drumlins and flutings have a northwest alignment. The till is a pale brown, calcareous boulder loam in which Norfolk (Bois Blanc) limestone is the dominant material, although grey or pale brown dolomite is also abundant. A group of kames around Lakeside in the northwest corner of Oxford county are included in mapping this region. The valleys that cut across this till plain are at present occupied by the headwaters of the Thames River. These are obviously misfit streams, too small for the valleys they occupy. They emanate from a little clay plain southeast of Tavistock where the drainage divide is poorly defined. Trout Creek, which leads to the North Branch of the Thames at St. Mary's, and the South Branch of the Thames take rise in the same swamp about four miles west of Tavistock. The ancient rivers, by cutting away the overburden from the bedrock, have made quarrying possible, particularly at Beachville where a great deal of limestone is processed. Some parts of the old spillway system contain great quantities of gravel which is very useful for road construction and as concrete aggregate in spite of a sprinkling of chert.

In much of the area, the dominant soil type is Guelph loam, a typical Grey-Brown Podzolic, developed under a maple-beech forest. Because of the gentle slopes, good drainage, loamy texture, neutral reaction, and lack of extreme stoniness, this is a good soil. Stoniness varies probably because of differences in the depth of the superficial deposit of silt which is found throughout the region. Large areas of very gently sloping land are classified as London loam, while there are occasional flat areas with poor drainage which have a Parkhill soil profile.

LAND USE AND SETTLEMENT

During an earlier period there were many small farms on the Oxford till plain but a considerable consolidation of holdings has taken place, bringing the average size of farm to more than 120 acres. Change to other uses than agriculture has been very small in Oxford county. Statistics show that 85 per cent of the occupied land is improved and 58 per cent is devoted to crop production, 23 per cent to seeded pasture, 6 per cent to woodland, and 8 per cent is classed as other unimproved land. The crop pattern in terms of occupied farm land averages 21 per cent in hay, 19 per cent in oats, 7 per cent in mixed grains, 7 per cent in corn, 2 per cent in wheat, and small areas of many other crops. In recent years the trends are toward decreases in wheat and mixed grains and unimproved pasture, and increases in oats, corn, hay, and improved pasture.

Within the area there are notable variations, however; toward the south cornfields and silos have always been more numerous, with few farmers growing wheat; toward the north dairying seems to have been a later development, for many of the farms still have the wide German barns of the former beef-raising period, although they may be used to stable Holstein cows. It is toward the north also that one notices the modern tendency toward more beef cattle and hogs.

Although a few settlers were found in this area in the latter part of the eighteenth century, the great wave of settlement came after the Napoleonic wars when large numbers of Scots immigrants took up land in the bush. Scottish traditions and people of Scots descent are still dominant. Population increased rapidly, most of the townships reaching their population peak by 1871. The county of Oxford, in fact, reached a population peak in 1881, declining slowly thereafter until 1921. In 1951 it reached the same level as in 1881 and is still gaining. Farm population is still dropping although the average per farm has risen perceptibly in the last few years. There are no large cities in the area under study, nor in the whole of Oxford county; thus the growth of population in the future is likely to be slow in comparison to some other parts of Ontario.

The commercial and social focus of the area, in fact the seat of the county of Oxford, is the city of Woodstock (21,000) located on the southern bank of the Thames river. It has dairy plants, feed mills and other services required by the surrounding agricultural region. Woodstock also has a few secondary industries, but its population, and therefore its market potential, is not increasing as fast as the productive capacity of the surrounding farm lands. Ingersoll (7,000) is located in the Thames valley about nine miles downstream from Woodstock. It is the focus of a local trading area and a small manufacturing centre long noted for cheese and other dairy products. Half-way between the two larger centres is the village of Beechville (900) near which there are large limestone quarries in the floor of the valley. Thamesford (1,200) and Embro (550) are associated with the valley of the middle branch of the Thames. Innerkip, Hickson, and Bright are small centres in the north-eastern part of the area.

The undulating or gently rolling landscape of the Oxford till plain is most pleasing. Roads and farm lanes are commonly lined with maple trees and it is unfortunate that modern improvement has made it necessary to severely prune them or remove them from the highways. There are many well constructed houses and barns now seventy-five to a hundred years old, seeming almost to blend with the natural landscape. There are also many signs of modern progress—large new silos and circular cribs for the expanding corn crop and new metal-covered buildings to accommodate a greater wealth of livestock and new machinery.

MOUNT ELGIN RIDGES

Between the Thames valley and the sand plain of Norfolk and Elgin counties lies a succession of ridges and vales which are called the Mount Elgin ridges. Including the southeastern portion of Middlesex county, southern Oxford, and small adjoining parts of Elgin and Brant, the area consists of about 563 square miles or 360,000 acres. The general elevation of the lowlands is between 800 and 900 feet, rising gradually towards the northeast. The crests of the ridges are at 950 to 1,000 feet giving a local relief of about 100 feet with few slopes greater than 10 per cent. The ridges are moraines of pale brown calcareous clay or silty clay, while in the vales it is common to find alluvium of gravel, sand, or silt. The region is named after the little village of Mount Elgin which is built on the crest of a typical ridge, on No. 19 Highway.

The divides between the Thames River system and several small streams that run south to Lake Erie are found within this region. There are numerous examples of basins with no visible drainage outlets, ranging from small kettles such as Mud Lake, Walker Pond, and Whittaker Lake to the wide swampy hollows of the old spillways, as for example near Dereham Centre.

The northern slope of the Ingersoll moraine drains directly into the Thames. The Thames system also includes Reynolds and Dingman Creeks which provide drainage for two sections of the flat-floored trough south of the Ingersoll moraine. Deep loams, sand, and gravels in that valley bear witness to the former much larger stream which occupied it in glacial times.

South of the Westminster and St. Thomas moraines the country drains to Lake Erie by means of tributaries of Kettle, Catfish, and Otter Creeks. The broad Belmont vale is occupied by the main branch of Kettle Creek which, working headward from Lake Erie, has entrenched itself about 50 feet into the till. This stream has been dammed to provide a reservoir for the city of St. Thomas. The St. Thomas moraine divides the Kettle and Catfish drainage areas. The long narrow vales on either side of the Culloden strand of the Norwich moraine are drained by tributaries of Catfish Creek. At Delmer, however, Stony Creek breaks through both the Brownsville and Culloden ridges into the central depression to capture a little drainage for Otter Creek. Farther east several other branches of Otter Creek cross the Norwich moraine into the hollows south of Mount Elgin. Big Creek rises in the moraine near Oriel while further north the ridges are drained by Kenny Creek and Horner Creek, both tributaries of the Grand.

The two major landform components of this region provide obviously contrasting soils. The ridges are well drained while imperfect and even poor drainage characterize the hollows. The ridges are formed from clay till similar to that of the Port Huron moraines and the Stratford plain, and the soils naturally belong to the Huron catena. Huron clay loam is commonly found but in some places there has been enough superficial silt deposited to create the Huron silt loam. Silt deposits are more widespread and usually deeper in the lowlands, Perth silt loam being mapped over wide areas. There are also considerable areas of sandy deposits with varying degrees of drainage as well as a number of undrained basins with peat and muck soils.

The advantage of well-drained, well-aerated soil is nicely illustrated in this region. Farms on the ridges are generally more prosperous than those in the vales, in spite of the erosion that has taken place on the hillsides. The tendency, even on the same farm, is to use the low-lying land for pasture and cultivate the rolling fields. Perhaps at some future date the status of the upland and lowland soils will be reversed, if progressive erosion accelerates deterioration on the slopes and the tile drains are installed in the flats to improve the warmth and aeration of the lowland soils.

From all appearances this is one of the most prosperous dairy and livestock regions in Ontario. Farm premises are neatly kept and in good repair. Paint is used more liberally, perhaps, than in most parts of the province. Corn fields and silos are almost universal and of larger capacity than in most other areas. Holstein cattle predominate in the lush pastures. There are also thousands of well-fed beef animals, but comparatively few beef-cow herds. Buildings for the housing of swine and poultry are numerous. Most of the farms still consist of the original one-hundred-acre lots, but smaller farms are being eliminated and the average size (1961) is about 113 acres. While the total area of farm land has fallen slightly in recent years the percentage of improved occupied land has risen to about 86 per cent with 60 per cent in crop and 20 per cent in seeded pasture. In terms of total farm land the various crops line up as follows: wheat 4 per cent, oats 19 per cent, mixed grains 2 per cent, hay 19 per cent, corn for grain 6 per cent, and corn for ensilage 5 per cent; other crops 5 per cent. Unimproved land totals about 14 per cent, about half of which is in woodlots. In comparison with a previous period the acreages of oats, corn, and wheat have increased, while mixed grains and hay have decreased. Small acreages of soybeans and other minor crops are grown also. The average farm in this area carries about 35 animal units of livestock including 14 milk cows, 15 other cattle, 18 pigs, and 250–300 poultry. The actual livestock pressure on the land appears to have risen by over 20 per cent in three decades, thus maintaining the region as one of the better farming areas of the province.

Despite its physical uniformity and its agricultural excellence, the region has no social or commercial unity. Politically it is divided amongst three different counties. It has, of itself, no large town which might serve as a nucleus. Instead it is flanked by London (190,000), Ingersoll (7,000), Woodstock (21,000), Tillsonburg (7,000), Aylmer (5,000), and St. Thomas (23,000), each of which includes a portion of the area within its own umland. The construction of Highway 401 along its northern boundary has given it more rapid access to distant metropolitan markets and is bound to exert an influence on the land use pattern. The small villages such as Glanworth, Belmont, Mount Elgin, Norwich, Burgessville, and New Durham, which originally were market, service, and supply points, now find their business influence on the decline, even though some of them may now be gaining in population.

At this point perhaps one might take a quick economic inventory of agriculture in Oxford county. In 1961 there were 3,743 farms with average capitalization of over $40,300 per farm: land and buildings, $28,000; machinery and equipment, $5,500; livestock and poultry, $6,800. The average value of cash sales per farm in 1960 was $10,400, of which 30 per cent was derived from crop sales, 28 per cent from dairy; 16 per cent from sale of cattle; 14 per cent from hogs; and 12 per cent from poultry and eggs. However, it is necessary to point out that most of the crop sales came from about 500 tobacco farms, averaging $20,000 per farm, and that these do not belong to the till plains and ridges, but to the adjoining Norfolk sands. If we try to eliminate these, the remainder have a total sales average of about $8,600, composed of: dairy 36 per cent, cattle 22 per cent, hogs 20 per cent, poultry and eggs 16 per cent, and crops about 6 per cent. While Oxford is still the outstanding dairy area in Ontario, both in production and in number of dairy farms, it is obvious that there is now much more diversification than formerly.

CARADOC SAND PLAINS AND LONDON ANNEX

In the neighbourhood of London there is a series of small plains which differ from the adjacent moraines and clay plains in that they are covered with sand or other light-textured, waterlaid deposits. Together they comprise about 300 square miles or 192,000 acres in which the soils are conducive to specialized agriculture.

Immediately surrounding the city and extending several miles eastward there is a basin lying between 850 and 900 feet above sea level. Into this basin the earliest glacial spillways discharged muddy water, laying down beds of silt and fine sand. Later, when standing water had retired westward to lower levels, gravelly alluvium was spread over the lower parts of the basin. These gravels continue along the Thames to Komoka where high-level terraces now appear. Later, when the standing water had lowered to the level of Lake Whittlesey the early Thames River cut through the Komoka terraces and built a delta which covers most of Caradoc township.

The sands of Caradoc thin out towards the west until eventually the underlying clay appears on the surface. The surface is nearly level except near Mount Bridges where some old fixed dunes and other sandy ridges appear. Another variation worth noting is the channel extending from Komoka towards Strathroy, one section of it being occupied by a bog. In an area west of Falconbridge a peculiar layer of clay several inches deep appears near enough to the surface to affect the soil.

The main part of the Caradoc sand plains in Caradoc township has been characterized by three soil types on the Middlesex soil map. Fox fine sandy loam appears on the finer sands which are deep and well drained, while the main type in those areas with a shallow layer of sand over clay, and having wet subsoil, is classified as Berrien sandy loam. On the old fixed dunes and other sandhills, Oshtemo sand appears. The Burford gravelly loam of the terraces along the Thames is a well-drained productive soil. The shallow silts and fine sands of the London basin are not freely drained and the incompletely developed profile of London loam is found. When tile-drained, this is an excellent soil.

In the early days the system of farming on the light soils was not very different from that on the clay soils around them. Poor drainage, of course, prevented some land from being taken into cultivation and as time went on a good deal of poor land was retired from cultivation and used for pasture. For many years the farmers of Caradoc had more land in pasture than in cultivated crop, although each farm averaged only about 15 head of cattle. Hay, oats, and wheat were the most generally grown crops, although Caradoc was long known as a potatoes-growing area. There is a moderate concentration of apple orchards in the Mount Bridges area and canning crops are grown for the factory in Strathroy. There are also a number of market gardens. In recent years a large number of tobacco farms has been established. Many of them have been able to set up irrigation systems by tapping the water-soaked sands a few feet below the surface. Livestock farming has shown a slight increase, but fewer potatoes are grown than formerly.

The gravelly terraces along the Thames from Delaware to London were characterized by orchards and market gardens. In recent years much of this land has been taken over by suburban development.

The London basin, to the east of the city was at one time dominantly a dairy region, although fruits and garden products were grown. During the Second World War, Crumlin airport was established in the centre of the little plain, signalling the end of the agricultural period.

The city of London (180,000) is expanding rapidly and most of the London basin will soon have been put to urban uses. Some of its deposits of sand and gravel have been excavated for use in building roads, streets, and in concrete. London is the market and commercial centre not only of the small region under discussion but of most of south-western Ontario. Its situation is related to the physiographic surroundings in several ways. It was originally located at the forks of the Thames because the river was the early route of travel, and because the high alluvial terrace offered a good site on which to build. The underlying sands also offered a good water supply which is now proving too limited for the large city which has grown up. Water will have to be secured by pipeline from the Great Lakes as London grows larger.

EKFRID CLAY PLAIN

Below the Whittlesey delta, or west and south of the Caradoc sand plain, stratified clays appear in Ekfrid and parts of nearby townships. The surface is nearly level except where it is cut by gullies near the Thames and Sydenham rivers. Here and there, knolls or low smooth ridges of sand and gravel are superimposed on the clay. The clay beds are thinnest between the Thames River and St. Thomas where the boulder clay often comes to the surface. The area of silt around Fingal is included, making an area of 360 square miles in all.

The sediments in the central part of the Ekfrid plain no doubt were imported from higher up the Thames valley. Those in the Fingal area likely came from farther east. They are all pale greyish-brown in colour and strongly calcareous due largely to limestone from the Norfolk formation. In Ekfrid township and the Caradoc Reserve the clay is fine in texture. More silty beds appear near Kerwood and south of Alvinston.

The silty sediments give rise to particularly good soil, being fairly pervious and easy to till. Slow drainage is their main limitation. Some of the better-drained clay lands around Melbourne are used only for pasture. The common dark-surfaced clay loam is a good soil when drainage is secured, better than that on the bevelled till plain of Lambton county. Most of this soil is tiled.

Beef cattle, hogs, and poultry are the mainstay of agriculture on this little clay plain. Winter wheat is grown, partly as a cash crop, and sugar beets are seen on a few farms. However, about half the land is in pasture, indicating the predominance of cattle in the economy. The plain is highly cleared with only 6 per cent of land taken up by woodlots.

BOTHWELL SAND PLAIN

The Bothwell sand plain is the delta of the Thames River in glacial Lake Warren and stands at an altitude of between 600 and 700 feet above sea level. The sands were spread thinly over the clay floor, covering some 700 square miles. With only three or four feet of sand, more or less over

78. *Two aerial photographs of the sand plain west of Glencoe. The one on the top shows some blow-outs and the soil is more variable than in the area shown below.*

clay, water invariably collects above the clay because the rain readily infiltrates the sandy surface but percolates slowly through the clay below. With a water table so close to the surface the depressions are swampy, or at least moist. Near Bothwell and Newbury the surface exhibits a series of knolls and swales, and the variation in the colour of the surface soil gives a mottled appearance to vertical aerial photographs. Other sections of the sand plain have a fairly smooth surface and the aerial pictures are correspondingly uniform. The contrast between the two may be seen on the accompanying pair of photographs (Figure 78).

This sand plain is cut in two by the Thames River. Between Aldborough and Mosa townships the river has cut a valley 50 to 100 feet deep and meanders over a flood plain that is up to a mile wide. Farther downstream where the channel is shallower, it overflows periodically. Another departure from the generally level topography appears near the shore of Lake Erie where gullies dissect the promontory above the shorecliff.

This is a region of low-grade soil, in contrast to the fertile Chatham flats to the west or the clay plain of Ekfrid township. In its wetter parts a high proportion of the land remains in woods. This is on the Watrin and Granby soils as named on the county soil maps. The Berrien sandy loam is soil that would become valuable for truck crops or canning crops if the demand for them arose.

Within this area some natural gas is found, particularly in Mosa township. A few oil wells have operated but on the whole with disappointing results. Some of the drinking water is tainted with sulphur.

ST. CLAIR CLAY PLAINS

Adjoining Lake St. Clair in Essex and Kent counties and the St. Clair River in Lambton county are extensive clay plains covering 2,270 square miles. The region is one of little relief, lying between 575 and 700 feet, except the moraine at Ridgetown or Blenheim which rises 50 to 100 feet higher. However, there are minor variations in levelness that have had a great effect on the vegetation and soils which will be discussed separately. There is a deep cover of overburden on the bedrock except near Amherstburg where a dome of limestone comes to the surface. Wells are frequently drilled 100 to 200 feet, often through a fairly uniform bed of clay, before striking solid rock. Limestone underlies Essex county and the adjacent part of Kent, while the remainder of the region is underlain by a black shale. This shale is an important component of the clays but limestone flour is also abundant.

Glacial Lake Whittlesey, which deeply covered all of these lands, and Lake Warren which subsequently covered nearly the whole area, failed to leave deep stratified beds of sediment on the underlying clay till except around Chatham, between Blenheim and the Rondeau marshes, and in a few other smaller areas. Most of Lambton and Essex counties, therefore, are essentially till plains smoothed by shallow deposits of lacustrine clay which settled in the depressions while the knolls were being lowered by wave action. In general the levelling is better done in Essex than in Lambton. The very flat tract of land east of Lake St. Clair was submerged after the disappearance of Lake Warren in a correlative of Lake Algonquin and received a deeper covering of stratified clay and silt.

Essex clay plain. Essex county and the southwestern part of Kent county have a fairly uniform environment and may be discussed together as a subregion. Standing between the basins of Lake Erie and Lake St. Clair, the surface is, essentially, a till plain overlying the Cincinnati arch, which is a low swell in the bedrock of the area. The surface drainage of the plain is nearly all northward to Lake St. Clair, but the gradient is extremely low and the drainage divide near Lake Erie is rather vague. Although it is almost level, the clay plain has a faint relief so that it is better drained than the very flat, low-lying area along the western shore of Lake St. Clair. The prevailing soil type is Brookston clay loam, a dark-surfaced gleisolic soil developed under a swamp forest of elm, black and white ash, silver maple, and other moisture-loving trees. There are also numerous undrained areas where peat and muck have accumulated. Most of the Essex plain has such imperfect drainage that dredged ditches and tile underdrains have had to be installed in order to provide satisfactory conditions for crop cultivation. A better drained strip of land through Charing Cross and along Middle Road originally supported a forest of oaks and hickories; here the surface is grey, leached, and acid. Because it is so hard to work, this soil has largely been used for pasture.

Lying between the July isotherms of 71° and 73° mean temperature, the Essex plain has summer conditions similar to those of the northern fringe of the United States Corn Belt. Its situation also, between Lake Erie and Lake St. Clair tends to reduce the daily range of temperature and prolong the frost-free season.

The most important crops in this area are corn, which is mostly grown for grain, soybeans, soft winter wheat, oats, and alfalfa hay. This is also the area in which most of the early potatoes in Ontario are grown. Sugar beets, white beans, tobacco, barley, tomatoes, cucumbers, melons, and many other vegetables and canning crops are also found.

While the one-hundred-acre farm is commonly found over much of the Essex plain, and larger farms are not uncommon, actually more than half of the operating units have less than 70 acres, with 50 acres a very frequent size. This area is not notable for its livestock production, each hundred acres of farm land supporting about 12 livestock units, which is only three-fifths of the Ontario average and about one-quarter the strength of the best livestock areas. However, the presence of Windsor with its large demand for milk makes it possible for more than 10 per cent of all commercial farms to be dairy operations. Only about 6 per cent rely on the production of beef cattle and hogs as compared to over 35 per cent who specialize in small grains and other field crops.

Essex and the southern part of Kent also have a number of areas which are somewhat atypical of the region. One of these is the Blenheim moraine, a modest ridge of rolling clay land, rising 50 to 100 feet above the surrounding plain. It is fringed by gravel terraces, which meet at Blenheim and continue to the southwest through Cedar Springs to the shore of Lake Erie. The clay land on top of the ridge is used for general crop and livestock farming, with a considerable acreage in pasture. The gravelly soils of the terraces have lent themselves to specialization: peaches, cherries, small fruits, beans, and tobacco are the chief crops. A particularly attractive horticultural area is based on the silt loams between the Blenheim ridge and Rondeau, while the Rondeau marshes themselves are devoted mainly to onion production. All in all, the southeastern corner of Kent county has gained a good reputation for its special crops.

The continuity of the clay plain is broken near Leamington by a small morainic hill, standing about a hundred feet above the general level. Comprising a good deal of sand and gravel in the first place, it was smoothed by the action of Lakes Whittlesey and Warren, and surrounded by a series of beaches. Down the slope towards Lake Erie are broad aprons of fine sand and silt. Being in the warmest part of the province these well-drained soils warm up early in the spring, and it is here that the earliest truck crops such as tomatoes, strawberries, sweet corn, and cucumbers are grown. The long growing season makes it possible to grow water melons and other southern crops. Tender fruits are also important, especially peaches.

In the lee of the long sand spit of Pelee Point there is marshland which has been drained and developed, like the Rondeau marsh, for the growing of onions.

A third area of sandy soil is found on a slight elevation around Harrow. At 625 feet it was not high enough to form an island in Lake Warren and may be regarded as a group of shallow-water sandbars. Harrow was once an important flue-cured tobacco centre, but it has now turned to the culture of asparagus, tomatoes, sweet corn, and other truck and canning crops.

The flatness of the Essex clay plain is further broken by a low gravel ridge extending through Essex, Cottam, and Maidstone, while similar gravel stands are found at Pelton and shallow gravel deposits occur near Edgar, Gesto, and Comet.

Lambton clay plain. Lambton county also contains a large area which might be termed a bevelled till plain, often having a shallow veneer of lacustrine clay over the underlying till. Over extensive areas it has the faint knoll and sag relief, typical of ground moraine. Vegetation and soil development reflect a somewhat better natural drainage than was found in the Essex plain and, as a consequence, municipal ditches and tile drains are not nearly so prevalent. In some areas near Courtright and Corunna, where the till comes to the surface in low crooked ridges, the cultivated soil contains many small pieces of chert or flint.

There are two main soils found on the Lambton clay plain, Brookston clay and Caistor clay; both are imperfectly drained.

Agriculture does not seem to be so intensively developed in Lambton as in Kent and Essex. While originally, and indeed up to 1930, every surveyed lot had been occupied, the census farms of 1961 included only about 85 per cent of total land area. Farm size increased from about 105 acres to 145 acres while the number of farms decreased by 35 per cent. The total area of both improved land and crop land remains practically the same while hay and pasture have been somewhat reduced. An average land use pattern for the area is approximately as follows: hay 12 per cent, wheat 11 per cent, corn 10 per cent, oats 10 per cent, soybeans 7 per cent, other cereals 2 per cent, and a large number of minor crops taking 10 per cent, to make the total crop area equal to about 62 per cent of the farm land. About 20 per cent of the land is in seeded pasture, 9 per cent in natural pasture, and 9 per cent in woods. The average hundred acres of farm land carries about 22 livestock units, including about 18 cattle, 10 hogs, 3 sheep, and 165 poultry. In this respect the area is about equal to the average for Ontario farm land. Most of the farms depend upon the sale of meat animals for the greater part of their incomes. There are a few dairy farmers, mainly shipping to Sarnia, a few depend exclusively upon the sale of field crops, and there is also a fair number of specialized poultry farms. Lambton county is the most important turkey-producing area in Ontario.

The Chatham flats. The flat land east of Lake St. Clair in Dover and Chatham townships covers a 256-square mile area. In the north it extends into Sombra township where it is sharply bounded by a low but distinct bluff just above the 600-foot contour. Beyond the bluff the clay plain has very little relief but it lacks the deep lacustrine clays and extreme flatness of the lower lands. In Chatham township a shallow surface layer of sand on the clay is prevalent. Immediately south and east of the city of Chatham beds of silt appear. In Dover township the stratified clay or loam comes right to the surface except where a few sand spots occur, or, in two good-sized tracts, where black muck has accumulated. The southern and eastern boundaries of the area are indefinite but roughly follow the 600-foot contour.

In flowing across this plain the Thames River has a fall of barely one foot to the mile. Its channel is not deep enough to contain the waters in times of freshet, and recurrent floods have spread clay on the land adjacent to the river. This alluvium has a reddish cast which is in contrast to the grey clay of the lake bottom.

In its virgin state this area supported a swamp forest of elm, black ash, white ash, and silver or red maple, except on the fringe of Lake St. Clair where there was meadow. Its development was delayed until artificial drainage was established by the municipalities; now, the big open ditches along the roadsides form an integral part of the landscape. The main canals end in pumping stations designed to elevate the water a few feet into Lake St. Clair, thus lowering the water in the ditches below the outlets of the tile drains. Most of the farms are systematically tiled.

The Chatham flats is a small region of highly fertile soils. Climatically it is favoured with a frost-free period of 155 to 160 days. The warm summers are indicated by a July normal of 70° F., while in winter the snow blanket seldom stays long on the ground. Perhaps its most serious climatic handicap is the frequent summer droughts. On the other hand spring floods often delay spring planting and impair the soil.

This area is given to the production of cash crops rather than livestock, although beef cattle and hogs are of considerable importance. While some hay, barley, and oats are grown for feed, the main emphasis is on corn, sugar beets, soybeans, winter wheat, white beans, black or burley tobacco, tomatoes, peas, or other canning crops grown for sale. At Jeanette Creek a little tract of deep, lack sandy loam is noted for its onions. Near Wallaceburg, cereal grass for dehydrating is produced and some locally grown alfalfa is dried and ground into meal in the same plant.

The secondary place of livestock in the agriculture of the Chatham

79. *Chatham flats, seen on the Bear line. Black tobacco is being harvested (centre) and on the left are sugar beets.*

flats shows up in the small, often poor barns. In fact the houses and grounds all too frequently appear neglected, even on good farms. The high farm values which prevail in this area are not due to the buildings, but to the highly productive and durable soils.

From the foregoing it is evident that within a broad clay plain such as that around Lake St. Clair and east of the St. Clair River there are elements of similarity and of contrast. The plain illustrates in striking fashion the effect of small differences in drainage and parent materials on the character of the resulting soils. Some of the best soil in Dover township sells for more than ten times the price per acre paid for land in parts of Lambton.

In describing other regions we have occasionally commented on the need of controlling soil erosion in order to insure a permanent agriculture. The St. Clair clay plain has no erosion problem, instead, the problem to be met and solved is that of regularly renewing the organic content and maintaining the granular structure of the soil. Indeed it must be increased because the scarcity of livestock and the predominant place of corn and other row crops has already seriously reduced the perviousness, workability, and fertility of the soil.

St. Clair delta. The delta at the mouth of the St. Clair River is a splendid example of a bird's-foot delta; the several channels drawn on a map produce the fanciful resemblance to a bird's foot. Most of the delta lies on the Canadian side of the South Channel, 64 square miles of it in all. It consists of a series of islands; Walpole Island, the largest one, is flanked by Squirrel Island and St. Anne Island, while within it is a smaller island produced artificially by canals which is called Potaminie Island. The whole delta is marshy with the outer border being a meadow.

The only settlement on the delta is in the north where the water table is lowest. The national topographical maps show roughly 200 buildings. The area is an Indian Reserve.

PELEE ISLAND

Pelee Island is a detached part of Essex county, the distance from Point Pelee to Lighthouse Point being eight and a quarter miles. To the west and south are several tiny islets north of the international border. Middle or Boundary Island off to the south of Pelee is the southernmost land of Canada. It is a bit of limestone of less than ten acres, shaped like a tadpole with its tail to the west where the tip comes within 100 yards of the international boundary. A recreation club is its only tenant. The half dozen islets to the west are all smaller than Middle Island.

Pelee Island covers 15 square miles or 10,000 acres. Roughly rectangular in shape, it is five miles from north to south and three from east to west. The somewhat rectangular shape is due to limestone ledges which form the shore at the northwest and southeast corners of the island. Most of the shore consists of a sand beach backed by a thin line of low dunes. Lighthouse Point has a series of protecting sand bars. The southwest section of the shore is a low bluff, part of which is protected from erosion by concrete groynes. From the southwestern corner a gravel spit juts into the lake for a mile or more, curving eastward at the tip. The foundation of Pelee Island is a low dome of limestone, rising at its highest only 38 feet above the mean level of Lake Erie. Bedrock comes to the surface in several parts of the island and forms certain parts of the shore. Thus, approximately 15 per cent of the area has shallow drift with no more than

80. *Two views of Pelee Island: on the left, soybeans on level clay; on the right, a pheasant hunter and blackish soil*

a few inches of clay or stony loam on the limestone floor. The remainder is a level clay plain lying barely above lake level. The drainage canals and deep ditches that facilitate under-drainage, without which the soil would be of little use, are now an established part of the landscape. The lowest tracts are, in fact, below high water mark and would at times be inundated except for the dykes which are built along the shore. The canals are sealed at the ends and equipped with pumps to hoist the water into the lake. In this way the water level in the ditches is lowered below the outlets of the tile drains which are installed in the fields.

Over most of the island the clay is of moderate depth. The rock is exposed in the bottom of the main east-west drainage canal on the eastern side of the island. A similar covering of about ten feet is common in the southern-central part of the island. The deepest drift is on the western side, 40 to 50 feet of boulder clay being recorded in several wells, although towards the northwestern corner one well is reputed to extend 96 feet in clay without hitting bedrock. Incidentally, a few dug wells derive their water from a gravelly layer just above the rock but most of them are drilled into the limestone.

The clay plain of Pelee Island, as in most of Essex county, is a bevelled till plain. A search for stratified sediments proved fruitless; although there must be some reworked material on the surface it is shallow and the stratifications are largely destroyed by plant roots or by weathering.

The till is of fairly constant composition, fine in texture and containing very few big boulders. Pebbles appear in variable numbers. Several counts of the various rock types showed about 70 per cent of limestone, flint and a little dolomite, 18 per cent of brown to black shale or sandstone, and an estimated 12 per cent of granite, etc. of Precambrian age. The grey and pale brown limestones give the till its dominant colour, while its content of shale and sandstone are responsible for its extreme susceptibility to rusty mottling.

The soils of Pelee Island were described and mapped several years ago as part of Essex county. The black-surfaced, wet clay in the lowest sites is a type of exceptional fertility and durability. It was developed under a swamp forest of elm, ash, and soft maple, but this may have been preceded by meadow. The Brookston clay is a little less poorly drained but requires tiling also before it can be used. Its surface layer is shallower and contains less humus. There is some shallow muck but it is not extensive and does not appear on the map. The remainder comprises the shallow soils where the limestone comes nearly to or to the surface.

With an average frost-free period of 179 days, Pelee Island boasts the longest growing season of any part of Ontario, longer in fact than that of most of Ohio extending 150 miles to the south. However, because it is isolated from the markets, truck crops are not grown commercially. The clay soil also limits the choice of crops to be grown. For these and other reasons the present-day agriculture of Pelee Island can be fairly well covered by one word—soybeans. Over 85 per cent of the cropland is in soybeans, the remainder being sown to seed corn, black tobacco, winter wheat, and alfalfa. A change in the relative price of soybeans, corn, and tobacco and available labour for tobacco culture, might cause a shift from soybeans to either tobacco or corn; but soybeans are favoured because they are a reliable crop, require no special fertilization, and, above all, are easily handled. Moreover, they provide feed for the pheasants for which the island is noted.

The environment of the island must approach an optimum for pheasants, as the population in autumn in recent years has exceeded 20,000 birds. Each fall 1,000 to 1,500 hunters flock to these 15 square miles and take up to 14,000 birds. For their hunting privilege they pay fees which amount to a considerable proportion of the municipal taxes. The mild climate with its limited snowfall, the abundance of food, and the cover provided by the waste land on the island, no doubt combine to make it a good place for pheasants.

ERIE SPITS

From Haldimand county westward the shore of Lake Erie takes the form of three big scallops with as many big sand spits at the apexes. The centre of the scallops are marked by high bluffs which are contsantly shifting inland. Conversely the spits are gradually extending farther out into the lake using the sand made available by the erosion of the shore.

It is not surprising to find the longest of these spits built off the shore in Norfolk county opposite the great Norfolk sand plain; Long Point, as it is called, is built from the west at the eastern apex of the biggest scallop. A smaller spit built from the east is growing out to meet it, threatening to confine a section of the lake that even now is called Inner Bay. At the other apex of the big Port Stanley scallop is the Rondeau spit which terminates in Point aux Pins, and this one is built from the east. Here, also a smaller bar built from the opposite side virtually closes off Rondeau harbour.

The most westerly of the three great spits is Point Pelee. It juts straight out from the shore and both sides are built up, indicating that sand is being added from both east and west.

The conformation of Long Point is clearly apparent from a map. It is fabricated loosely from a succession of sand bars which run at an angle to the long axis of the spit. Marsh or open water appears between the bars. Point Pelee also includes several small lakes or ponds.

Much of these spits is covered with meadow, except in the case of Rondeau Park. The surface is at lake level or a few inches below so that cat-tails and other water-loving plants grow rather than trees. The spits are favourite feeding grounds for wild ducks.

Apart from Pelee Island the Erie Spits have the most temperate climate of any part of Ontario. In this respect they are exceeded in Canada only by small areas on the west and east coasts. Trees, such as hackberry, which are scarce inland, grow in groves on Point Pelee and wafer ash is plentiful. Also, from those who study them, we understand that a great variety of snakes live on Long Point.

A limited area on Point Pelee is used for fruit and truck crops, but there

81. *Beach at Turkey Point on Lake Erie*

is no farming on the other spits. Their major value is recreational, as the sandy beaches attract summer cottages. Here, at least, there is no encroachment of the bluffs on the shore properties as there is along most of Lake Erie.

NORFOLK SAND PLAIN

The Norfolk sand plain is wedge-shaped with a broad curved base along the shore of Lake Erie and tapers northward to a point at Brantford on the Grand River. It includes the greater part of Norfolk county, the eastern end of Elgin county, southern Brant, and a small corner of Oxford, with an area of 1,215 square miles or 775,000 acres. The plain declines southward from about 850 feet to the level of Lake Erie (572 feet) or in the west to the top of the shorecliff 100 feet or more above the lake. In good-sized sections of the plain the slope is only a foot or two to the mile, while a noticeable break in the slope occurs five to ten miles from the shore of Lake Erie.

The sands and silts of this region were deposited as a delta in glacial Lakes Whittlesey and Warren. A great discharge of meltwater from the Grand River area entered the lakes between the ice front and the moraines to the northwest, building the delta from west to east as the glacier withdrew. Thus it covered most of the area west of the Galt moraine. This and other moraines to the west are partly buried by sand; some sections are entirely covered whereas others still stand up 50 to 75 feet above the general level. From observations of exposures in river valleys and along the bluffs of Lake Erie we have records of sand beds up to 75 feet, but usually silt or clay strata or beds of boulder clay occur within 30 feet of the surface.

The drainage is through small rivers flowing directly to Lake Erie, except in a small area in the north which is tributary to the Grand River. These streams, the largest of which are Otter Creek and Big Creek, have cut deep valleys across the sand plain, often being incised 75 to 100 feet. The valleys, since the earliest days of settlement, have been the cause of frequent gaps in the road net. Dissection of the sand plain is quite incomplete, however, and drainage conditions vary accordingly. Near the streams and their laterals the water table has been lowered so that good or even excessive vertical drainage operates. In the intermorainal sections in the north and other interfluvial parts of the plain there are large tracts of bog or wet sands.

One of the favourable attributes of the Norfolk sand plain is that of

82. *Tobacco farm at Cathcart*

abundant well water. Frequently, a good supply may be obtained by driving in a sand point. Infiltration is rapid into the sandy soil, while finer sediments below hold up the water table. It is also a fact that summer rainfall is greater in this area than in most of Southern Ontario.

Norfolk was the first county in Ontario to have a soil survey, the work being finished in 1928. It came at an opportune time to guide in the selection of soil for tobacco. The dominant type of the well-drained areas was identified as Plainfield sand, in which the coarseness of the material and free drainage has produced a rather featureless profile and a thin surface horizon lacking in organic matter. Once cleared and cultivated it is readily blown about by the wind. Elsewhere finer sand has permitted the development of normal grey-brown earths which were classified in the Fox series.

Settlement began early in the Norfolk sand plains. The land was taken up rapidly after the townships were opened in 1792 to 1812, except in the wet areas. The light-textured soils, however, could not stand up to the regular cropping. As the original humus became exhausted, productivity declined and wind erosion increased so that farm abandonment became common. A programme of reforestation was urged and a forest nursery was established at St. Williams to provide the young trees. Whereas wheat, rye, corn, and peas were important field crops in the 1890's, the tendency towards canning crops and orchards increased during the first quarter of this century. Then, tobacco-growing started and it has come to occupy the leading role, the glass starting houses and box-like curing kilns now being part of the landscape. Tobacco-growing occupies most of the warm, well-drained sandy soils and is a good example of fitting a crop to specifically adapted soils. The Fox soils are valued more highly than the Plainfield sand because they contain more moisture and fertility, but the latter type is also used extensively.

Total population of the sand plain reached its lowest ebb at the end of the First World War. Much land was lying idle while efforts were being made to have large areas put back into forest. The discovery that successful cultivation of flue-cured tobacco was possible, changed the pattern of land use very rapidly. By 1931, about 16,000 acres of tobacco were being grown by more than 500 farmers in what was known as the "new tobacco belt" (the "old belt" being the areas in Essex and Kent county which had grown mainly Burley and dark tobaccos). By 1931, also, the total population on the sand plains had recovered to a level slightly higher than it was at the turn of the century and it has been rising ever since. Farm population has fallen slightly (perhaps 10 per cent), occupied land has also decreased slightly while the size of farm has increased, but the average is still only about 100 acres per farm, much smaller than in most agricultural areas in the province. In 1961 the Norfolk sand plain had more than 3,200 tobacco growers with a tobacco crop area of more than 103,000 acres, giving an average of about 32 acres of tobacco per farm. According to statistics issued by the Ontario Department of Agriculture the average yield per acre has increased from less than 900 pounds per acre in 1931 to more than 1,800 pounds per acre in 1963. In fact, in the early 1960's the production of flue-cured tobacco had reached such a level that mandatory acreage reduction has to be enforced. This caused a noticeable switch to other crops, the acreage in husking corn in Norfolk county for instance, increasing from 10,000 to 30,000 acres in three years. The other important field crops are rye, oats, wheat, and hay. Rye is commonly grown in alternative years on tobacco land. Between one-half and two-thirds of the commercial farms on the Norfolk sand plain are tobacco farms. The other specialties are dairy, fruits and vegetables, livestock, and poultry.

The urban centres of the region follow a peculiar distribution pattern.

There is a string of small ports and fishing villages along the Erie shore, including Port Dover (3,100), Port Rowan (800), Port Burwell (800), and Port Stanley (1,500), but they are not directly connected with each other by either railway or Provincial Highway. Both these facilities are forced inland a distance of ten to twenty miles from the lake by reason of the entrenched river valleys. The major centres of population are found at the points where the major traffic arteries cross the larger streams. Thus St. Thomas (23,000) is on Kettle Creek, Aylmer (4,800) on Catfish Creek, Tillsonburg (6,800) on Otter Creek, Delhi (3,500) on Big Creek, Simcoe (9,000) on Lynn River, and Waterford (2,300) on Nanticoke Creek. Among the smaller centres which might be mentioned are Vittoria, Teeterville, Courtland, and Vienna. Brantford (60,000) dominates the northern part of the region, but certainly has had other reasons for existence and growth. There are a few villages, much closer to the soil, including Scotland, Oakland, Burford, and Princeton.

While other smaller areas in Ontario also grow tobacco, there is no doubt that the Norfolk sand plain presents the observer with one of the most distinctive landscapes in Southern Ontario. It is striking, for instance, where one breaks over the low moraine from the east to find the cattle, pastures and hayfields of the Haldimand clay plain suddenly disappear, replaced by a new assemblage of tobacco, rye, groups of curing barns, glass-covered starting houses, and water tanks. The old farm barns remain; some tobacco farmers also feed cattle and hogs; corn remains, or, rather, has again come to be a significant crop.

Population of	Total	50,475	100%
NORFOLK COUNTY	Urban (four towns and one village)	18,253	36.1
1961	Rural	32,222	63.9
	farm	14,718	29.2
	non-farm	17,504	34.7
	unincorporated villages (9)	3,181	6.3
area 634 sq. mi.	small hamlets (32)	3,187	6.3
density: 16.6	scattered	11,136	22.1

Perhaps the outstanding feature of the plain, however, is the present distribution of the population. In the old days of rural Ontario there were many mill sites and road corner settlements. With declining agriculture these reached a low ebb by the end of the First World War. Most of them have now revived to a point far beyond their previous peak, while in addition thousands of people now live in scattered rural houses along the concessions and sideroads. Norfolk county does not include the whole region but the population analysis of the county is revealing.

Should one relate this density of population to the underlying physiography? Forty years ago when the total population of the county was only about 26,000, the towns had 8,000 people, and most of the remainder lived on farms. The lack of prosperity of the county was then freely attributed to the poverty of the sandy soil. Now under different technology, must its prosperity not also be attributed to its physiographic pattern and soil development, which have permitted a new system of agriculture that, directly or indirectly, has given work to thousands of new people? The very nature of the land, the ease of excavation, the accessibility of water, and the porous soil which permits the operation of septic tanks, also are major factors in the settlement pattern. One must also mention the availability of road gravel in the old moraines and deltas, which enable the new tax dollars to be used effectively. Finally one might comment favourably upon the use of abandoned but water-filled gravel pits, such as those at Waterford, for recreational purposes.

HALDIMAND CLAY PLAIN

Lying between the Niagara escarpment and Lake Erie, thus occupying all of the Niagara peninsula except the fruit belt below the escarpment, the Haldimand clay plain has an area of about 1,350 square miles. Although it was all submerged in Lake Warren the till is not all buried by stratified clay; it comes to the surface generally on the low morainic ridges in the north. In fact, there is in that area a confused intermixture of stratified clay and till. The northern part has more relief than the southern part where typical level lake plain occurs.

The underlying rocks consist of a succession of Palaeozoic beds dipping slightly southward under Lake Erie. The vertical cliffs along the brow of the escarpment are formed of Lockport dolomite and this formation underlies a narrow strip of the plain to be succeeded southward by the Guelph dolomite. Overlying these hard dolomites is a series of softer rocks which include shale members. They have been worn down to form a broad depression in front of the Onondaga escarpment. The latter is a low north-facing scarp formed of hard limestone. Small areas of bare rock appear along its crest; otherwise the change in bedrock makes little difference in the clay plain.

The overburden on the rocks near the Niagara escarpment is generally less than 50 feet deep, but increases in depth as one goes southward. In the wide vale north of the Onondaga escarpment the depth of drift may

83. *Two views of Haldimand plain near Caistorville: above, the grey soil and the pine, oak, hickory, and elm trees are typical; below, a field of wheat*

reach 150 feet. As noted before, the brow of this escarpment is often bare, but the overburden increases southward again and is 10 to 50 feet in depth along the Lake Erie shore.

The Haldimand clay plain can be described as falling into a series of parallel belts. The first is the highest ground adjoining the Niagara escarpment, where recessional moraines were built by the ice lobe that occupied the basin of Lake Ontario. It varies in elevation from 600 feet to 750 feet except on the gravelly "Short Hills" which reach an altitude of about 850 feet above sea level. The sand and gravel of the Fonthill vicinity are in strong contrast to the heavy boulder clay of the moraines and adjoining ground moraine. It is a shaly till derived in large measure from the red and grey beds below the Niagara escarpment. As moraines, these ridges have a much subdued relief due to having been built under water. The intervening troughs are floored with lacustrine silt or clay.

The drainage of this belt is controlled by the modest ridges which directed it eastward in several parallel streams. In the north, some of them have been captured through deep notches in the escarpment. Twenty Mile Creek, Forty Mile Creek, and the Welland River are the most important of these streams. They have cut shallow channels but many undrained depressions remain on the higher ground. The wet sloughs may be seen on the ground, and the pattern in which they occur is clearly visible on aerial photographs.

In the central part of the plain, the Grand, being a larger river, has cut a deep valley in the clay and silt below Brantford. Consequently there has been much dissection by tributary drainage. Near Caledonia an odd variation is provided by a scattered group of drumlins partially buried in moderately dissected clay beds. The lower course of the Grand River is almost without gradient and apparently controlled by the Onondaga escarpment, until a low saddle is reached east of Dunnville.

The southeastern part of the peninsula might almost be considered as a separate subregion, characterized by levelness and poor drainage. The main part of Welland county comprises heavy clay, while the lowest part of the plain lies in the southern portion of that county. Here the watershed is provided by the Onondaga cuesta which, though quite low and lying close to the shore of Lake Erie, nevertheless forces the drainage to the north and east. Large undrained areas remain, covering several square miles in which the Wainfleet and Humberstone peat bogs have formed.

Besides the outright bog, there is a large area of poorly drained sand, silt, and clay in Moulton and Canborough townships. These deposits undoubtedly represent a delta of the Grand River built into a higher stage of Lake Erie. Remnants of its abandoned shoreline may be seen near Lowbanks.

A few other beaches deserve mention. Beaches of Lake Warren are found upon the upper levels of the Short Hills. The ridge at Lundy's Lane is a beach formed upon a faint moraine. The gravelly streaks at Ridgeway and Ridgemount are also lacustrine beaches. Another is followed by the road from Crystal Beach to Sherkston, while still more are found capping high ground near Crowland. These gravel bars are seldom over six or eight feet deep, and since the upper four or five come within the weathered soil profile it is only the deepest of them that provide gravel for road metal.

South of the Onondaga escarpment the drainage is directly to Lake Erie by way of Nanticoke Creek, Sandusk Creek, Hemlock Creek, and other shorter streams. East of Port Maitland, morainic hills appear to form the headlands on the Lake Erie shore called Lapp Point and Kinnard Point, and a more extensive area of till lies between Ridgeway and Fort Erie.

The shore of Lake Erie is low and appears to be controlled considerably by the surface of the limestone, resistant beds of which form headlands. Besides the morainic headlands east of Port Maitland there are several lines of active dunes and a wide sand beach at Crystal Beach.

The outstanding characteristics of the soils of this region, the heavy texture and poor drainage, have already been inferred. Unevenness in drainage must also be mentioned, areas of better drained soils being dotted with wet depressions of irregular size and shape. The better-drained soils have thin surface horizons immediately underlain by brownish-grey horizons which have lost most of their lime and phosphorus. The greyish soil of the higher ground contrasts strongly with the darker soil of the depression and it presents a bold pattern on aerial photographs.

The better-drained soils are represented by the Oneida clay loam which is found in the dissected area along the Grand River around Caledonia where the sediments are somewhat silty, and by Haldimand clay loam on the bevelled plain north and south of the Grand. Caistor clay loam, mapped in the Welland River drainage area, is an association of imperfectly drained soil and wet, often swampy spots. Farther east in Welland county where the topography is more uniformly level there is the heavy intractable soil called Welland clay. These soils are all lacking in lime and phosphorus.

Of the light-textured soils which break the uniformity of the Haldimand clay plain the deep sands and gravels at Fonthill are more prized for fruit and truck crops. The other gravelly loams on the old beaches are also valuable for peaches, cherries, and vegetables. The wet sandy loams of the Dunnville area are gradually being improved for canning crops and general agriculture.

SETTLEMENT AND LAND USE

The Niagara peninsula has the function of a geographic corridor. Traffic between Canada and the United States flows to and from the gateways at Fort Erie, Niagara Falls, and Queenston. Much of it descends the Niagara Escarpment to traverse the fruit belt below, but the remainder crosses the Haldimand clay plain. Because of the falls over the escarpment the Niagara River does not provide an open corridor for shipping between Lake Erie and Lake Ontario; this has been artificially supplied beginning with William Merritt's first Welland Canal in 1829 and represented by the Welland Ship Canal of the present day, with its Twin Flight Locks on the face of the escarpment. While Niagara Falls (55,000) and Fort Erie (10,000) are the modern phases of the old gateways, the later development of the canal bridge points and terminals have given rise to St. Catharines (85,000), Thorold (11,000), Welland (36,000), and Port Colborne (16,000). Smaller centres include Chippawa (3,500), Fonthill (2,500), Welland Junction (1,100), Ridgeway (1,900), Ridgeville (400), Stevensville (600), and Crystal Beach (1,900). The latter is not strictly of the clay plain, but is a resort centre on Lake Erie. St. Catharines, of course, is not nucleated on the Haldimand clay plain, but in the area below the escarpment, yet, as the largest urban centre of the canal area, it has considerable influence on human affairs in the clay plain.

The power plants at Niagara Falls provide the key to much of the industrial activity in that vicinity. And the falls themselves are one of the foremost tourist attractions on the continent. Each year recently nearly 3 million tourists visit them on the Canadian side. Tourist establishments catering to these visitors dominate the scene in the city and along the highways in the outskirts. No doubt Niagara Falls has earned the title of Honeymoon Capital.

While the eastern end, or Frontier Zone, of the clay plain is highly urbanized, the western and far larger area is almost strictly rural. Haldimand county has no large towns or cities, its largest centres being Dunnville (5,500), Caledonia (2,300), and Cayuga (1,000), on the Grand River, and Hagersville (2,100), and Jarvis (800), in the extreme southwest. Smaller rural service centres which might be mentioned include Fisherville, Smithville, St. Ann's, Binbrook, Wainfleet, and Wellandport. The existence of the industrial city of Hamilton (400,000) on the northwest corner of the region overshadows the whole area.

In Ontario, clay plains of low relief are usually general farming regions with emphasis on livestock. While this is true of the Haldimand clay plain, the development is by no means uniform. Only 45 per cent of the area of Welland county is occupied by farms, and only 55 per cent of the farms are regarded as commercial; this explains the observable fact that much land in Welland county lies idle. Dairying prevails along the brow of the escarpment and along main roads leading to Hamilton. A second, somewhat less concentrated area of dairy farming is found on the better drained soils of Haldimand county. The most important crops of the clay plain are hay, oats, wheat, and corn. Husking corn is not extensive in spite of the favourable climate.

The plain to the north of the escarpment is well known as the most intensively developed horticultural area in Canada, and it is natural that some of this activity should spread onto the upland. Consequently, orchards and vineyards appear on the "mountain," especially on the slopes of the Vinemount moraine. Generally, the poorly drained vale of Twenty Mile Creek forms an effective barrier to this development, but in the Binbrook area some areas of good silt loam are planted to grapes, pears, and apples.

The light soils around Fonthill form a horticultural "island" in the clay plain. Here practically all the crops of the Niagara fruit belt may be grown, including peaches, sweet cherries, and vegetables for the early market. A smaller area with similar development is found at Ridgeway on the soils of the old beach ridges. The sands north and east of Dunnville are being drained and used for vegetable crops, as are some marginal areas of the bogs. Areas of truck farming are found near Niagara Falls and Welland, although in both cases they are being overwhelmed by urban expansion.

The heavy clay soils of the plain pose a number of problems. They are deficient in lime, phosphorus, and organic matter, and many areas have poor drainage as well. In some areas there are numerous wet "sloughs." Much of the land is very weedy. Wild carrot, chicory, and daisy have taken possession of many meadows and roadsides. The beauty of these weeds is, unfortunately, poor compensation for the lack of grass, the dense fibrous roots of which are needed to impart good tilth.

SAUGEEN CLAY PLAIN

This small clay plain is situated in the drainage basin of the Saugeen River, north of the Walkerton moraine. It is underlain by deep stratified clay deposited in a bay of Lake Warren and is thus distinguished from the more extensive plain of Lake Warren along the shore of Lake Huron, which has only a shallow surface layer of clay.

The clay is pale brown in colour and highly calcareous, being no doubt largely derived from the limestones and dolomites of the local bedrocks. These sediments were brought into the basin by the ice border drainage channels from the east, the forerunners of the present branches of the Saugeen system. The largest area of lacustrine clay lies between the Singhamton (Walkerton) and Gibraltar (Chesley) moraines, and it would seem that the Chesley moraine marked the ice front in Lake Warren during most of its lifetime. The shoreline around the "Paisley Island" and the clay to the north of the ridge indicate that there was a post-Chesley stage of Lake Warren. The next continuous low ridge of till, which occurs about five miles north of Chesley, makes a convenient boundary between the clay plain and the rolling landscape of the Arran drumlin field.

The original relief of this plain was flat to undulating, somewhat similar to that of the Schomberg clay plain, but the Saugeen River, Teeswater River, Deer Creek, and North Branch have cut deep valleys in the clay beds. The main valley of the Saugeen near Paisley is 125 feet in depth. Consequently there has been considerable lateral gullying with the development of "river breaks" such as are found along the Grand River southeast of Brantford or along the Conestoga near Glen Allen.

The Teeswater and the Saugeen emptied into Lake Warren separately, but their deltas coalesced to form a sand plain between Eden Grove and the Ellengowan. A small part of this sand plain lies on the east flank of the Saugeen valley while another small tract of sand is found just north of Vesta. The latter seems to be associated with a group of kames along the border of Elderslie township. Another group of kames appears along the road between Gillie Hill and Salisbury. Although the dominant landscape of the area is that of a clay plain, there are thus a number of variant minor topographic elements. Drainage varies greatly, for while most of the southern part of the plain is well drained because of stream dissection, the northern area along Snake Creek has very poor drainage and was for decades known as the Elderslie swamp.

The uplands of this area had an original forest cover consisting largely of sugar maple, beech, and hemlock, while the lowlands carried an association of elm, ash, soft maple, and cedar. The clay soils have been placed in the Saugeen catena by the Ontario soil survey. The well-drained profile is that of Saugeen silty clay loam, that of the imperfectly drained areas, which are actually more widespread, is Elderslie silty clay loam, while Chesley silty clay loam is found in the poorly drained flats. In large part these soils, developed on highly calcareous clays, exhibit characteristics of the Brown Forest soil group. They have relatively shallow profiles (12 to 18 inches) with either a very slight development of the A_e horizon or apparently none at all. Under conditions of poor drainage the brownish B horizon is replaced by a grey, massive, plastic glei. On the sand plain, the deeper beds near the river afford warm dry soil which is classified as Fox sandy loam while there are also areas of the more poorly drained Brady series. Much of the sand plain, however, is underlain at shallow depths by clay, and the internal drainage is imperfect, producing a Berrien profile.

SETTLEMENT AND LAND USE

First settlement in this area took place in the years 1851–55, following the building of the Durham road and the Elora-Saugeen Road which opened up the area west of the Saugeen River. Later, the eastern and larger portion of the clay plain was also opened by main roads. Around 1870 both the eastern and western parts of the plain received rail service with the opening of the Toronto, Grey, and Bruce and the Wellington, Grey, and Bruce railways. During this period and for some years thereafter, lumbering was a dominant industry as the land was cleared for farming. The northern part of the plain, the "Elderslie swamp," required drainage in order to make it fit for agriculture and township ditches were dredged.

As in the rest of Bruce county, the agriculture of the first half-century of settlement was directed largely towards wheat production. Today the emphasis is placed upon pasture and hay for beef production but a few farmers still grow some wheat along with oats and mixed grains. There has been considerable consolidation and enlargement of farm holdings, the average farm in 1961 comprising about 190 acres. The clay plains of Bruce county have seen the abandonment of many farm homes and a great reduction in total population.

The Saugeen clay plain contains no large towns. The trade of the southern portion is divided between Walkerton (4,000), and Hanover (4,500), the two larger towns of the upper Saugeen basin. Chesley (1,700), on the north Saugeen, is the centre for the northern part of the clay plain with Paisley (750) on the western margin. Smaller places such as Elmwood (400) on the eastern edge, and Cargill, Chepstowe, and Pinkerton on the Teeswater River are grist-mill and local supply points.

HURON SLOPE

Occupying an area of about 1,000 square miles along the eastern side of Lake Huron, the land between the Algonquin shorecliff and the Wyoming moraine slopes gently upward from 600 feet to 850 or 900 feet above sea level. It is essentially a clay plain modified by a narrow strip of sand, and by the twin beaches of glacial Lake Warren which flank the moraine. Below the Warren beach the surface has been bevelled, but the deposition of lacustrine clay seldom amounts to more than three or four feet, and the till often comes to the surface. Deeper clays of a silty nature are found between Clark Point and Ripley. Above the Warren levels, the surface is without the planing effect of the stratified clays, and is similar to that of the Stratford till plain. The till is formed from a brown calcareous clay, containing a minimum of pebbles and boulders. The sheet of till is only six to ten feet in depth and it rests on stratified clay of the same colour. There seems little doubt that the matrix of clay in the upper till sheet is merely reworked lacustrine material similar to that of the beds below.

The sandy strip mentioned above runs parallel to the Warren beach, but usually separated from it by a mile or so of clay or loam. It consists of a shallow deposit of sand spread over the clay, probably as a sort of offshore apron created by wave action. Under such conditions a perched water table occurs on the surface of the clay with the result that the sandy soil has a wet, cold subsoil. The surface of the sand is unevenly ribbed and swampy streaks occupy the depressions. There are two small deltas where the Bayfield and the Ausable emptied into Lake Warren thus creating other sandy areas. The Ausable delta around Sylvan is the larger of the two. In both cases the beds are deep enough to form dry sand plains.

Farmers on this slope generally emphasize the raising of livestock, grazing is featured, and grass farms with unoccupied houses are common. Oats and hay are the usual crops. South of Goderich, wheat, beans, and corn appear. The sandy strip is rather poor for agriculture, even for grazing, as the cattle fail to thrive on it. Most of the cattle belong to the beef breeds, but some dairying is found near Goderich and Kincardine. Cheese is produced at Ripley. The summer rainfall in the lee of Lake Huron is fairly reliable, enabling the production of good crops of oats and mixed grains which are fed to hogs.

The Algonquin bluff at the lakeward border of this slope, or the shorecliff of Lake Huron where it has undercut the earlier cliff, have a drop of about 75 feet. With such a head the streams, large and small, have cut deep gullies, thus making a much-dissected fringe just about the bluff. It is this that forced the Bluewater Highway (No. 21) away from the shore and, indeed, mostly out of sight of the blue water of Lake Huron. These gullies also present problems of control to the farmers whose land is being spoiled by their growth. Both Bruce and Huron counties, in which the Huron Slope lies and which comprises a very continuous body of agricultural land, are rural counties, yet the area contains a number of small towns and villages. Some of them, of course, have lakeside sites and are connected with the Huron Fringe, but they draw much of their support from the general farming area and will be mentioned here. Goderich (6,500) at the mouth of the Maitland River, is the county seat of Huron county, an important port, a mill site, and a salt mining centre. Clinton (3,500), twelve miles inland from Goderich is an old agricultural trading centre and, since the days of the Second World War, the site of an R.C.A.F. training base. Zurich (750) is centrally located in the southern part of the area. Bayfield (400) is a small lakeside centre some distance south of Goderich. Lucknow (1,000), actually in the spillway between the moraines, serves adjacent cattle raising country in both Huron and Bruce counties. Kincardine (2,900) is the largest town on Lake Huron north of Goderich. Ripley (500) is centrally located in the best dairy section of the region under discussion. It is famous because, allegedly, the natural fluorine content of its water supply has given its inhabitants remarkable freedom from dental caries. Tiverton is a small village a few miles north of Kincardine.

HURON FRINGE

A narrow fringe of land along Lake Huron from Sarnia to Tobermory needs to be distinguished from the clay plain adjacent and above it, because it is so very different. It comprises the wave-cut terraces of glacial Lake Algonquin and Lake Nipissing with their boulders, gravel bars, and sand dunes. Although it is narrow, the strip is approximately 200 miles long and its total area amounts to about 435 square miles.

On the Bruce peninsula this Huron Fringe is a scoured belt of limestone just above the lake level, back of which are either a series of beaches, or sand dunes, sometimes interspersed with swamp. These are the work of Lake Nipissing or the present lake. Lake Algonquin covered practically the whole of the peninsula north of Wiarton and consequently left few shorelines. A few miles south of Wiarton, back of Sauble Beach, the boundary of this region swings inland around a sand plain. Sauble Beach itself is flanked by a range of sand dunes, and blow-sand is common on the adjacent sand plain. A massive gravel beach skirts the northern border of the Arran drumlins, while along the western border bluffs cut in the sides of drumlins occupy most of the shoreline north of the Bluewater Highway. Below the bluffs lie wide boulder pavements. The French Bay Road follows a line of sand dunes, and about half way between this road and the lake the Nipissing beach may be recognized. Beyond it are old beaches, one after another, to the shore of Lake Huron.

Across the mouth of the Saugeen valley, Lake Algonquin built a massive beach of sand and gravel. Behind it was a lagoon in which fine sand and silt were deposited to a considerable depth. Delta sands were spread outside the beach, also, ending at a distinct bluff about half a mile from the present shore. The terrace below the bluff is ribbed with gravel bars built by Lake Nipissing and, as is the case along so much of the shoreline, the waves have washed most of the overburden off the bedrock on the lower or Nipissing terrace.

Between the big barrier beach at Port Elgin and the barrier of sand dunes south of Grand Bend the Huron Fringe is bordered by a shore-cliff, from 50 to over 100 feet high. From Point Clark to Grand Bend there is little or no terrace below this bluff. The few strips of sandy beach are eagerly sought by summer cottagers.

Grand Bend takes it name from the hairpin turn of the Ausable River as it doubled back to flow the length of the Ipperwash sand dunes to Port Franks before emptying into Lake Huron. This route has been used less since a channel was forced to the lake at Grand Bend, and a canal cut directly inland behind Port Franks to the Ausable, shortening the course of drainage by about 20 miles. The dunes here cover a belt about a mile and a half wide. Behind them is a lagoon with a marly floor, including a big marsh and Smith Lake. Some of the marsh nearest to Thedford has been drained and developed for the production of celery, onions, and other special crops.

The well-known Kettle Point is a shelf of bedrock which has not yet been worn away but stands uncovered so that big round concretions, called kettles, are exposed. These peculiar forms are like cannon balls of stone; the term "kettle" seems to be something of a misnomer.

Adjoining the Kettle Point Indian Reserve is a small plain ending at Gustin Grove below the Algonquin bluffs. This protective wall to landward is 50 or 60 feet high and the little area also has some sandy soil suitable for fruit growing. Even peaches have been grown commercially.

84. *Fissures in limestone, Bruce peninsula*

The most southerly ten miles along Lake Huron from Brights Grove to Point Edward has sandy beaches along the shore protecting a marshy lagoon. There was once a shallow lake in this lagoon, but it was emptied by a drainage canal in a scheme to reclaim the marshland for the culture of sugar beets and truck crops. It might be well to record that the lowest of the marshlands are frequently inundated in the spring due to freshets, or plugging of the outlet at the lake, or both. As in the Thedford marsh area, the Blackwell marsh and the flats around are underlain by marl or marly silt and clay.

The Huron Fringe contains the resort areas of the Lake Huron shore. Such places as Port Franks, Grand Bend (1,000), Bayfield (400), Goderich (6,500), Bruce Beach, Kincardine (2,900), Inverhuron, Port Elgin (1,650), Southampton (1,850), and Sauble Beach are well known. Much of the clientele is drawn from London and other cities in southwestern Ontario, but there is a goodly number of American visitors as well.

Another use for relatively isolated shorelines is demonstrated by the location at Douglas Point, just north of Inverhuron, of the Ontario Hydro-electric Power Commission's first large-scale nuclear-powered generating station.

BRUCE PENINSULA

The Niagara cuesta near the escarpment is generally a zone of scour and the Bruce peninsula section is no exception. Apart from the silt beds of Eastnor township and a few drumlins, gravel bars, and sand dunes, the Bruce peninsula has only a little overburden scattered on the grey dolomite. The surface of the rock is more irregular than that of the limestones of central and eastern Ontario and many wet swampy basins and lakes appear. The dip of the rock strata is towards the west; the surface rises gradually from the water's edge on the Lake Huron side towards the escarpments on or near the Georgian Bay shore, the highest bluffs on Georgian Bay being well over 200 feet in height. Thus the latter shore is one of rugged beauty, while the opposite shore is low, with boulders, gravel, and sand bars, and intervening strips of wet ground extending some distance inland.

The greater part of the Bruce peninsula has very shallow soils, with much bare rock exposed. Classified by the Ontario Soil Survey as the Breypen series, this land type covers an area of about 400 square miles. There is a small plain near Lion's Head in Eastnor township which has deep but poorly drained lacustrine deposits. Ditches have been dug to improve the silt or clay loam soils and considerable land has been

85. *Rough bouldery pasture, Keppel township*

reclaimed by burning peat from the surface. This is the main farming area north of Wiarton, and at one time it enjoyed a good reputation for the production of peas. South of Wiarton, in Amabel and Keppel townships, there is much rough stony land with soils similar to those of the moraines and drumlins farther south.

SETTLEMENT AND LAND USE

The Bruce peninsula was opened for settlement in 1850 and, for a period, lumbering was the chief activity. Most of the good stands of pine, spruce, and hardwoods have long since been removed, but limited cutting of second growth is still carried on. The farmer followed the lumberman but in the northern townships there was little land suitable for crops. In keeping with the characteristics of the land there are two land-use districts, north and south of Wiarton. In the northern townships only 45 per cent of the land is occupied, and farms average over 340 acres, some indeed being very large. The scattered farms on the shallow soils are mainly beef cattle enterprises. In summer the cattle graze over wide areas. Toward the north, however, agriculture becomes less and less important. St. Edmunds township in 1961, had only 14 farms, and only 1.5 per cent of their land was improved. In Eastnor township on the better soils, both the total land holding and the improved land have recently been slightly increased although there are fewer farms. The southern area, occupying the neck of the peninsula west of Owen Sound, is somewhat of a transition zone. About 70 per cent of the land is occupied, the farms averaging more than 200 acres in extent, while only 25 per cent of the farm area is in crop. Beef production is still the most important livestock enterprise but dairy cattle, hogs, and poultry are much more numerous than they are in the northern area.

The characteristics of population distribution deserve some notice. There are several Indian reserves on the peninsula with an almost static population of about 900. The population of the municipalities has decreased from about 15,800 in 1901 to about 10,500 in 1961. The city of Owen Sound, which is not here regarded as part of the peninsula, has grown in the same time from 8,800 to 17,500. It is a port of some consequence, and the commercial centre of wide areas in both Bruce and Grey counties, as well as being county seat of Grey.

The following table is a population analysis of peninsular municipalities, as delineated above.

Population of the BRUCE PENINSULA 1961			
	Total	10,518	100%
	Urban (1 town, 3 villages)	3,252	30.9
	Rural	7,266	69.1
	farm (964 farms)	3,882	37.0
	non-farm	3,384	32.1
	unincorporated villages (2)	572	5.5
area: 670 sq. mi.	small hamlets (8)	666	6.2
density: 16.6	scattered	2,146	20.4

As one might expect, more than half of the scattered non-farm population is to be found in two townships, within a few miles of Owen Sound. All in all, the Bruce peninsula is one of the least densely populated areas in southern Ontario.

However, permanent population reveals only one side of the character of the Bruce. Approximately 60 miles in length, its greatly indented shoreline provides 180 miles of water front, ranging from the superb sands of Sauble Beach to the commanding cliffs of Cape Dundas and Lion's Head. The trail from Dyer Bay to Cabot Head provides a particularly impressive array of cliff formations and cobble and boulder storm beaches. The peninsula is fringed by many offshore islands, including the famous Flowerpot Island National Park. There are, also, at least thirty inland lakes of varying size, occupying depressions in the bedrock created by glacial erosion. Summer brings a great, even if uncounted, influx of temporary population. In addition many people use Highway 6 to or from Tobermory, because of the ferry service to Manitoulin and the Huron north shore. Many small settlements are busy in the tourist season. Large cottage colonies are found at Sauble Beach, Oliphant, Red Bay.

Howdenvale, Pike Bay, and Stokes Bay on the western side of the peninsula. In addition to its transportation function, Tobermory is the home port of a small lake fishing fleet and a yachting centre of some importance. Motels, resorts, and private cottages are numerous. Cottage colonies are found on the eastern side of the peninsula at Dyer Bay, Lion's Head, Barrow Bay, Hope Bay, and Colpoys Bay. Physiography may not have been very favourable to agricultural development in the Bruce peninsula, but it offers some compensation in the attraction of cottagers and other summer visitors.

MANITOULIN ISLAND

Because of its geographic individuality Manitoulin Island may be taken as a unit in this series of physiographic regions. It lies in the northern part of Lake Huron in line with the Bruce peninsula, Cockburn, Drummond, and St. Joseph Islands. It is eighty miles long, from three to

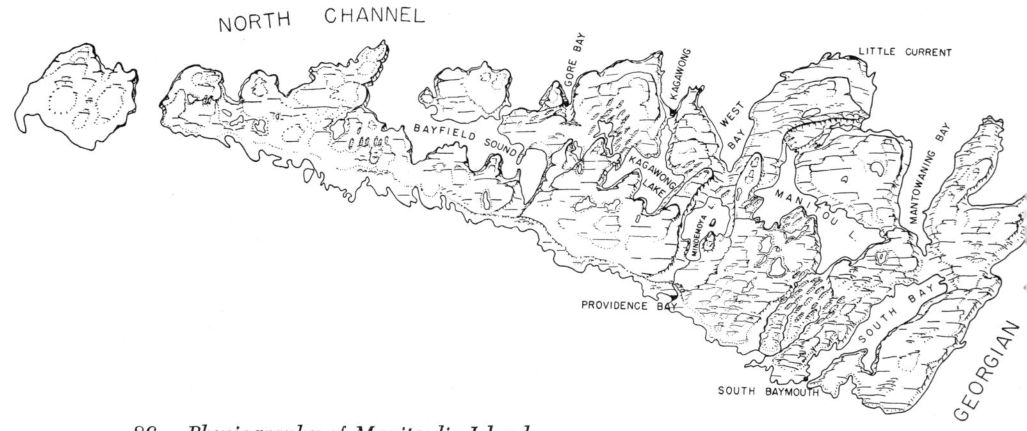

86. *Physiography of Manitoulin Island*

thirty miles wide, covers 1,588 square miles, and is said to be the largest "fresh-water" island in the world. It contains more than one hundred lakes, the three largest of which, Kagawong, Mindemoya, and Manitou, together cover more than 65 square miles. Consisting mainly of limestone tablelands tilted towards the southwest or south, Manitoulin is basically similar to the Bruce peninsula or Prince Edward county.

The island is part of the Niagara cuesta, the rim of the great dolomitic saucer that underlies the Michigan basin. If it were not for a few low spots this rim would divide Lake Huron in two and a long land arc would extend from the Bruce peninsula to northern Michigan.

Lockport dolomite covers the southern two-thirds of the island. Along the south shore the rocky floor dips gradually under the water, but the northern edge of this formation is seen in perpendicular cliffs. The headland projecting into Bayfield Sound, and High Hill between Manitou Lake and West Bay are outstanding examples. The latter stands over 500 feet above the level of Lake Huron or more than 1,100 feet above sea level and is one of the two highest points on the island. In both places the dip of the cap-rock is clearly evident.

Below and north of the Lockport dolomite, the Manitoulin dolomite forms widespread rock plains. In places it too terminates in steep cliffs which in turn overlook Ordovician shales and limestone, probably to be correlated with the Dundas, Collingwood, and Trenton series in Southern Ontario. Between the two dolomites lie the reddish shale of the Cabot Head member, but its outcrops are very limited. In this part of the Niagara escarpment one misses the red Queenston shales which form the lower slopes of the escarpment farther south.

Only a small proportion of the island is covered with deep drift; on the contrary the bedrock over large areas was washed clean of a scanty overburden by the waves of glacial Lake Algonquin. These rock outcrops bear strong evidence of erosion by the Pleistocene glaciers, often exhibiting gouges or flutings running 20 to 40 degrees west of south. Striae are often well preserved when covered with a few inches of soil. South of Gore Bay, where dense strata of Manitoulin dolomite appear on the surface, the rock is smooth and polished. From one site, striated slabs are cut from the upper stratum for use in hearths and mantels.

For the most part the striae point about 40 degrees west of south, agreeing with those on the Bruce peninsula. Towards the west end of the island, however, the direction is 15–20 degrees west of south. They have the same alignment as the elongated ridges of dolomite and the drumlins. Traces of another earlier set of striae trending about 60 degrees east of south were seen on Barrie Island.

Just north of Shequiandah there appears a ridge of white quartzite, of Precambrian age, protruding up through the Palaeozoic strata.

Manitoulin Island has nothing to compare with the big drumlin fields of Southern Ontario, but there are several groups of these hills. They are mostly of the long thin type, some being actually cigar-shaped; and their crests are often bouldery from wave action at the time of Lake Algonquin. The drumlins around Tehkummah contain a good deal of acidic Precambrian rock, those near Ice Lake contain red shale which is not evident elsewhere on the island, and the several drumlins near

Silverwater consist almost entirely of dolomite and limestone. As in drumlins generally, the till is loamy in texture.

In the Tehkummah vicinity and in the valleys of Manitou River and Blue Jay Creek varved clay occupies the hollows and encroaches on the sides of the drumlins. Clay flats are also associated with the little group of drumlins in Billings township, south of Kagawong.

With the possible exception of the crest of High Hill, all of Manitoulin Island was submerged by glacial Lake Algonquin. Gravel bars are common on the island since Lake Algonquin had several stages at as many levels. The waves also washed the surface of the rock, gathering the sediment that finally settled into the depressions. The larger stones and boulders which resisted the waves are strewn over the limestone floor in varying numbers throughout the island. The best tracts of stratified fine sand, silt, or clay, are in the valleys of Black Creek and the

87. *Flutings in bedrock, Grimsthorpe, Manitoulin Island*

Manitou River, in the lowlands both north and south of Mindemoya Lake, the Spring Bay area south of Kagawong Lake, and the Burpee and Evansville flats near Wolsey Lake.

The effects of Lake Nipissing are also evident, not so much along the precipitous north shore where in places there are gravel beaches from 30 to 50 feet above the water level, but along the south shore where a belt one to three miles wide was under Lake Nipissing. The old shoreline is marked by low bluffs, sand dunes, or sand and gravel beaches. Much of this shoreline belt is a barren rock plain from which the forest has been cut or burned. In the Providence Bay and Poplar vicinities it is apparent that silt and clay beds left by Lake Algonquin were washed off the rock by the waves of Lake Nipissing.

Pedologically, Manitoulin Island is very much of a transition area. Typical Podzols and typical Grey-Brown Podzolic soils occur, but in many cases the profiles are weakly developed. Many profiles have characteristics similar to those of the grey wooded soils of western Canada. Others found on highly calcareous parent material resemble the Brown Forest soils of central Ontario while some, on somewhat less calcareous materials, resemble the Brown Podzolic soils of Parry Sound district. It is also fairly common to find a Podzol or Brown Podzolic development superimposed upon what was a Grey-Brown Podzolic soil. Many areas on Manitoulin Island are imperfectly or poorly drained and have varying development of a glei horizon at the base of the profile. Fair-sized areas, also, have a covering of peat or muck. However, by far the most obvious and most important characteristic of the soil cover is its shallowness. Almost two-thirds of the island, most of it classed as Farmington, is too shallow for cultivation and is suitable only for woods or rough pasture.

SETTLEMENT AND LAND USE

White settlement on Manitoulin dates from 1862 when a treaty was concluded with the Indians of the western and central parts of the island, setting aside certain reserves and throwing the remainder open. The Indians of the eastern end of the island, however, refused to sign and their territory remains on the map as "Manitoulin unceded." By 1900 the island had over 11,000 people of whom about 2,000 were Indians. Since then the white population has increased. Manitoulin is still a rural area, its two small towns, Gore Bay and Little Current, having about 2,250 people. There are also two small unincorporated villages and about ten rural hamlets. About one third of the white population lives on some 700 farms with a total area of 280,000 acres.

Population of MANITOULIN ISLAND 1961			
	Total	10,692	
	Indian reserves	2,765	
	Other areas	7,927	100%
	Urban (2 towns)	2,243	28.5
	Rural	5,684	71.5
	farm (703 farms)	2,682	33.7
	non-farm	3,002	37.8
	unincorporated villages (2)	756	9.6
	small hamlets (10)	776	9.7
	scattered	1,440	18.5

There is a common saying in Manitoulin that the island is made up of one-third good land, one-third rock, and one-third water. Actually about 40 per cent of the island is occupied by farms, but the area annually cropped is less than 50,000 acres. The average size of farm is about 400 acres, which is almost three times the provincial average. Each farm has about 70 acres in crops, the most important being hay, oats, and mixed grains. A small amount of corn is grown for ensilage. The agriculture of Manitoulin consists mainly of extensive grazing, the latest statistics (1963) disclosing 25,500 cattle and 14,000 sheep on the island.

Many farmers have achieved a practical organization which utilizes to advantage both the good deep soils and the shallow rocky land. They live on small lowland farms on which they grow their crops and keep their herds of breeding stock. In addition they have several hundred acres of "range" on which they pasture growing animals during the summer. There are many farms of 500 to 1,500 acres on the island.

Most of the island cattle are of the beef breeds, Herefords being dominant because they adapt best to range conditions. The island has for years been free of bovine tuberculosis and the range animals are in good demand in southern Ontario where they are finished for market. Livestock sales are held annually at Little Current.

Dairying is of minor importance, less than 10 per cent of the commercial farms being classified as dairy enterprises. There are creameries at Mindemoya and Manitowaning, but cheesemaking never became established here as it did on the shallow soils of eastern Ontario. Manitoulin gained some fame in the production of turkeys. In 1930 about 10,000 birds were marketed; some years later the number had increased to 50,000, but in recent years there has been some decline in the face of competition from other areas. There are also a few large fur farms.

A great deal of Manitoulin still remains in forest. At one time it was an important source of saw-logs. Later it produced quantities of spruce and poplar pulpwood. At present this activity is much reduced but some reforestation has been carried out.

Apart from agriculture, summer recreation is the most important activity. Lodges, cottages, camps, and private cottages are widespread but they are especially attracted to the shores of the large inland lakes. The south shore, facing the wide open expanse of Lake Huron, is almost uninhabited except for a few spots such as South Baymouth and Providence Bay. Little Current, the nearest approach to the mainland, is a popular yachting centre.

In many ways, land use and human activity on Manitoulin Island resembles that of the Bruce Peninsula, with which it also shares so many physiographic characteristics.

ST. JOSEPH AND COCKBURN ISLANDS

In line with the Bruce peninsula and Manitoulin Island on the Niagara cuesta are the three islands of Cockburn, Drummond, and St. Joseph. Drummond Island is part of the State of Michigan and therefore beyond our province, but to some extent it resembles Manitoulin. St. Joseph and Cockburn Islands bear little similarity to Manitoulin and are distinctly reminiscent of the Penetang peninsula. In particular, they are almost all covered with deep drift derived from Precambrian rocks, whereas Manitoulin Island consists largely of limestone rock plains. With an area of 150 square miles and a population of 900 people, St. Joseph is much the more important of the two islands. Cockburn Island is about 67 square miles in extent and supports less than 100 persons.

The landform pattern on the two islands is fairly similar. It consists of morainic hills near the centres which stood as islands in glacial Lake Algonquin and which are given prominence by high bluffs that encircle them. They stand about 500 feet above Lake Huron. Below these bluffs there are broad bouldery terraces, gravel beaches, and plains of deeper sands. A group of drumlins with bouldery surfaces add variety to both islands. Glacial moraines which are prominent in the northern part of St. Joseph have smaller counterparts north of the big hill on Cockburn Island. On St. Joseph Island a good-sized area of Algonquin and Nipissing clay occurs in the northwestern section, while only a very small clay area occurs on Cockburn Island. Limestone comes to the surface along the north shores of both islands, but there are no large areas of shallow soil.

On Cockburn Island farming is no longer carried on. The northern part of St. Joseph is a general farming area, strawberries, sweet corn, raspberries, and other vegetables grown on the island are noted for quality in the Sault Ste Marie market. Even a few apples are grown.

The forests on these two islands contain mostly hardwoods with sugar maple the dominant species. Yellow birch, white birch, and beech are fairly abundant, along with some white and red pine, white spruce, balsam, white cedar, and aspen. Limited stands of black spruce are seen in the valleys of St. Joseph. There is some regeneration taking place on abandoned farmland, chiefly of hardwoods. On Cockburn Island some spruce has recently been planted, pointing the way for better future care of the forests which occupy over 90 per cent of the area.

OAK RIDGES

Oak Ridges interlobate moraine stands out as one of the most distinctive physiographic units of Southern Ontario. Its general altitude is about 1,000 feet above sea level and it extends from the Niagara escarpment to the Trent River, forming the height of land dividing the streams of the Lake Ontario drainage basin from those flowing into Georgian Bay and the Trent River. Over 100 miles in length, its width varying up to eight miles, the total area is approximately 500 square miles. In three places it narrows almost to the point of extinction and, whether by accident or design, these saddles are crossed by three north-south provincial highways: 27, 7, and 28. It is also crossed by Highways 50, 11, 48, 47, and 35, where it is higher and wider.

The surface is hilly with a knob-and-basin relief typical of end-moraine. The highest parts are at the western end where, along the brow of the escarpment, the sandy hills have an elevation of about 1,400 feet near Mono Mills and Sleswick. A few miles east, Wolfe Mountain rises to 1,200 feet. The triangulation station north of Lake Marie in King township is above the 1,175 foot contour, while the Uxbridge triangulation station is above 1,325. The highest point in Clarke township is over 1,300 feet while several nearby hills are nearly as high. The Haldimand triangulation station, situated on the highest point south of Rice Lake, has an altitude of over 1,175 feet.

While, for the most part, these hills are composed of sandy or gravelly materials, this is not always the case, some of the highest ridges being formed of boulder clay which protrudes above the outwash. The depth of this sandy material is not completely known over the whole area but in King township, where grading has been done for the dual highway from Toronto to Barrie, deep cuts expose till within 30 feet of the surface. On the other hand there are areas where stratified deposits are known to be 100 feet or more in depth. The northern border of the morainic area is deeply indented by swamp-floored valleys, along which many outwash terraces are found.

In general this upland is to be regarded as the source area for many streams which drain the till plains on either side of it. However, as Carman (10) pointed out some years ago, there is in much of the moraine, itself, a virtual lack of streams. The water drains vertically through the sand and gravel, moving laterally only when it reaches less pervious beds and reappearing as springs along the slopes of the moraine. An important exception is to be noted in the case of the Humber River which cuts deeply into the sandhills northwest of Bolton. In that case the source of the water is not so much in the sand hills themselves as in the Niagara escarpment which adjoins them to the west. This lack of streams puts the area at great disadvantage for recreational purposes and also limits its usefulness as pasture land.

The original vegetation of the area was a mixed forest of pine and hardwoods. The valuable white pine was early sought out and very few trees of any size are left in the present wood lots. Among the hardwoods, hard maple, beech, and red and white oak were most important and a few good stands may still be seen.

On the Oak Ridges moraine there is much hilly sandy soil which is subject to blowing. Blow-outs are frequent and it is common to find the original surface horizon of the soil in quite level areas buried under a few inches of calcareous sand. Some nearly level topography is provided by sand and gravel outwash, or occasionally fine sandy loam.

The characteristic soil of the kames is droughty and unstable under cultivation or even pasture. Attempts have been made to fix the blow-sand by reforestation here and there on the ridges, as for instance in the York county forest near Vivian, the Durham forest east of Pontypool, the Northumberland forest south of Fenella, and many smaller private plantations. They provide splendid demonstrations of the effectiveness of pine plantations on such soil. Furthermore, this land serves as the headwater area of the streams which take rise in springs on the shoulder of the moraine, and forest is needed to maintain these springs (10).

On the gentler hillsides and the outwash aprons the soils are more useful. They may be pastured or cropped, depending on the slope, and the texture of the sand. Cattle seem to thrive on the sparse pastures, perhaps because of the lime and phosphorus which the soil inherits from the local limestones. The livestock economy is supplemented by potatoes and considerable rye is grown. A few tobacco farms have been established in recent years near Pontypool.

88. *Oblique aerial views of the Oak Ridges moraine, above, south of Uxbridge and, below, near Pontypool*

The Pontypool series is found upon the hilly topography and the surface soil may be sandy loam, sand, or gravelly sand. The Brighton series is developed upon the almost flat or gently undulating outwash aprons which are here and there associated with the moraine. While the coarser-textured materials predominate, yet there are many areas in which the soils are classified as sandy loams, and which may be expected to be more retentive of water and therefore more useful soils.

PONTYPOOL SAND

A_h 2–3 inches, light grey brown sand; low in organic matter, frequently stony; pH 6.8.
A_e 10–30 inches, yellowish sand, sometimes brownish and indicative of a slight development of a B horizon; pH 6.4.
C grey, coarse, poorly sorted, calcareous sand with numerous cobblestones; pH 7.4.

BRIGHTON SAND

A_h 3–4 inches, grey brown sand, low in organic matter, stone-free; pH 6.6.
A_e 12–20 inches, greyish yellow sand; pH 6.4.
B 2–4 inches, brown loam with incipient blocky structure, sometimes extending deep tongues into the parent material; pH 6.6.
C grey, stratified, calcareous sand; pH 7.6.

Following the exploitation of the timber, practically the entire area was occupied as farm land and, despite the abandonment of the poorer hillier farms, much of it is still farmed. Only one township can be said to be entirely within the area, namely Uxbridge, and for various reasons it cannot be considered strictly representative. From the 1961 Census of Agriculture, however, the following information was obtained. Sixty per cent of the township is occupied by farms, the average size of which is 120 acres. Both farm land and farm population are only about two-thirds of what they were thirty years ago. Of the farm area, 73 per cent is improved land, 47 per cent is under crops, 21 per cent improved pasture, 12 per cent woodland, and 15 per cent in natural pasture and waste land. The outstanding fact here is not so much a changed pattern in the use of land in farms but that the area of farm land has decreased by 30 per cent in two decades.

The Ganaraska River which has a large part of its area in the inter-lobate moraine was the subject of the pioneer study of conservation and river development in Ontario (115). It recorded the land use for a contiguous area of over 22,000 acres, showing 39 per cent crop land, 35.5 per cent pasture land (both improved and unimproved), 18.9 per cent woodland, and 6.4 per cent idle land. Problems of land use on the inter-lobate moraine were discussed in some detail. The primary problem was to decide what land should be reforested and to make plans for a program of tree planting. For much of the remainder, good sod culture was seen to be the key to management of this light land. However, it is of interest to note that some of the sandy soil has been used for tobacco culture, despite the short frost-free season of these uplands.

Declining agricultural land use and the extension of forest area usually means a declining rural population. This is not the case throughout most of the interlobate moraine, and particularly in the western portion which has become increasingly accessible to the people of the metropolitan area through the building of improved highways. In thirty years farm acreage has declined by 20 per cent and farm population by 26 per cent, but rural non-farm population has increased more than six-fold. Much of this is located on scattered roadside lots, but there have also been efforts to create scenic rural or woodland subdivisions.

Much of the land which has been taken out of agriculture has been put to recreational uses. There are a number of kettle lakes, such as Wilcox Lake and Musselman's Lake which have attracted summer resorts and cottage settlements. Ski resorts have been built in some of the hilly areas. There are some excellent golf courses. The Metropolitan Conservation Authority has developed a number of areas for a comprehensive, year-round program of activities, including the Albion Hills Conservation School. Thus the varied characteristics of the land which were unfavourable to agricultural development have been used to advantage.

Agriculture itself has some special characteristics. Many of the occupied farms are in actuality rural residences where only a nominal amount of farming is carried on by people whose livelihood is earned elsewhere. There are also many hobby farms owned by persons whose wealth is derived from industry or other business in the city. Increasingly, the Oak Ridges will come to be considered a non-agricultural area. It should be regarded as a reserve of open space for amenity purposes which might well be directed thither in order to conserve good agricultural land in other areas.

PETERBOROUGH DRUMLIN FIELD

Lying between the Oak Ridges moraine and the area of shallow overburden on the Black River limestone there is a rolling till plain with an area of about 1,750 square miles. Extending from Hastings county in the east to Simcoe county in the west, and including the drumlins south of the moraine in Northumberland county, this belt contains approximately

89. *Oblique aerial photograph of a drumlin near Lake Scugog*

3,000 good drumlins in addition to many other drumlinoidal hills and surface flutings of the drift cover. The name of Peterborough has been used to designate the whole group because the city occupies its geographical centre and because the drumlins are most typical in form and most densely distributed in Peterborough county. The Peterborough sheet, National Topographic series, depicts the contours of at least 1,000 drumlins. Because it was built upon drumlins, Peterborough was once dubbed "City of Seven Hills," but it has since spread its development upon several more. Ashburnham Memorial Park, the City Hall, the Hospital, the golf course, and the university all make use of drumlin sites.

For the most part the rock underlying this region is Trenton limestone, a somewhat softer and less massive formation than the Black River limestone. It is also highly fossiliferous and easily disintegrated. The beds slope slightly towards the southwest and the edges of overlapping strata face the north, thus providing the advancing glacier with good loading conditions. The general level of the bedrock plain in the east is around 600 feet above sea level but it rises to about 800 feet in the west. The drift varies considerably in depth, being rather thin on the borders of the limestone plains to the north and thicker towards the moraines on the south.

While the general orientation of the drumlin axes in this field is from northeast to southwest, there are local variations worth noting. In Hastings county the direction of the ice movement seems to have been 20–30 degrees west of south. Around Campbellford in Northumberland county, it was only 10 degrees west of south; west of Rice Lake drumlins are found pointing 40 degrees west of south and in the Lake Simcoe area as much as 60 degrees west of south.

The drumlins throughout are composed of highly calcareous till but there are local differences. In Victoria county, particularly in Mariposa township, the bedrock is seldom far from the surface and the till contains great quantities of angular limestone rubble. Here, also, the drumlins are more scattered and not so well formed as in other areas. Farther east in Peterborough, Northumberland, and Hastings counties, the till contains much less small rubble. Instead, there is a much greater occurrence of "hardheads," boulders of Precambrian origin, many of them being two to three feet in diameter. These are often noticed to be more numerous on or near the surface than in deeper excavations.

Toward the south, along the border of the interlobate area, and in the western part of the field, the till is somewhat more sandy. Many drumlins near the moraine have shallow coverings of nearly stone-free silt and fine sand which is probably wind-blown material. Its depth, on the average, is less than two feet and hence it is all within the weathered soil profile; but in extreme cases a depth of six or seven feet of this material has been noted.

The Peterborough drumlin field is notable for its eskers as well as its drumlins. These gravel ridges afford poor soils but they are valuable as sources of road metal since other gravels of good quality are rather scarce. Good examples may be seen near Cannington, Cameron Lake, Omemee, Norwood, Stirling, Tweed, and Marlbank.

The Cannington esker is a rather narrow and extremely crooked gravel ridge which crosses the highway just west of the village, ending in a large kame from which much gravel is removed. Across concession XII of Brock township its crest is followed by a farm lane. In concession XIII the Beaverton River has cut away a short section. For the most part it is a single ridge about 25 feet in height. Between Cannington and Manilla the county line crosses a much larger esker which in places is over 60 feet in height. It is much segmented, its sections ranging from one-tenth to two miles in length. It can be followed definitely from Blackwater to a point about two miles north of Manilla, a distance of eight miles. Southwest of Blackwater the deep swampy valley of the Beaverton River extends for ten miles, almost to Uxbridge. Rising from the floor of this valley in various places are unmistakeable esker segments. The esker near Cameron Lake follows the valley of Martin Creek and is more or less in line with it. The distance from Blackwater to Cameron Lake is 35 miles, but it is not definitely proven that these ridges all belong to the same esker train.

The Omemee esker is crossed by Highway 7 about one mile west of the village. Known as the "Hogsback," it is from 50 to 75 feet in height. It is cut through by the Pigeon River, and for a short distance south of this a road has been built along its crest. A massive gravel ridge, it can be identified without difficulty from a point in the valley of Fleetwood Creek near Bethany to the shore of Pigeon Lake north of Fee's Landing, a distance of twelve miles. A mile or so northwest of Omemee, it is joined by another esker which can be followed northward through Downeyville for about six miles. Both northern branches are broken into many segments.

The Norwood esker is also a very massive ridge, some parts of it being over 75 feet in height and the angle of slopes about 26 degrees. In places the esker is compound, two or three ridges lying side by side. This esker can be followed for about twelve miles.

Between Stirling and West Huntington the Ridge Road follows a medium-sized esker for several miles. Another large esker can be followed for about twelve miles in the vicinity of Thomasburg and Tweed.

The longest of these gravel ridges is the one which can be seen near Marlbank, Frankford, and Codrington. It trends roughly northeast-southwest and can be followed, with some interruptions for nearly fifty miles. In the vicinity of Marlbank its crest is followed by a highway for several miles. For the most part its course seems to be controlled by preglacial limestone valleys, or, at least, it follows them very closely. Beaver Lake, four miles northeast of Marlbank, is almost divided into two by this narrow gravel ridge. To the northeast it was traced across the boundary between the Precambrian and Palaeozoic rocks. On the Shield the stones in the esker are all of Precambrian origin, but about five miles to the southwest they are about 90 per cent of Palaeozoic limestone.

While the eskers are perhaps the most striking features of the plain, apart from the drumlins themselves, they are not as important in respect to soils as the deposits of clay which lie between the drumlins in some areas. This "drumlin and clay flat" landscape is particularly noticeable near Stirling, Bailieboro, Lindsay, Bradford, Schomberg, and other places flooded by the old glacial lakes. More clays are found in the Trent River region where Lake Iroquois covered all the lowland as far upstream as Peterborough. In other places such as to the south of Lake Simcoe we find drumlins or drumlin uplands rising from sand plains. In the flooded areas also many of the drumlins have been wave-washed sufficiently to leave a bouldery surface, while in some cases the hillsides have been undercut and over-steepened as well.

Another important landscape variation is due to the size and shape of the drumlins themselves, and these characteristics are fairly strongly correlated with the depth and abundance of the drift. Along the northern edge of this belt where the drift is thin, drumlins are small, scattered, and often imperfectly formed. On the other hand, along the southern edge of this group there are many very large drumlins. In the townships of Percy and Alnwick some outstanding specimens exceed 200 feet in height and 25 per cent slopes are common. The typical drumlins of the central region around Peterborough are less than a mile in length, one-quarter mile or less in width, and 75 feet in height. They are closely spaced, averaging four or five to the square mile which leaves very little intervening space. Here also some of the drumlins are compound, with double crests, double tails or distinct secondary ridges, and flutings along the sides. In places where the hills are more widely spaced, swamps may intervene but there are not many lakes. On the upland just west of Peterborough, the drumlins are of a wide oval shape, sometimes almost round; but around Lindsay, particularly in Mariposa township, they are poorly developed, low, elongated swells in the till plain. Further west, near Lake Simcoe, they are of somewhat more typical shape with many swampy areas intervening.

A description of the area is not complete without mention of the series of deep valleys leading northward which dissect this till plain. They contain such streams as Black River, Pefferlaw Creek, Uxbridge Brook, Beaverton River, Nonquon River, Eastcross Creek, Pigeon River, and Fleetwood Creek. Occupying a pair of similar valleys, Lake Scugog is caused by the damming of the Scugog River. All of these valleys have wide swampy bottoms traversed by sluggish streams which have not been able to keep pace with the northward upwarping of the earth's crust since the ice age. The valleys, themselves, are deep enough to provide excellent drainage to the adjacent uplands.

The forest on the higher, well-drained sites consists largely of maple and beech with a fair admixture of white pine and hemlock. Precipitation is around 33 inches annually with summer droughts a regular occurrence. The zonal soils belong to the Grey-Brown Podzolic group. Owing to the high limestone content of the till, however, the profiles are often shallow, the A_e horizons being thin so that these soils may be more properly classified as Brown Forest soils.

Otonabee loam, recognized extensively in Hastings, Peterborough, and Victoria counties is the best representative of the Brown Forest soils. The following is a generalized description of the profile often seen in drumlin roadcuts.

OTONABEE LOAM

A_h 3–4 inches, very dark greyish brown, soft crumb-structured, moderately stony loam; pH 7.0.

B 6–12 inches, very dark brown, firm, finely nut-structured, moderately stony loam; pH 7.4.

C grey or greyish brown, loamy, hard, stony, highly calcareous till; pH 8.0.

In uncultivated areas uprooted trees and rodents have mixed much calcareous material with the surface soil. In cultivated areas the plough turns up the calcareous lower horizons so that the surface soil almost invariably gives a positive test for carbonates. Many stones have been gathered from the surface and dumped into piles or built into stone fences. No doubt the stoniness of this soil has helped to retard erosion, but considerable topsoil has been lost from the steeper slopes. Frequent patches of greyish subsoil show that the soil has been removed. Fortunately the uniform slopes of most drumlins are amenable to contour cultivation and strip cropping. To a certain extent some farmers have long followed such practices, keeping bands of steep land along the drumlin sides in sod or in forest. These precautions and working the gentler slopes on the contour will increase, and the sooner the better, because the continued productivity of this soil depends in the first instance upon the control of erosion.

BONDHEAD LOAM

A_h 3–5 inches, dark greyish brown, finely granular, friable, slightly stony loam; pH 6.6.

A_e 12–18 inches, light greyish to yellowish brown, finely granular, friable firm, slightly stony, loam; pH 6.2.

B 6–8 inches, dark brown, somewhat blocky, firm but friable, slightly stony loam or clay loam; pH 6.8.

C light greyish brown or grey, stony loamy, calcareous till; pH 7.8.

Bondhead loam, a zonal Grey-Brown Podzolic soil, occurs in the western and southern parts of the area where the drumlins generally have a more moderate relief. In comparison with the Otonabee, the Bondhead has a deeper profile, while the surface horizons are more sandy and have fewer stones; it is thus a more desirable soil for agricultural purposes. Bondhead fine sandy loam also occurs in many places. Another soil of some importance is the Dundonald series; it occurs where the basal till is overlain by deposits of fine sand to depths of two to four feet.

LAND USE AND SETTLEMENT

The utilization of intensely drumlinized areas is under the triple handicap of stoniness, steep slopes, and wet swampy hollows. Throughout most of this area a second series of difficulties is imposed upon the farmer by reason of the cadastral survey. In common with the rest of Ontario the land was surveyed into townships, concessions, and lots with base lines parallel to the shores of the Great Lakes. This means that in the Peterborough drumlin field the roads and farm lines make angles of about 45 degrees with general trend of the drumlin axes. Consequently there are a great number of triangular and diamond-shaped fields and many areas of wasteland just too small or too awkward to be worked successfully.

The average size of farm, on the basis of six representative townships, is about 170 acres—considerably larger than the average for Ontario. In the last thirty years the area has lost one-third of its farmers, while the average farm has increased in size by more than 35 per cent. The percentage of improved land varies somewhat around an average of 60; about 13 per cent remains in woods leaving 27 per cent for natural pasture and wasteland. About 37 per cent of the farm land is cropped each year and about 20 per cent is in improved pasture. Altogether, about 40 per cent of the farm land is used for grazing. Of the crop land slightly more than half is used for hay, the most important cereal crop is oats; winter wheat is grown in some areas while the use of ensilage corn is increasing.

On the average the farms of the drumlin belt support about 22 animal units per hundred acres, or 37 per farm. That this is lower than in many other regions is not surprising in view of the high proportion of unimproved land. The composition of the livestock population varies somewhat, however; in Hastings county dairy cattle and swine are important, Victoria county specializes in beef cattle, and in Peterborough there is a mixture of beef cattle, dairy cattle, and swine, with poultry becoming important in some areas.

A widespread area such as this must necessarily have market points. Peterborough (50,000) is the largest and most centrally located. Its meat packing, dairy, and milling plants provide local markets. Peterborough, however, has outgrown the local phase of development, becoming an independent manufacturing centre of some consequence, using materials from many sources and selling in a national market. Its initial impetus was physiographic, its advantage being the power developed from the rapids of the Otonabee River upstream from the city. Now, however, power is also brought in from other sources by the Ontario Hydro-electric Commission. Other centres of importance are Lindsay (12,000), Campbellford (3,500), Lakefield (2,200), Frankford (1,700), Norwood (1,060), Hastings (900), and Stirling (1,400). There are numerous interesting smaller villages such as Sunderland, Oakwood, Little Britain, and Omemee, as well as many road corner hamlets. We are unable to match the population census with the estimated area of the Peterborough drumlin field, but we can derive some information from the same six townships which we used for land use analysis. Here we see that only 40 per cent of the rural population lives on farms while 60 per cent lives in rural hamlets and in scattered rural residences. Again there is a certain physiographic influence. A residential site overlooking a drumlin landscape has an aesthetic appeal. Odd corners on many sloping farms have much more value as building lots than they could ever have as agricultural land, while township councils have been slow to impose zoning restrictions. This is particularly evident in the environs of Peterborough and less so in more remote areas. Even the smaller towns, however, show some development of the rural-urban fringe.

SOUTH SLOPE

Between Lake Ontario and the interlobate moraine the area has been divided into three regions. The Iroquois lake plain occupies the lowest land, under 400 feet in the west and 600 in the east. The Peel plain occupies a central position in the expanded western portion and is separated from the Iroquois shoreline by the Trafalgar moraine and a strip of fluted till plain. The South Slope is the southern slope of the interlobate moraine but it includes the strip south of the Peel plain. It rises to the line of contact with the moraine at 800 to 1,000 feet above sea level. In other words it rises three or four hundred feet in an average width of six or seven miles. Extending from the Niagara escarpment to the Trent River it covers approximately 940 square miles.

The central portion, in Ontario and Durham counties, is drumlinized, although drumlins are scattered and are of the long thin type pointing directly up the slope. The streams flow directly down the slope; being rapid they have cut sharp valleys in the till. In addition, numerous gullies have been cut by intermittent drainage so that the east-west sideroads

90. *Gullied till plain north of Bowmanville, showing bare eroded slopes*

cross a succession of valleys. Bare grey slopes where soil is actively eroding are common in this area, and the extension of gullies into otherwise unbroken fields is a critical problem. It is a particularly important problem because the land affected is otherwise of high quality.

In Hope township between Kendal and Rossmount there are several large hills standing two or three hundred feet above surrounding lowlands. They have steep sides but their crowns are moulded into drumlins. The south face of the moraine in this section appears farther north and is rather abrupt. Because of the gap in the moraine opposite the southwestern tip of Rice Lake there is also a gap in the South Slope.

The eastern portion of the slope in Northumberland county is thickly covered by big drumlins pointing to the southwest, which have diverted the streams diagonally down the slope. The side of the Murray hills, which constitute the eastern extremity of the South Slope, is an abrupt slope over 200 feet in height, steepened near the base by the shorecliff of Lake

Iroquois. On a clear day this imposing formation may be seen by looking north from No. 2 highway west of Trenton. A much closer view may be obtained from Highway 401.

The western portion of the south slope of the interlobate moraine lies north of the Peel plain, but the Trafalgar moraine and adjacent till plain to the south of the Peel plain is also included. East of Maple the slope is smoothed, faintly drumlinized, and scored at intervals by valleys tributary to the Rouge, Don, and Humber river systems. West of Maple the surface is morainic, most of it a ground moraine of limited relief. South of the Peel plain in Scarborough township there is a gently rolling till plain exhibiting bold flutings running about 30 degrees west of north, and low drumlins. Toward the west in York, Etobicoke, and Toronto townships this type of surface fades out in favour of ground moraine with its irregular knolls and hollows. West of the Credit River the Trafalgar moraine provides subdued morainic topography, while a narrow belt above the Iroquois shorecliff is planed and fluted. In the Sheridan area the flutings appear to be worn in the shale bedrock. All the rivers in this area have cut through the boulder clay and into the shale, the valley-walls in the shale often being almost perpendicular.

The South Slope lies across Trenton limestone, the grey Collingwood and Dundas shales, and the reddish Queenston shale. The material in the drift is related to the underlying rock, allowing for some importation by the glacier. East of Oshawa the till is highly calcareous and the cultivated soils often contain free lime carbonates on the surface. Although shale appears in the till west of Oshawa, it is not until Scarborough township is reached that acid soil is found and it is only slightly acid. The shale content increases west of Toronto until a till consisting nearly all of red and grey shale is reached west of the Credit River.

Reference should be made to the marly clay till in Durham and Ontario counties. Containing few stones, it probably consists of lacustrine clay and silt reworked by the glacier.

In Northumberland county fine sand and silt is found on the surface of the till up to a depth of six or eight feet. This wind-blown deposit is generally less than two feet thick in Durham and Ontario counties and unimportant west of Toronto.

The South Slope contains a variety of soils, some of which have proved to be excellent through more than a century of agricultural use. They are developed upon tills which are more sandy in the east and clayey in the west while the slopes are often steeper in the east than in the west. They are highly calcareous east of Pickering; west of Oshawa a sprinkling of shale appears in all the tills and the shale content increases westward.

Black and grey shales are found east of Brampton, while red shale characterizes the area west of there, particularly beyond the Credit River. Mostly, this slope is drumlinized, the main exception being the Trafalgar moraine between Streetsville and Nelson with its moderate knob-and-kettle relief. The soils on the sandier tills of Northumberland have been classified in the Dundonald Series. The Bondhead, Darlington, and Woburn soils are loams in texture, the latter two containing some shale along with the limestone. The King clay loam also contains a blend of shale and limestone while the main soils on the more shaly tills west of Toronto are called Chinguacousy clay loam and Oneida clay loam. The Oneida soil was developed on the reddish tills of the Trafalgar moraine under a forest dominated by oak, hickory, and white pine.

Of all these soils, the Woburn loam on the gently rolling plain of Scarborough township appears to be the best, although the fine farms of this township are now largely urbanized. The Bondhead and Darlington loams are also very desirable soils. The Oneida and Chinguacousy clay loams are acid, harder to work, and more limited in their adaptation to various crops, but they cannot be regarded as poor soils. Lime and available phosphorus contents in these latter shaly soils are lower than in the soils east of Toronto, but potash content is higher. Thus the physical and chemical properties and productivity of the soils of this slope illustrate nicely the effect of bedrocks and landforms.

SETTLEMENT AND LAND USE

Lying behind the lakeshore areas of first settlement in Upper Canada, the South Slope was colonized by the "second wave," composed largely of British immigrants after the close of the Napoleonic wars. Toronto (York) was founded in 1793, Markham, Whitby, and Bowmanville in 1794, but the interior of Peel and Haltan counties was not laid out for settlement until 1819. The interior of Ontario and Durham counties was also settled later; names like Enniskillen, Kendal, Kirby, Tyrone, and Raglan undoubtedly give some clues concerning the origin of the settlers.

A mixed, subsistence agriculture was undoubtedly the rule in the early settlements but grain soon began to be exported from the little lake ports. This was increased by the repeal of the Corn Laws in Britain and by reciprocity with the United States; several of the main haulage roads were improved, and some were made into plank roads. In the mid-fifties the railways appeared. The period of grain growing was a period of prosperity, as the stony soils were cleared to make them suitable for the use of horse-drawn machinery. Ready cash and handy building material resulted in many fine fieldstone houses throughout the region.

Wheat growing declined, to be replaced by commercial mixed farming in which beef cattle, hogs, and dairy butter were the chief sources of income. The erosional scars of continuous grain growing can still be seen on some slopes but for the most part they have come under the control of the sod cover of long-term pasture. The expansion of the Toronto milk shed displaced the beef cattle and hogs over a large part of the region but much of the sod remained. Finally, of course, large areas, west, north, and east of Toronto have come under the "urban shadow" and are rapidly becoming part of the built-up metropolis.

Population of FOUR TOWNSHIPS OF THE SOUTH SLOPE (EAST)			
	Total	48,893	100%
	Urban (3 towns, 1 village)	27,406	56.0
	Rural	21,487	44.0
	farm (1,491 farms)	6,453	13.2
	non-farm	15,034	30.8
	unincorporated villages (10)	4,643	9.5
area: 425 sq. mi.	small hamlets (16)	1,792	3.7
density: 115	scattered	8,599	17.6

The eastern part of the South Slope seems to have preserved its rural character to a greater extent than the western part. Let us take a closer look at the area comprised of the four original townships, Darlington, Clarke, Hope, and Hamilton, which are now part of the combined counties of Northumberland and Durham. Until 1930, 99 per cent of this area was still in farms which averaged almost exactly 100 acres each. In 1961 only 80 per cent was still in farms, farm population had dropped by one-third, and the farms were now over 140 acres in area. The area in crops had decreased by 32 per cent, the area in pasture had remained almost the same, while the total pressure of livestock populations had slightly increased, but had been reconstituted to give considerable preference to beef cattle. It is well to point out that the physical differences between the South Slope and the Iroquois Plain in this area are not sufficient to promote much contrast in agricultural practices. Orchards and canning crops were grown both near the lake and on the slopes, although both of these specialties seem to be of less importance than formerly, while a general livestock and crop pattern seems widespread. The whole slope between the lakeshore and the Oak Ridges may be considered as one geographic unit served by a series of strategically placed lakeshore towns.

The combined counties of Northumberland and Durham reached a population peak in 1880 declining to a stagnant level which changed little between the two world wars. Since then, although farm population has continued to decline, total population has increased by almost 50 per cent. The incorporated towns, now well served by land transportation have had a new season of growth but, as may be seen from the accompanying table, rural non-farm population is of considerable importance.

PEEL PLAIN

The Peel plain is a level-to-undulating tract of clay soils covering 300 square miles across the central portions of York, Peel, and Halton counties. The general elevation is from 500 to 750 feet above sea level and there is a gradual and fairly uniform slope toward Lake Ontario. Across this plain the Credit, Humber, Don, and Rouge rivers have cut deep valleys, as have other streams such as the Bronte, Oakville, and Etobicoke creeks. There is, therefore, no large undrained depression, swamp or bog in the whole area, although in many of the interstream areas drainage is still imperfect.

The underlying geological material of the plain is a till or boulder clay containing large amounts of Palaeozoic shale and limestone. In much of the Peel plain this has been modified by a veneer of clay which, when deep enough, is clearly seen to be varved. When tile drains were laid during the building of the original Malton Airport, deep beds of stratified clay were seen. More recently during the widening of the Seventh Line road in Oakville, deposits of varved clay ten feet in depth were found near Drumquin. The clay is heavy in texture and more calcareous than the underlying shaley till, having, presumably, been brought by meltwater from the limestone regions to the east and north, and deposited in a temporary lake impounded between the higher land and the ice lobe in the Ontario basin. The plain extends across the contact of the grey (Dundas) and red (Queenston) shales; consequently the clay to the southwest of the Credit River is reddish in colour and somewhat lower in lime than the clay in the eastern end of the plain.

There are exceptions to be noted to the general heavy texture of the soil. In various places the stream valleys are bordered by trains of sandy alluvium. This is true of the Credit below Norval and of the Humber near Nashville and Thistletown. A small isolated sandy tract lies in concessions VIII and IX of Trafalgar township in Halton county. Near Unionville there is a flat sandy strip, which, like the others, is being occupied by market gardeners. North of Brampton, there is a partly buried esker which serves as a local source of road metal and, even more important, provides an aquifer supplying water for the town.

The water supply of the plain is somewhat of a problem. Except in the central part of York county, the overburden is not deep, the till is dense,

91. *Peel plain near Malton*

and there are few thick beds of sand to serve as aquifers. The high degree of evaporation from the deforested clay surface also militates against adequate recharge of underground water supplies. Further difficulties arise from the fact that the underlying shales are not good aquifers; even when water is obtained by deep boring it is often found to contain salt.

Although now almost completely deforested there is evidence that this plain carried a hardwood forest of high quality and great wealth of species. In the better-drained parts grew sugar maple, beech, white oak, hickory, basswood, and some white pine. The depressional areas carried elm, white ash, and white cedar. Below is a description of Peel clay, the imperfectly drained but more commonly encountered member of the catena. The well-drained member, Cashel clay, found on similar materials with slightly more relief has a profile from 30 to 36 inches deep.

PEEL CLAY

A_h 5–6 inches, dark brown, crumb-structured, stone-free clay loam; moderately high in organic matter; pH 7.00.
A_e 4–5 inches, brownish grey to yellowish brown, slightly blocky, clay loam.
B 9–12 inches, dull brown to brownish grey, blocky clay.
C brownish grey containing a few stones; free carbonates present.

SETTLEMENT AND LAND USE

Settled during the early part of the nineteenth century, the fertile clay soils were cleared rapidly. Once the pioneer stage was passed the plain become a noted wheat growing area which, besides supplying the growing city of Toronto, produced quantities of grain for export to the United

States through various lake ports such as Oakville, Port Credit, and Whitby. Later a mixed type of crop and livestock farming developed with its chief market in Toronto. When alfalfa was introduced into Ontario it was found that this area not only produced abundant crops of hay but it was also ideal for the production of seed and, for a time, this crop was the source of a fleeting prosperity. Since 1928, however, the alfalfa seed crop has generally been a failure. Being within easy trucking distance of Toronto, and having a good mileage of improved highways the Peel plain rapidly became a well-developed portion of the Toronto milk shed. Nearness to Toronto also resulted in the development of a number of gentleman's farms and in the establishment of some orchard, small fruit, vegetable, and poultry farms.

Until the Second World War practically all of the land was used for agriculture. Farms averaged about 100 acres in size, slightly larger in the western part, and slightly smaller just north of Toronto. About 87 per cent of the land was improved, 66–70 per cent was under field crop, 10–20 per cent in pasture, and only 6–7 per cent remained as woodlots.

A number of small urban centres had developed within or near the borders of the clay plain. While they might have had small industries they were still, until the Second World War, mainly farmer's towns. Among them are Milton, Streetsville, Woodbridge, Richmond Hill, and Markham all with less than 2,000 people. Brampton, the most important supply point of the area and centre of a large greenhouse industry, had over 6,000.

In many ways, of course, the Peel plain is simply the central part of a land use area which also includes the western arms of the South Slope. Urban land uses have taken over Scarborough, North York, and Etobicoke which in their northern reaches once exemplified some of the best of the Peel plain. The same will soon be true of Toronto Township, as well as Trafalgar and Nelson, both now incorporated with the lakeshore urban municipalities. In these areas 3,500 farms have been erased to accommodate 750,000 urban folk.

The northern part of the Peel plain, as well as much of the South Slope is contained in the next rank of townships, stretching from Markham west to Esquesing. In 1931 these were not touched by urbanization and, indeed, very little by 1941. This area had about 3,200 farms averaging about 92 acres and carrying about 20 livestock units per farm. There was apparently no vacant land. In 20 years an area of 44,000 acres (19 per cent) was subtracted and the remaining farms consolidated to less than 1,900 units of 127 acres, each carrying about 37 livestock units. Horses and sheep almost disappeared to be replaced by beef cattle. The number of dairy cows hardly changed at all but they are now kept in larger herds and yield more milk per cow. Under these changed conditions the area of pasture has been somewhat increased while the area of crop has been considerably reduced. Of course, there are still a number of special farms with small areas which means that the dairy and livestock farms are correspondingly larger.

Population of the NORTHERN PEEL PLAIN 1961			
	Total	104,032	100%
	Urban (4 towns, 3 villages)	59,152	56.7
	Rural	44,080	43.3
	farm (1,866 farms)	7,304	7.0
	non-farm	37,576	36.3
area: 474 sq. mi.	unincorporated villages (15)	14,499	14.0
density: 220	small hamlets (23)	2,411	2.3
	scattered	20,666	20.0

92. *Erosion of clay loam near Schomberg*

What happened to the 55,000 acres lost from agriculture in the thirty-year period? It became living space, and to some extent work space, for over 70,000 extra people; the population expanded from less than 34,000 to more than 104,000, the distribution of which is shown in the adjoining table. From the municipal directory we learn that the 59,000 urban people occupy nearly 15,000 acres, about 4 per acre. This is a low density, but even so, the amount of land taken to accommodate them is a minor part of the total. Large areas, of course, were taken for railway yards, industrial areas, and unincorporated subdivisions, but a major factor seems to have been the land required by the scattered non-farm group. A few years ago, Crerar (31) estimated that each thousand persons added to the population of the Toronto–Hamilton Metropolitan region during the intercensal period 1951–56 required 382 acres of farm land. His definitions are more precise than ours, and our sample area lies outside the official metropolitan areas. Nevertheless it is realistic to examine land use changes over a long period. In 1931 this area, outside the small towns, was almost completely occupied by farms. Seventy thousand people have moved in, and 55,000 acres have been lost, each thousand added has meant the sterilization of about 785 acres, which is more than twice Crerar's estimated rate. Possibly there is a land use principle involved here, which either permits or demands more space for population at greater distances from large urban nuclei. At any rate, more than half of the Peel plain, one of the most productive dairy and livestock areas in Southern Ontario, is now under the "urban shadow."

SCHOMBERG CLAY PLAINS

A number of topographic basins along the northern slopes of the Oak Ridges moraine contain deep deposits of stratified clay and silt. Located near Schomberg, Newmarket, and to the north of Lake Scugog, the three larger areas, taken together, cover about 475 square miles, and are included under the name of the Schomberg clay plains. In the first two areas the surface under the clay is that of a drumlinized till plain. The smaller drumlins are completely covered, but many of the larger ones escaped complete burial although the clay may occur well up the slopes of the hills. The average depth of the clay deposit seems to be about 15 feet, but deep deposits are known. The construction of the highway near Holland Landing in 1935 resulted in the exposure of nearly 50 feet of varved clays. Since the rolling relief of the underlying till plain has not entirely been eliminated these areas are not so flat as many lake plains.

The Scugog area on the other hand overlies a flat till plain and has a more normal appearance for a lake plain. Even here, however, a few drumlins occur. In the area along the Holland River between Newmarket and Holland Landing considerable dissection has taken place giving rise to rough topography. Elsewhere, the beginning of soil erosion is indicated by light-coloured surface patches on knolls and gentle slopes.

The Schomberg sediments are typically varved clays with annual layers of two, three, four or more inches in thickness. The summer band makes up three-quarters to four-fifths of the thickness, is somewhat more silty, and is grey in colour. The winter band is slightly denser and is brownish grey. Both portions of the varve are decidedly calcareous and small fossil shells are sometimes found in them. Chemical analysis shows the clay to contain about 50 per cent calcium and magnesium carbonates. Mechanical analysis by the Bouyoucos hydrometer method indicated about 50 per cent clay and 40 per cent silt, but its behaviour is more like that of silt than clay. It is very slippery when wet and inclined to be mealy when dry. It is probably composed of freshly ground rock flour rather than weathered clay minerals.

The soils of the Schomberg clay plains have been placed in the Schomberg catena by the Ontario Soil Survey. It contains the well-drained Schomberg silty clay loam, the imperfectly drained Smithfield silty clay loam, and the poorly drained Simcoe silty clay and silt loams. In its natural state the well-drained member of this catena is one of the better clay loam soils of the province. Furthermore, in the Schomberg and Newmarket districts, the well-drained soil predominates, occupying at least 75 per cent of the area. Tile drains have been installed in most of the depressions so that whole fields may be cultivated at the same time, avoiding the patchiness seen in some clay plains.

The original vegetation was hardwood forest, the well-drained sites supporting sugar maple, black maple, beech, ironwood, and basswood; the areas of imperfect and poor drainage were dominated by elm, ash, soft maple, and white cedar.

Since almost no virgin forest remains, especially on well-drained sites, virgin soil profiles are hard to find. However, it is apparent that the dark brown surface horizon of this soil is deeper than usual for the region, and it possesses a good granular structure. The light yellowish brown leached (A_e) horizon is quite shallow; in fact it is sometimes not apparent at all. Thus the soil might better be regarded as a Brown Forest rather than a Grey-Brown Podzolic soil. Free carbonates are commonly present on the surface in uncultivated areas, and are invariably present in the ploughed layer of the cultivated soil. From the standpoint of chemical nutrients the most serious limitation of all members of the Schomberg catena is a deficiency of available phosphorus. Phosphorus deficiency symptoms may often be seen in corn, especially if the soil is low in organic matter and in poor tilth.

LAND USE AND SETTLEMENT

Being associated with well-drained upland soils of drumlinized areas, such as the Bondhead series, and being fairly easily accessible to colonization routes from York, these clay plains were well settled and thoroughly cleared during the first half of the nineteenth century. Little forest cover remains except in the wettest places. Mixed farming was the rule with a dominance of grain in the cropping program. The suitability of the land for wheat was such that for many years the concentration of this crop was greater than in any other part of Ontario except the clay plains of Kent and Essex. Oats and barley were plentiful also, being largely used as feed for hogs. In the imperfectly drained Scugog area, wheat growing was not so successful and alsike and red clovers were commonly used in the crop rotation and often matured for seed. In this latter area there was a great tendency to put down crop land to grass, while a good deal of land also remained in natural or unimproved pasture. All three areas have long been noted for the raising of good beef cattle while in an earlier period sheep were also fairly numerous. With the extension of paved roads these areas come within the range of the Toronto milk shed and some of the farms became fluid milk suppliers. Since these clay plains are small and are highly integrated with the adjoining till plains, and moreover since they are influenced differently by transportation and by other cultural factors, they can hardly be said to have developed a cultural pattern of their own. Yet some trends are apparent. Farms have decreased in number and increased in size, but considerable areas of land have been lost from agriculture to other uses. In the Newmarket area especially, but also in the Schomberg area, rural non-farm population and, consequently, rural subdivision of the land, have increased greatly.

In agricultural land use, itself, wheat remains almost as important as formerly, but other small grains have decreased. Hay and pasture have increased both in area and in yield. There are about the same number of dairy cattle but they produce more. On the other hand, beef cattle have increased by at least 50 per cent as well as being of better quality, hogs have increased more than 50 per cent, and the number of poultry has been multiplied several times.

These areas contain a number of towns and villages, such as Bradford

(2,400), Schomberg (600), and Newmarket (9,000), to which they must have contributed much in their early periods, but of course they serve wider areas as well. Most important, in recent years, is the fact that they are directly connected with the metropolitan area by four or five provincial highways.

SIMCOE LOWLANDS

The lowlands bordering Georgian Bay and Lake Simcoe may be termed the Simcoe lowlands. They fall naturally into two major divisions separated by the uplands of Simcoe county. To the west are the plains draining into Nottawasaga Bay, mostly by way of the Nottawasaga River; it is therefore called the Nottawasaga basin. To the east is the lowland surrounding Lake Simcoe which will be referred to as the Lake Simcoe basin. These two basins are connected at Barrie by a flat-floored valley and by similar valleys between the upland plateaux farther north. Both the lowlands and transverse valleys were flooded by Lake Algonquin and are bordered by shorecliffs, beaches, and bouldery terraces. Thus they are floored by sand, silt, and clay. The lowest parts of the Nottawasaga basin were flooded by Lake Nipissing, which also covered a small area near Coldwater but did not enter the Lake Simcoe basin. The boundary of these lowlands is generally the upper Algonquin beach, although in the Penetang peninsula, which was mostly within the Algonquin bed, the terrain resembles, and will be classed with, the Simcoe uplands. They lie between 725 feet and 850 feet above sea level. The total area of the Simcoe lowlands is about 1,100 square miles.

The Nottawasaga basin. Named for the river which drains it, this basin is limited to the broad flats bordering the river and does not include the considerable areas of upland which help to swell the area drained by the river to about 1,145 square miles. For the most part, this restricted basin was at one time part of the floor of Lake Algonquin and its surface beds are therefore deposits of deltaic and lacustrine origin and not glacial outwash.

The southern portion of the basin, lying mainly in Tecumseth township, represents a bay pretty well separated from the main basin by moraine uplands. Into this area, which we may call the Tecumseth flats, the upper Nottawasaga and its tributaries transported enormous quantities of sand and silt from the dissected outwash terraces of the Hockley valley. Through these level plains, the present streams have cut only shallow

93. *Map of the Simcoe lowlands*

channels and the drainage generally is poor. Near the head of Innisfil Creek in Innisfil township there is a bog of some 2,000 acres, while another of 1,200 acres is found near Randall station farther down the stream. A third large bog of about 2,000 acres lies along Bailey Creek in Tecumseth and Adjala townships. In addition to the bogs, this valley contains large areas of imperfectly drained and poorly drained soils which would be useful for special crops if properly drained. These areas have been mapped by the Ontario Soil Survey as Alliston sandy loam, Smithfield silt loam, and Simcoe silt loam.

The sands of Camp Borden and of the Lisle and Tioga areas were brought into the basin by the Pine and Mad rivers. These loose, coarse textured materials have been well drained by the entrenchment of the rivers of the area. Classified as Tioga sand or sandy loam, these soils are poor and droughty. They have never satisfactorily sustained an agricultural economy and should all be reforested.

The smaller area adjacent to the confluence of the Boyne and Nottawasaga which we have called the Essa flats, has soils of better quality and it is unfortunate that these sandy loams and fine sandy loams should also be classified in the Tioga series. Apart from a few wet areas the soils are mildly acid Grey-Brown earths with an abundance of lime in their parent materials and a fair moisture-holding capacity. These are the soils which gave the Alliston vicinity its reputation for potato growing and which today support the most important area of potato production in the province. Numerous tobacco farms have also made their appearance on the warm sandy loams of the Essa flats.

The Minesing flats represent an annex of the Nipissing lake plains, a small basin shut in by the Edenvale moraine. Part of it, the Minesing swamp, is still undrained and filled with deep peat. For the most part, the swamp is forested with elm, black ash, and soft maple, but in the centre there is an open bog several hundred acres in extent. The remainder of the basin is floored with calcareous clay, some marl and, in the south, with the sandy delta of the early Nottawasaga. This low area is at times the scene of severe floods caused by the overflow of both the Mad and the Nottawasaga rivers. The wettest lands are used for pasture, while general farming with wheat as a cash crop is practised on the areas which are somewhat better drained.

Within the basin there is one important hill, the Minesing "island," upon which the hamlet of Minesing stands. Its soils are formed partly of calcareous clays and partly of overlying sands, both under good drainage conditions.

The Stayner clay plain, despite its low relief, is one of considerable complexity. In part it is a bevelled till plain with pebbly till appearing at or near the surface. Other parts of the area are floored with deeper beds of calcareous clay, while in other places the clay is covered with one to several feet of sand.

The morainic ridge north of Edenvale provides a strip of better-drained soil between poorly drained clay areas to the north and south. Its crest is smoothed by deposits of clay, sand, and gravel. The Nottawasaga River has cut a sharp valley nearly one hundred feet deep through the ridge thus draining most of the lake which once occupied the Minesing basin. Toward the east, the lower Algonquin shorelines are marked by an area of sandy boulder pavement which has mostly been left in woods.

North of the Edenvale moraine, and behind the Wasaga sand dunes lies the small, flat, Jack Lake basin, containing two small lakes with a normal water level of 589 feet above sea level. This is well below the elevation of the Nipissing beaches and not far above the normal level of

Georgian Bay. The lowlands adjacent to Jack Lake are subject to seasonal flooding by the Nottawasaga River. Like the Minesing basin, this depression contains marl and marly silt deposits.

Northeast of the Jack Lake basin lies the Elmvale clay plain which resembles the Stayner plain. Around Crossland and Allenwood boulder clay is found at the surface and the soil is pebbly. In places there are shallow surface layers of sand. East of Elmvale the stratified clay is deep and marly. Lying, as it does, between the Nottawasaga drainage and

94. *Minesing flats: Nottawasaga River in the foreground, and marlbeds in the bank*

the streams flowing to the northeast, the whole plain is either imperfectly or poorly drained. About three miles northwest of Elmvale there is a flat area about 2,000 acres in extent, formerly occupied by the shallow Cranberry Lake which overflowed to the Wye River. It has been artificially drained but the depression is still subject to floods. East of Wyevale, the Wye River and its tributaries have cut deeply into the clay beds, providing better drainage for the adjoining lands. It is, perhaps, stretching a point to include the upper part of the Wye drainage area in the Nottawasaga basin, but in such a featureless plain there is no clear line of demarcation.

Along the slopes of Nottawasaga Bay to the north of Wyevale, the Algonquin shorelines are marked by considerable areas of sandy boulder pavement, while below the Nipissing bluff a wet bouldery strip has remained largely uncleared. The bluff itself gains prominence toward the north where it is more than seventy-five feet in height, thus serving as a clear-cut boundary between the lowlands and the uplands.

As an integral part of the Nottawasaga basin the beaches on Nottawasaga Bay must be mentioned, especially Wasaga Beach. The latter is a splendid sandy beach extending for six miles, backed by sand dunes. The dunes constituted a barrier in front of a lagoon in the Jack Lake area in Nipissing times, and they are still actively forming. Hundreds of summer cottages are built on the dunes behind this beach, and during July and August the summer colony amounts to a veritable city. On the east shore of Nottawasaga Bay, Woodland Beach, Wimblewood Beach, and Balm Beach are also lined with summer cottages owned mostly by Toronto citizens.

The Nottawasaga basin exhibits some notable contrasts in agricultural patterns of land use which must in part be correlated with physiographical differences. In both the sand plains and the clay plains, numbers of farms, farm population and occupied farm land have considerably decreased in the past thirty years. In both areas the average farm size has risen from 120 to 150 acres while crop area per farm has risen from 60 to 80 acres per farm. Both areas show a considerable decrease in cereal production, particularly in wheat and barley. The clay plains show a much greater increase in hay and pasture than do the sand plains. On the other hand, the Essa sand plains in particular show a great increase in cash crops, including potatoes, tobacco, and corn. Livestock raising is more important than formerly. There are about the same number of dairy cows but individual herds are larger and milk production is higher. The greatest change is in the number of beef cattle, the clay areas tending toward larger cow herds in keeping with their greater pasture areas while the sand areas specialize more in finishing cattle for market.

While actual farm population has fallen fairly uniformly by about 25 per cent in three decades throughout the basin, total population has risen. Much of the increase has no connexion with agriculture but is due to the presence of Camp Borden in the approximate centre of the basin. A smaller number have also been attracted by the development of summer recreational areas along Nottawasaga Bay. There are two small towns; Alliston (3,000), the service centre for potato and tobacco producers, and Stayner (1,800), the farm centre for the northern part of the basin. The villages: Tottenham (800), Beeton (850), Cookstown

(700), Elmvale (950), Angus (1,200), Midhurst (350), and New Lowell (200) share in the farm service trade.

Wasaga Beach, with a permanent population of less than 500 is a summer recreation centre for thousands of people, drawn thither by the attraction of the sandy margin of Nottawasaga Bay. The smaller summer colonies at Woodland Beach and Balm Beach farther north along the bay also have few permanent residents.

The Lake Simcoe basin. The eastern portion of the Simcoe lowlands is termed the Lake Simcoe basin since about half its area is occupied by the waters of Lake Simcoe. Along the northern and western shores of the lake the lowland consists of a narrow bouldery terrace for the most part confined by a low bluff cut by the highest stage of Lake Algonquin. On the south and east there are broader plains which may be further described.

From the southern end of Lake Simcoe, known as Cook Bay, a broad valley extends southwestward for 15 miles between high morainic hills. Once a shallow extension of the lake, the floor of this valley is now a marsh of 20,000 acres through which the Holland River meanders sluggishly to Lake Simcoe. Near the upper end where it is crossed by the new Barrie Highway, from six to eight feet of peat was removed in order to base the highway foundation upon the sand and clay below. Towards Lake Simcoe the peat is considerably deeper. The central portion of the marsh supported a vegetation of sedges, cat-tails, and other reeds, while the margins had swamp forest consisting mainly of white cedars.

During the first century of settlement the marsh remained uncleared, being crossed by a single road now known as No. 11 Highway. Since 1925 an area of 7,000 acres south of the highway has been dyked and drained as garden land. In 1935 a number of natives of the Netherlands came to the eastern side of the marsh, building a new village known as Ansnorvelt. Since then practically the whole area within the drainage scheme has been put into such crops as onions, lettuce, celery, spinach, carrots, potatoes, and other vegetables. Development is beginning in the area north of the highway and it is estimated that about 13,000 acres may eventually be reclaimed. The marsh gardens have brought considerable prosperity to the village of Bradford (2,400) which has become the location of vegetable storage and packing plants as well as the trading centre for a new population of about 1,600 people on the marsh. Although

95. *Aerial view of vegetable farms on the Holland marsh—one of the most intensively cultivated areas of the province*

a great deal of the produce is sent direct to Toronto markets, a considerable amount is shipped in refrigerator cars to all parts of Canada and even to the United States.

East of the Holland Marsh the Algonquin lake plain consists of level plains based on deep deposits of sand and silt. Between the marsh and Holland Landing the soils are sandy and have cold wet subsoils. Though partially cleared they have not supported successful general farming; but since the development of the marsh, part of this sand plain has also been occupied by market gardeners. Farther north in what is sometimes called the Queensville flats, the soil is silty in texture and highly calcareous. Much of it has been cleared and occupied by general farmers, who operate under the double handicap of phosphorus deficiency and poor drainage.

Directly south of Lake Simcoe a low, swampy, sandy plain covers most of Georgina and part of adjoining townships. Black River and Pefferlaw Brook are the most important streams but they have failed to provide good drainage. Extending upstream for several miles along each of these streams and their tributaries, Zephyr Creek, Mount Albert Creek, and Uxbridge Creek are long swampy valleys a mile or so wide which may be considered as southern extensions of the lowland. It is to be noted that several areas of drumlinized till, which were islands in Lake Algonquin, break the continuity of this plain. On certain exposures the sides of these hills are marked by shorecliffs and barrier beaches, as for example at Little Hell Hill.

About half of this area is swamp or wet sand which has not been cleared of forest, and more of the poorly drained sandy soil has never been fully developed. On the other hand a few very good farms are found on the loamy soil near Virginia. There are no large towns in the vicinity to provide markets, Sutton, with 1,470 people according to the 1961 census, being the largest village. The summer resort population creates demand for dairy produce and fresh vegetables, yet most of this demand is supplied from outside sources.

A good deal of Thorah township is similar to the wet sand plain just described, modified by several tracts of nearly bare limestone. Fortunately, the better silt loams along the Beaverton River and north of Beaverton make up a higher proportion of the area than in Georgina township.

Northeast of Lake Simcoe in Mara township an interesting variation is met in the form of a clay plain dotted with drumlins. The clay is less marly than that occurring south and west of Lake Simcoe. The drumlins are elongated and consist of calcareous till and, though stony, the soil is good for general farming. The clay land between the hills is more difficult to handle.

Another clay plain without the drumlins occupies the southern part of Orillia township. The continuity of this flat plain is broken by outcrops of limestone, glacial hills, and patches of sand and gravel. In the north the clay extends to the Precambrian rocks and that boundary is not easily defined.

The western shore of Lake Simcoe has many sandy beaches which are almost completely occupied by summer cottages of Toronto people. This littoral area is poor agriculturally, but it includes Barrie (22,000) at the head of Kempenfelt Bay and Orillia (16,000) at the narrows between Lakes Simcoe and Couchiching. Barrie is the county town while both places are centres for manufacturing, local trade, and the tourist traffic. The eastern shore of the lake has fewer sandy beaches but summer colonies exist at Atherley, Brechin, Beaverton, and Port Bolster. On the south shore there are fine beaches at Jackson's Point, Roches Point, Keswick, and at several intermediate points.

On the whole, the Lake Simcoe basin is a poorer farming district than the Nottawasaga basin, its major handicaps being the extensive areas of bogs and wet sand. However, these latter soils may become useful as population grows, since both could be drained and developed for vegetables.

SIMCOE UPLANDS

The Simcoe uplands comprise a series of broad curved ridges separated by steep-sided, flat-floored valleys. They are encircled by numerous shorelines, indicating that they were islands in Lake Algonquin. Included, also, is a portion of the broad upland south of Barrie, which has certain similarities, and the high ground north of Alliston. These till plains stand about 200 feet above the adjoining lake plains at an average altitude of about 1,000 feet above sea level. In northern Oro and Medonte township a range of sandhills stands up above the general level. On the Penetang peninsula the uplands were submerged in glacial Lake Algonquin with the result that boulder pavement, sand, and silt appear on the surface. They are included here because they are elevated and have rolling topography. The total area of these uplands is estimated at 400 square miles.

The origin of the broad ridges and valleys of Simcoe county is still in doubt. We know that deep wells have been drilled in the valleys without striking bedrock, but have no evidence of rock cores in the uplands. In spite of this the best interpretation which can be suggested is that the surface form follows the bedrock topography; that is, the valleys reflect stream valleys in the bedrock. A second suggestion is that the ridges originated as moraines because of an ice lobe in the Georgian Bay depression. In any event their surfaces were finally planed by a glacier advancing toward the southwest.

The till in these uplands differs from the till found east of Lake Simcoe; it consists mainly of Precambrian rock rather than limestone. Its texture is a gritty loam, becoming more sandy towards the north, and it is also bouldery. Some heavier, more calcareous till occurs near Lake Simcoe and near Midland. Several drumlins appear near Orillia.

The morainic sandhills of Oro are similar in structure to the Oak Ridges moraine but consist of less calcareous material. At least half of this belt is still wooded and, at that, clearing has been carried too far. It is not good pasture land, and only selected areas are good enough for potatoes and a rotation of crops.

The surface of the Innisfil uplands south of Barrie is smoothed by shallow sand and gravel deposits. North of Barrie also there are gravel streaks which appear to be shoreline deposits of water levels 200 feet or more above Lake Algonquin.

Between Orillia and Barrie and also south of Barrie certain areas are covered by the crooked ribs of loose sandy till and outwash which Deane calls "ice-block ridges" (33). Most of them are roughly parallel while in some cases they encircle little depressions.

The original forest consisted of hardwoods, chiefly sugar maple and beech, together with white pine. Yellow birch, basswood, and hemlock were also common. The white pine groves have largely disappeared, having been cut out during the lumbering period. Beech are more plentiful here than anywhere else in the province.

The dominant well-drained soil on the glacial till uplands north of Kempenfelt Bay has been placed in the Vasey series. It is a sandy loam which in some areas is rather steep and stony. Because there is often not much limestone in the till its profile differs considerably from the normal Grey-Brown soil of the drumlin fields of southern Ontario and has certain relationships with those of the Brown Podzolic soils on the Canadian Shield immediately to the north. The surface soil has a low organic matter content and is moderately acid. Weathering often extends to a depth of about three feet but the horizons may be weakly differentiated. The surface is undulating but there are few badly eroded slopes.

Swamps are less numerous than in a drumlin field; indeed the sandy till allows downward percolation of water to such an extent that streams are rare on the crowns of the uplands. Springs issue from part-way down the slopes, however, to feed the permanent streams in the lowlands.

The Simcoe uplands also include the sandy ridges of the Oro moraine, the Hendrie forest, the heights overlooking Thunder Bay, as well as many smaller areas. These are also extremely pervious soil areas, sometimes with dry depressions many feet in depth. The loose sandy texture of the surface soil is conducive to wind erosion when the vegetation has been unwisely removed.

The agriculture of the Simcoe uplands may be classified as mixed farming based on a variety of products including cream, beef, veal, hogs, eggs, and poultry. The region is on the northern edge of the winter wheat zone and the acreage of this crop per farm decreases from about six acres in the south to about one acre in the north. Another trend of the same nature is the decrease in fodder corn toward the north. It is grown twice as frequently in the south, perhaps because of better soil conditions as well as a more favourable climate. The Lafontaine area in the northern part of the peninsula has long been known for the excellence of its potato seed stock, in part because the slightly acid soils inhibit the development of the potato scab organism.

From several points of view agriculture seems to be receding; there has been great abandonment of land in recent years, the 1961 census showing only 70 per cent of the area as occupied by farms. The farms that remain, however, have increased in size to about 150 acres. Only about 60 per cent of the farm area is improved, about half of the remainder being in woods; about 16 per cent is in seeded pasture and about 40 per cent in crop. The important crops, in order, are hay, oats, mixed grains, corn, wheat, and potatoes.

Each hundred acres of farmland carries approximately 21 livestock units which means about 32 per farm. The average farm has about 27 cattle, 13 hogs, 4 sheep, and 80 poultry. About 25 per cent of the cattle are listed as milk cows. One gets the impression that greater emphasis has been placed on beef-raising in recent years. In comparison with the adjoining lowlands of the Nottawasaga basin the uplands exhibit considerably less specialization and a lower level of prosperity.

The upland areas have not developed any market centres of their own, beyond a few crossroad hamlets. They are connected by good highways to both Barrie and Orillia, the growing urban centres of the Lake Simcoe

96. *View in the Simcoe uplands near Wyebridge (overleaf)*

basin. They are also in close proximity to the small port and manufacturing towns of the Georgian Bay shore, including Midland (9,000), Penetanguishene (5,500), Port McNicoll (1,100), Victoria Harbour (1,100), Coldwater (750), and Waubaushene (600). These settlements also serve as centres for the Georgian Bay summer recreation region.

CARDEN PLAIN

Between the Kawartha Lakes and Lake Couchiching there is an area of 225 square miles of limestone plain with very little overburden. It is named for the township of Carden which occupies the central part of the area and where conditions are typical. While the physical conditions are much like those of the Napanee plain farther east, there are certain differences. This area was under Lake Algonquin and some beaches and offshore sand deposits are found. The original forest appears to have had some very good pine stands, but most of the present tree growth is of hardwoods.

There was considerable settlement during the lumbering era. The population peak was reached about 1881 when the limestone plain in Victoria county had about 3,300 people and 470 occupied farms. The latter averaged about 135 acres each with 31.5 per cent improved land. In 1941 there were only about 1,700 people in the area and less than 280 farms. Farms averaged 315 acres in size with 13 per cent improved land. By 1961 there were only 160 farms with an average size of over 420 acres, with 17 per cent improved, 10 per cent in crop, mainly hay, and 6 per cent in seeded pasture. The improved land has actually declined from more than 20,000 acres at the height of settlement to about 11,000 acres (1961). More than 60 per cent of the occupied land is accounted for by rough pasture, much of it being held in lots of 1,000 acres or more. Twenty-five per cent of all farm land is reported as woodland. It is probable that a large part of this region was cleared for its timber without any intention of making arable land. Following fire, of which burned stumps give ample evidence, the land gradually came into use for grazing. However, the dairy economy so noticeable on the Napanee and Smiths Falls rock plains never was established here, the land being used mainly as range for beef cattle. It would seem reasonable that some sections should be reforested.

While numbers of farms have continued to decline, and farm population has dropped to about 500, the total population of the area has remained fairly stable since the Second World War. There are a number of small unincorporated but nucleated settlements including Coboconk (500), Kirkfield (225), Burnt River (125), and Victoria Road (125), as well as numerous isolated residences along highways and county roads. The presence of a number of lakes has attracted summer visitors and numerous waterside cottages have been built. Physiographic reasoning would indicate that this area should no longer be regarded as agricultural; it requires more than 12 acres to support one unit of livestock; obviously this is an area which in the long term should be returned to forest with full supplementary development of its recreation potential.

DUMMER MORAINES

The Dummer moraines constitute an area of rough stony land bordering the Canadian Shield from the Kawartha Lakes eastward. The typical landscape is to be seen throughout much of Dummer township and so the name is applied to the region. Other townships in the area are Belmont, Marmora, Madoc, Rawdon, Huntington, Hungerford, and Sheffield. The area is about 600 square miles.

The underlying bedrocks are sedimentary limestones, mostly of the Black River group although including some of the overlying Trenton. They form a plain which declines gently southward from an elevation of 600 to 800 feet above sea level. The limestone terminates on the north in an escarpment 25 to 75 feet in height. In some places there are several smaller north-facing escarpments, while in a few places the rock face is

97. *Carden plain: shallow soil on limestone*

98. *Pasture scene on the rough and stony Dummer moraine*

hidden beneath a morainic mantle. On the south there is an irregular boundary between the limestone moraines and the drumlinized till plain.

Lying on the Shield sometimes several miles north of the main escarpment are a number of limestone outliers. In effect, they are detached pieces of this region and agricultural communities of several square miles exist, surrounded by a wilderness of Precambrian rocks. The settlements of Oak Lake, Eldorado, and Cooper might be mentioned in particular.

Crossing this morainic belt are several streams which are tributary to the Trent or Moira rivers. Most of them follow preglacial valleys, entrenched up to 100 feet in the bedrock. A number of these valleys are blocked by glacial drift, thus creating long narrow lakes or swamps. The Kawartha Lakes and Moira and Stoco Lakes are prominent examples of this type.

The moraines of this area are characterized by angular fragments and blocks of limestone with many Precambrian rocks also present. The surface is extremely rough even though most of the morainic ridges are quite low. Bordering the escarpment, and here and there among the moraines are areas of shallow drift and even bare limestone. A few drumlins appear in Verulam township and large eskers are found at Norwood, Tweed, Marlbank, and Tamworth.

In the present forests sugar maple predominates, with white cedar occupying the wetter areas. Basswood, oak, beech, and butternut are also common. There is evidence that white pine and hemlock were more abundant in the original forest than at present. In burned-over areas the second growth consists largely of poplars.

LAND USE AND SETTLEMENT

A surprisingly large proportion of this rough land was occupied and put to agricultural use. Small, odd-shaped fields were laboriously cleared of stones and boulders, and used to grow hay, oats, and corn. Apart from the presence of stones the chief drawback of the soil is its droughtiness and light crops are frequent. Actually, only one-third of the occupied land was ever made fit for the plough, about 20 per cent was left in permanent woodland, and the rest, without being cleared of stumps or stones, was used as rough pasture. Also a large area of land, once cropped, was allowed to revert to permanent grass. The farm economy of this area is similar to that of the adjoining limestone plains, being based largely upon summer dairying for the production of cheese. Some large grazing farms were used for beef cattle while a few flocks of sheep were seen.

In this area agriculture has steadily declined since the beginning of the century. Since 1931, when the census began to record farm population apart from other rural residents, the number of farms has decreased by 42 per cent, and the number of farm people by a similar amount. The average size of farm, however, has increased from 165 to 245 acres, the amount of seeded pasture has nearly doubled, but the total area of improved land has decreased by about 13 per cent. Since very little is grown that is not marketed as livestock or livestock products, it interests us to note that in 1931 each hundred acres carried 15 livestock units (each farm therefore about 25), while in 1961 each hundred acres had about 13 and each farm 34. From the regional point of view, however, since much land has been totally abandoned, animal production has declined by more than 20 per cent in three decades.

There are no large towns or cities but, for an area of marginal or declining settlement, there are a surprising number of small towns and villages. Fenelon Falls (1,350), Bobcaygeon (1,200), Lakefield (2,200), Havelock (1,260), Marmora (1,400), Madoc (1,350), and Tweed (1,800), are all incorporated places. Most of them have a river or railway site which was important in lumbering days. Madoc once was, and Marmora now is, close to important mining development. However, all these settlements are now mainly farmers' towns and tourist supply centres. Most of them have shown only moderate growth in the past thirty years. Besides these there are small villages such as Tamworth (375), Marlbank (240), Thomasburg (110), Queensboro (200), and Warsaw (250).

The rough bouldery land of this area, interspersed as it is with swamps and tracts of bare limestone presents a problem in land utilization. Most important are the pastures, which are in desperate need of renovation if the livestock industry is to continue. In many cases where pasture renovation is too costly, reforestation is in order. On the other hand, with large machines now available to handle the boulders and to level the surface, some more land clearing might be attempted on the better sites. Even though this should happen there is bound to be further withdrawal of agricultural activity from this physically difficult region.

NAPANEE PLAINS

The Napanee plain is a flat-to-undulating plain of limestone from which the glacier stripped most of the overburden. Based mainly on the Black River limestone it is a counterpart of the smaller Carden plain and the larger Smiths Falls plain which is underlain chiefly by Beekmantown dolomite. Centring on the town of Napanee it covers approximately 700 square miles. While the soil is only a few inches deep over much of the region, some deeper glacial till occurs in the stream valleys and toward the north where this region borders on the limestone moraines. There are also a few scattered drumlins of the long, thin type. In the south, particularly, the depressions often have shallow deposits of stratified clay which also provide better soil.

Cutting the rock plain to a depth of 50 to 100 feet, the valleys of the Salmon and Napanee rivers provide the greatest elements of relief. In these valleys are found a variety of deposits, mostly of alluvial origin.

99. *Shallow soil on limestone near Napanee*

In the original forest sugar maple was probably the dominant tree, with white elm, silver and red maple, and white cedar occupying the low ground. Basswood, beech, and burr oak are also important trees, and there was some white pine, hemlock, balsam fir, and white spruce. White cedar now occurs in fairly pure stands on the dry shallow soil where it is invading old pastures. Hawthorne, hickory, burr oak, and black ash also withstand grazing. The characteristic plants of the pastures and roadsides are Canada blue grass, mulein, blueweed, and ground juniper.

SETTLEMENT AND LAND USE

The waterfront townships of the Napanee plain were among the earliest areas of Upper Canada to be laid out and settled. The pioneer occupants were United Empire Loyalists who made their homes here in the early 1780's. The traveller is reminded of them constantly by the many historical sites and commemorative parks. Particularly interesting are the old meeting house on Hay Bay and the park at Adolphustown. The occupation of the townships in the hinterland took place a few years later.

The geographical centre of the region, and the county seat of Lennox and Addington is the town of Napanee which, in 1961, had a population of 4,500. It was located at the head of navigation where pioneer industry utilized the power possibilities of the Napanee River, which also served as a means of transportation for logs from the interior. As well as for sawmilling, Napanee became a centre for grist-milling and other farm service industries for a wide area. While there were boggy spots, the pioneer roads over the flat rock plains were generally passable at all seasons. Belleville (32,000) to the west, and Kingston (65,000) to the east exert a considerable influence over the extremities of this plain, but under the conditions of pioneer transportation they were unable to suppress the development of a local trading centre approximately halfway between them. At any rate, the earliest land route lay along the shore through Bath (700) and Adolphustown, to the ferry across the bay to Prince Edward county which was also an area of early settlement. Napanee was thus secure in its position of pioneer outpost. Scattered over the plain were many local supply points. Some of those still surviving are Cataraqui (360), Camden East (225), Newburgh (600), Odessa (850), Roblin (135), Sydenham (800), and Yarker (320). Amherst Island has the little village of Stella (100), while Wolfe Island has the somewhat larger village of Marysville (400), as local supply points. These are important, since all transportation to the mainland involves ferry services.

The land was early and completely occupied by farms. In 1931, 99 per cent of the area in Lennox county was listed as agricultural land, while in 1961, it was still over 90 per cent and among the highest degrees of farm spread in the province. The number of farms meantime has been reduced by 42 per cent while the size of farm has increased from about 125 acres to 195 acres with a number of holdings over 1,000 acres. The average farm still has about 13 per cent of its area in woods and 38 per cent in other unimproved land. Only 32 per cent is devoted to crops of which hay and oats are by far the most prevalent, while wheat, corn and mixed grains, taken together occupy less than 5 per cent of the farm land. Grazing is thus on a very extensive scale. Roughly 60 per cent of the commercial farms of the area are dairy farms while 33 per cent depend mainly on beef, cattle, and hogs. It is noteworthy that beef cattle have increased in recent years while the population of dairy cows has declined, although the individual herd has increased in size and production. Swine, sheep, and poultry have notably decreased.

The average farm of 195 acres in 1961 was valued at $12,000, its equipment $3,800, and its livestock $4,600. It sells an average of $3,000 per year of which milk provides 55 per cent, cattle and hogs 42 per cent, with small amounts from many other enterprises. Not all farms are dairy farms, but on those which report dairy cows the average herd contains eleven cows and the average production is about 9,000 pounds per cow. By comparison, the average production in 1931 was only about 5,000 pounds per cow.

Averages unfortunately tend to obscure some basic realities. The census shows that only two-thirds of the farms in this area can be classified as commercial, that is, in 1961 they had gross sales of more than $1,200. But only about 25 per cent of these or about 16 per cent of all census farms sold more than $5,000 per farm. This is a much more realistic measure of the agricultural capabilities of the plain. Furthermore, approximately 20 per cent of all farm operators appear to hold a full-time job somewhere away from the farm. One cannot discount the fact that some men have achieved an agricultural adjustment with the environment, but the shallow soils of the plain are a distinct handicap.

Population of the NAPANEE PLAIN 1961			
	Total	20,415	100%
	Urban (1 town, 2 villages)	5,762	28.7
	Rural	14,553	71.3
	farm (1,356 farms)	6,406	31.3
	non-farm	8,237	40.0
	unincorporated villages (3)	1,388	6.8
	small hamlets (12)	1,169	5.7
area: 431 sq. mi.	scattered	5,680	28.5

We do not have exact data for population distribution over the whole of what we have designated the Napanee plain, but we can use a land area of 431 square miles, containing ten municipalities for which we have a complete census classification, to make a breakdown.

It is only fair to say that the southern part of the plain, through which run the main corridors of traffic, contains not only the incorporated places but also a much greater density of the scattered, non-farm rural residences. Total population density, in comparison with other Ontario areas, is sparse at 47.5 people per square mile. In the non-corridor township of Camden East, total density is only 24 per square mile, of whom 55 per cent are farm residents, 22 per cent live in small nucleated places, leaving a scattered remnant of 23 per cent. One can only conclude that the Napanee plain is a zone of difficulty and that a fair number of those who live on it do so for the sake of cheap space adjacent to a highway.

PRINCE EDWARD PENINSULA

Prince Edward county is a plain or low plateau of limestone projecting into the eastern part of Lake Ontario and almost separated from the mainland by the Bay of Quinte. The isthmus between the Bay of Quinte and Presqu'ile Bay at Carrying Place has been cut through by the Murray Canal, artificially creating an island. This isolation gives us our best chance in Southern Ontario to use a county as a natural region to which census data apply. Manitoulin Island is a closely related case, but there the administrative district includes a small section of mainland as well.

The area of Prince Edward county is 390 square miles, wholly underlain by Trenton limestone except for one small hill of Precambrian granite near Ameliasburgh. The highest point in the county lies between Picton and Glenora, reaching slightly more than 500 feet above sea level or more than 250 feet above the surface of Lake Ontario. Near this, the oddly located Lake-on-the-Mountain has an elevation of 415 feet above sea level, while most of the area of limestone plain has an elevation of about 350 feet or about 100 feet above Lake Ontario. Two other upland lakes worth noting are Roblin Lake near Ameliasburgh and Fish Lake near Demorestville. The shoreline is irregular because of a number of deep valleys dissecting the limestone and thus forming long bays or inlets. The surface of the plain has a slight gradient towards the southwest

100. *Nodular Trenton limestone*

and the western and southern shores are very low. The inlets there are often closed by transverse bars, thus forming lagoons such as West Lake, East Lake, and other smaller bodies of water. On the other hand, the northern and eastern shorelines are often precipitous rocky bluffs rising a hundred feet or more.

As a result of the irregular surface and the much indented coast line, the land survey of the county is also irregular. Shore roads were the natural original routes, and lots were laid out along them. Routes of rather irregular direction also follow the strips of deeper soil. Nevertheless the roads of the peninsula seem to be unnecessarily crooked although, as one resident wryly remarked, "they had to bend them or else they would have stuck out into the lake."

More than half of the county has shallow soils, that is, soils with less than 30 inches and mostly only a few inches of unconsolidated material over the bedrock. The profile development of these soils as well as their agricultural uses are conditioned by lack of depth and water holding capacity. A large area, approximately 55,000 acres according to the Ontario Soil Survey, must be classed in the Farmington series and regarded as similar to the soils of the Napanee Plain, Smiths Falls Plain, and other mainland regions of shallow soils.

In the townships of Hillier and Hallowell, however, the surface strata of the Trenton limestone are thinly bedded and have shale partings which have disintegrated to some depth. Consequently these soils provide better conditions for plant growth than those overlying more massive limestone formation. These soils are referred to as the Hillier series by the Ontario

Soil Survey. Among the farmers of Prince Edward county, they are referred to as "clay gravel" or "limestone gravel" depending upon the relative proportions of these constituents. They cover about 25,000 acres. In other areas, notably in Ameliasburgh, about two feet of till overlies the limestone. There the soils have been classified as Ameliasburgh loam and clay loam, the areas covered by each type being about 25,000 acres.

HILLIER CLAY LOAM

A_h	4–6 inches, dark reddish brown clay loam, frequent limestone fragments.
A_e	4–18 inches, reddish brown clay loam, frequent limestone fragments.
B	4–6 inches, heavy, brown rubbly, clay loam.
C	shattered limestone strata interbedded with frequent shale partings.

In the low areas bordering Big Bay, Muscote Bay, South Bay, and Prince Edward Bay there are extensive areas of deeper clay deposits forming clay plains of about 34,000 acres in extent. Profile development varies greatly according to drainage conditions and a number of soil series have been identified. In certain areas, more notably in the township of Sophiasburgh and Hallowell, there are a few areas of deeper drift with some low, but well-formed drumlins. The soils of these areas have been referred to the Darlington series. Notable, also, are the areas of sandy soils in the region southwest from Picton, between East Lake and West Lake. Sand dunes up to 75 feet in height are found on the barriers separating these lakes from Lake Ontario.

Prince Edward county contains over 13,000 acres each of marsh and muck. Marshes border nearly every lake and lagoon in the county but are especially notable near Muscote Bay, East Lake, and West Lake. Many small areas of muck are found in the shallow depressions on the rock plains such as those near Ameliasburgh and Hillier, but by far the largest area is that contained in the Big swamp which is about ten miles in length from east to west and from half a mile to a mile and a half in width. There has not, as yet, been any extensive development of muck gardens in the county. In all, twenty-eight soil types have been described. They occur in a tightly packed mosaic of small areas and have had considerable effect upon the diversification of agriculture within the county.

From the climatic standpoint Prince Edward is favoured with warmer summer temperatures and a longer frost-free season than the adjoining mainland. Its chief limitation is lack of enough rain in summer. Of course, droughts are particularly severe on the shallow soils. One of the greatest possibilities for development of this peninsula appears to be in supplemental irrigation.

101. *Active sand dunes in Prince Edward county on the shores of Lake Ontario*

SETTLEMENT AND LAND USE

Prince Edward county was early and almost fully occupied by farmers, most of them of United Empire Loyalist stock, and some very attractive old farmhouses still exist. After the pioneer stage, grain farming was tried by many and for a time a great deal of barley was shipped to the United States. Then the shallow lands of Prince Edward county, like those of the mainland limestone plains, went over to summer dairying for cheese-making. The protection from frost afforded by the lake and the excellence of some pockets of deep loamy soils encourage some farmers to set out orchards of apples and some other fruits. Peaches, however, are not grown. For the same reasons, also, canning crops, mainly green peas, sweet corn, and tomatoes, became popular and a number of canning factories were set up.

In recent years most of the commercial farms of Prince Edward county fall into four types, dairy farms, livestock other than dairy cows, fruit and vegetable farms, and mixed farming with livestock combinations. There are, of course, others (about 25 per cent) which are not commercial, usually being worked part-time. In addition, it may be pointed out that some farms have been abandoned as operating units and are now devoted entirely to summer grazing of growing stock.

The area of agricultural land in the county is 207,000 acres, about 82.5 per cent of the total area. During the Second World War some areas were taken for government purposes, and there has recently been some land taken for industrial use. The census of 1961 revealed 1,337 farms with an average area of 155 acres. For the most part they are of medium size, only about 30 being larger than 400 acres. About 68 per cent is in crops and 17 per cent in improved pasture. About 12.5 per cent of the farm area is woodland and there is considerable unimproved or natural pasture. The chief items of agricultural revenue may be listed as follows: dairy products 26 per cent, horticultural products 23 per cent, sales of cattle 16 per cent, poultry and eggs 15 per cent, and hogs 13 per cent, leaving about 7 per cent to be accounted for by field crops and other minor items. It is unfortunate that good agricultural land is so scarce in Prince Edward, otherwise its natural advantages of climate and position would make it one of the more productive areas in the province.

Prince Edward county reached a population peak in 1881, declining about 20 per cent during the next fifty years. Since the Second World War it has come back to its old level but shows little sign of advancing very rapidly. Its incorporated places, Picton (4,900), Wellington (1,000), and Bloomfield (800), are farm service and resort supply centres. Picton has recently become an iron ore port and the site of a cement factory. Canneries have long operated at Wellington, Bloomfield, and West Lake and there are a number of cheese factories in the county.

Population of PRINCE EDWARD COUNTY 1961	Total	21,108	100%
	Urban (1 town, 2 villages)	6,729	31.9
	Rural	14,379	68.1
	farm (1,337 farms)	5,482	26.0
	non-farm	8,897	42.1
	unincorporated villages (2)	568	2.6
area: 390 sq. mi.	small hamlets (12)	1,402	6.7
density: 54	scattered	6,927	32.8

Prince Edward's long shoreline, enclosed bays, sand beaches and sand dunes, are natural resources that have been put to good use. Resorts, cottage colonies, and provincial parks draw many people to this delightful area where lake breezes modify the summer heat.

IROQUOIS PLAIN

The lowland bordering Lake Ontario was inundated in late Pleistocene times by a body of water, known as Lake Iroquois, which emptied eastward at Rome, N.Y. Its old shorelines, including cliffs, bars, beaches, and boulder pavements are easily identifiable features, while the undulating till plains above stand in strong contrast to the lake bottom which has been smoothed by wave action or lacustrine deposits. The latter area is the Iroquois plain which is discussed in this section, excluding the areas to the east which were flooded by Lake Iroquois but which, because of shallow soils, are treated elsewhere. The Iroquois plain extends around the western part of Lake Ontario, from the Niagara River to the Trent River, a distance of 190 miles, its width varying from a few hundred yards to about eight miles. Then it extends inland to include a large area in the Trent valley. Conditions in the old lake plain vary greatly and it is convenient to divide it into a number of sub-sections for purposes of discussion.

Niagara fruit belt. The area commonly known as the Niagara fruit belt lies between Lake Ontario and the Niagara escarpment and extends eastward from Hamilton to the Niagara River. For the most part this lowland lies within the Iroquois plain but it also includes terraces adjacent to the escarpment. The plain is cut by no large streams, but a number of smaller ones cross it to Lake Ontario, including Twelve Mile, Sixteen Mile, Twenty Mile, Stoney, and Redhill creeks. All of these are drowned in their lower courses, producing lagoons or marshes cut off from Lake Ontario by a barrier beach.

From Grimsby east the lake plain contains areas of sandy soils which have enabled the area to become an outstanding fruit growing region. As the beds of sand are never very deep and often overlie clay at two to three feet, drainage is usually a problem. West of Grimsby the soil is largely developed upon red clay derived from the underlying Queenston shale. This soil is heavy in texture and of low permeability, hence sheds water easily and dries out rapidly. Acid in reaction, it has retained very little organic matter under the prevailing clean culture of orchards and vineyards, and is a difficult soil to work. From Stoney Creek to Hamilton there are broad gravel ridges upon which well-drained loams have developed.

The importance of horticulture is well shown by the fact that in Lincoln county, out of 2,128 commercial farms enumerated by the census of 1961, 1,353 (64 per cent) are classified as fruit and vegetable farms. Within the Iroquois plain itself, however, approximately 90 per cent of the commercial farms may be so classified. Approximately 70 per cent of the land is in crop, or an average of 23 acres per farm. The important crops are grapes and peaches but other tree fruits and small fruits are also grown extensively.

It is estimated that there were originally about 35,000 acres of well-drained light-textured soils in the Niagara peninsula suitable for the growing of peaches (88). However, the advantages of the area are primarily climatic (102) as there are considerable areas of soil elsewhere in Ontario which are just as suitable for fruit growing but are subject to greater climatic hazard. Unfortunately for fruit growing, the Iroquois plain in the Niagara peninsula is in great demand as sites for housing and industry as well as for transportation and other facilities that go along with them. The city of Hamilton has spread eastward to engulf much former fruit land while St. Catharines has annexed part of the township of Grantham thus effectively cutting the fruit belt in two. Nevertheless, fruit acreage in the Niagara area continued to increase at the expense of other crops until about 1956. Since then fruit growing has rapidly given way to non-agricultural uses and it has been estimated that by 1980 the fruit industry will almost have disappeared.

The Ontario lakehead. That portion of the Iroquois lake plain which lies at the head of Lake Ontario deserves attention as a special part of the Iroquois plain. Geographers mention many reasons why a lakehead location is likely to result in the building of a city. It is likely to be a port, a point at which transfer is made between water carriage and land carriage of persons and goods, a place of trade with a large hinterland, a place of processing goods brought thither by both land and water, and a place for the establishment of secondary manufacturing industries.

While it now has all of these characteristics, the city of Hamilton did not start out with them in mind. Dundas was at first much more the port and entrepôt, having built the Desjardins canal to exploit as fully as possible the advantage of water-carriage, and having most direct access by road to southwestern Ontario and the settlements of the middle reaches of the Grand River. Burlington on the north side of the lake was also better suited to the area developing behind it. Thus Hamilton was not founded as a village until 1812.

Its advantage seems to have been a focus of land routes. Later it grew downhill to the bay and developed its own port, quickly overcoming the lead of both Burlington and Dundas to become the most important lakehead community. Physiography, however, had denied it the supreme advantage, namely proximity to the natural connecting link between Lake Ontario and the upper lakes.

Lake Iroquois, however, had left it some advantages. The site, a narrow plain between the lake and the "mountain," had direct connections with Niagara by way of the old shoreline, the old bayhead bar provided a corridor to the north shore, and the old terraces inside the bar lead to an easy grade up the escarpment to Ancaster, thus bypassing Dundas in two directions. When the railways came, they also had to come around the head of the lake and the Iroquois bar became the natural route even though it involved a tunnel through the bar itself in order to reach the centre of the city. The old bar also furnished high, dry ground for the cemetery and, later at the northern end, its slopes provided the site for the rock garden. Iroquois terraces in almost all locations have furnished favoured building sites, among the more notable, perhaps, being the location of McMaster University.

Hamilton to Toronto. The next section is that occupying the northeast trending shore of Lake Ontario from Burlington to Toronto. Beginning with the great gravel bar which lies through Aldershot, the shoreline continues as a distinct bluff cut in red Queenston shale for about 18 miles. In the vicinity of Clarkson and Cooksville the old shore is cut in grey Dundas shale, while in some places as in Toronto itself, along Davenport Road, the Iroquois shoreline is cut in Pleistocene deposits.

Besides the Aldershot gravels the old lake built barrier beaches of varying size across Bronte Creek, Oakville Creek, the Credit River, Etobicoke Creek and the Humber and Don rivers in Toronto. Those at the Credit, Humber, and Don, especially, were noted sources of sand and gravel for many years.

The other boundary of the plain is the present shoreline where, as its water level has gradually risen, Lake Ontario has gradually cut back into the red shale from Burlington to Oakville, the grey shale at Clarkson, and the Pleistocene sands at Lorne Park. The mouths of the streams along this shore have all been drowned for some distance, but none of them has been cut off by a bar to form a lagoon like that of Jordan Harbour at the mouth of Twenty Mile Creek on the other side of the lake.

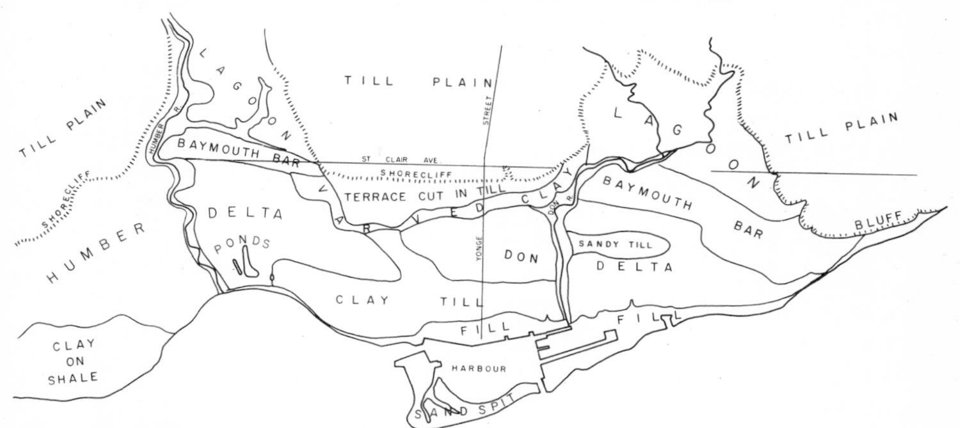

102. *Map of surficial deposits in Toronto*

Between the two shorelines, ancient and modern, the surviving portion of the bed of Lake Iroquois is a slightly sloping plain with an average width of about two miles. In some areas it is covered with stratified sands of varying depth, in others, the soil is formed directly on the wave-eroded surface of the red shale and, in still others, a shallow cover of till remains over the bedrock as in the Niagara. The sandy soils of Aldershot, Burlington, Bronte, Oakville, Clarkson, Lorne Park, and Humber Bay were preferred over the adjoining clay areas. This narrow belt with its discontinuous areas of good soil, being protected to a considerable extent from frost damage by its proximity to the lake, and quickly accessible to city markets by direct road and rail facilities, became an important horticultural area. Having just a little shorter season than the Niagara fruit belt very few of its farmers tried to specialize in peaches and grapes. Instead this area became famous for apples, pears, bush fruits, strawberries, and vegetables.

The Lakeshore fruit and vegetable district is, however, now almost a thing of the past. In 1941 there were about 1,000 fruit and vegetable farms in this area, with more than 15,000 acres of orchard, small fruits, and vegetables. There were also a number of general farms and rural estates. The three small towns, Burlington, Oakville, and Port Credit held a total population of 10,000 people. In 1964, having by this time annexed the adjoining rural townships and pretty well built up the Iroquois plain, Burlington and Oakville had a population of 100,000, while Port Credit and the urbanized part of Toronto townships within the Iroquois plain held a population of about 60,000. Within the area of Metropolitan Toronto itself, there were in 1941 three small urban communities, Mimico (8,070), New Toronto (9,504), and Long Branch (5,172), which now hold about 45,000, while the Iroquois plain in Etobicoke township must hold another 90,000. In a span of about two decades, horticultural activity on the interurban portion of the Iroquois plain has been engulfed by a flood of 250,000 urban people.

The gravels of the old beaches have been excavated for use in construction. The sand plains make excellent housing sites, often without the necessity of providing sewers, while any flat lake plain with bedrock at shallow depths can be used for industrial sites. The lakeshore has two added advantages: large quantities of water may be pumped from the lake and new port facilities may be installed. These advantages have brought thermal-electric power plants to Cherry Beach and Lakeview; oil refineries to Clarkson and Bronte, and cement, gypsum, and clay products industries to Clarkson. Worth mention also is the modern development of small-boat harbours in the drowned estuaries at Port Credit, Oakville, and Bronte.

Toronto. The growth and development of the city of Toronto may be assigned to the influence of many factors. Not the least of these, by any means, is the physiographic pattern of its site. Toronto was selected as a site for urban development because of its harbour. The harbour in its natural state was a phenomenon of the Lake Ontario shoreline and a product of the waves and currents of that water body. The land on which the first settlement was made and, in fact, the whole area in which the first century of growth took place was an inheritance from the realm of a previous water body, Lake Iroquois. The Iroquois lake plain, cut in previously deposited clay and till, is partly floored with sand deposits. It is about three miles in width, sloping gently northward from the level of Lake Ontario at a rate of 50 to 60 feet per mile. Along its northern border are evidences of the old beach and a steep bluff or shorecliff which, in places is about 75 feet high. Beyond this the gently rounded hills of the till plain stand at about 600 feet or more than 350 feet above the level at Lake Ontario. Both to the east and to the west of this shorecliff were the deep valleys of the Don and Humber rivers cut to pre-Iroquois base levels. Across these valleys the Iroquois built immense sand and gravel barriers, so that, in Iroquois times, Toronto had two harbours instead of one. When the Iroquois stage ended, the rivers resumed their down-cutting to create the Don and Humber valleys, which with the associated ravines, formed the eastern and western margins of the early city site. The site furnished not only space but a great deal of the material which went into the building of the city. The old bars furnished countless tons of sand and gravel, while in the Don valley and some of the ravines deposits of clay were exposed which were suitable for the making of bricks. The sand and gravel were deposited by Lake Iroquois, but the clay for the most part was of pre-Iroquoian age.

Eventually Toronto grew large enough to fill the area bounded by the harbour, the bluff and the two river valleys. It has crossed both rivers to occupy further segments of the Iroquois plain and expanded northward into the Peel plain. However, the downtown area, the "central business district," and most of those institutions which are the essential constituents of "the city" are still to be found in a few hundred acres of the Iroquois plain adjacent to the harbour.

Scarborough. At Scarborough the old shoreline lies very close to the present shoreline of Lake Ontario. Here are the famous Scarborough bluffs, standing about 350 feet above the lake at their highest point. These cliffs are cut in Pleistocene deposits composed of till, varved clay, and interglacial sands of various ages. Between the top of the cliff and the old shoreline is a wave-cut terrace floored by a few feet of sand laid down in Lake Iroquois. It is constantly being reduced in area, undercut by the waves of Lake Ontario, and is not important for agriculture. It is occupied by a golf course and other recreational facilities and, toward the east by residential subdivisions. As scenic attractions, the cliffs are unrivalled along Lake Ontario.

Scarborough to Newcastle. The Iroquois plain on the north shore of Lake Ontario from Scarborough to Trenton is an area of considerable complexity but one not easily divisible into well-marked geographical units. To a certain extent, however, the meanderings of the old shoreline, the soil pattern of the old lake plain, and the mosaic of present land use, enable us to distinguish certain areas.

First and largest is the area extending from Scarborough to beyond Newcastle. Eastward from Scarborough Bluffs the plain widens rapidly

103. *Scarborough bluffs: progressive erosion produces constantly changing shapes*

until at Greenwood the old beach is six and one-half miles inland from the present shore of Lake Ontario. Highland Creek and the Rouge River debouched sand into the old lake to build the sand plain in the southeast corner of Scarborough township and the adjacent portion of Pickering. Across Ontario county the Iroquois plain has a fairly constant pattern. The old shoreline is well marked by bluffs or gravel bars while immediately below it is a strip of boulder pavement and sandy off-shore deposits which varies from one half to three miles in width. Fairly level but often poorly drained this coarse sandy soil is not very productive and much of it is covered by cedar thicket. An exception, however, is provided by the dry sandy terrace north of Oshawa. The rest of the plain is a mosaic of till plains, drumlins, and areas of silty lacustrine deposits. The two most important soils of this area are Darlington loam and Newcastle loam. Both of these soils are well drained, but in some areas there are also imperfectly and poorly drained profiles developed on similar materials.

DARLINGTON LOAM

A_h 4–6 inches, dark greyish brown, crumb-structured, relatively stone-free loam; pH 7.4.
A_e 4–8 inches, pale brown, faintly blocky, almost stone-free loam; pH 7.0.
B 2–4 inches, medium brown, compact, nut-structured loam; pH 7.6.
C grey, stony, calcareous till containing some dark grey shale.

NEWCASTLE LOAM

A_h 4–6 inches, dark greyish brown, crumb-structured, friable, stone-free loam; pH 7.0.
A_e 8–12 inches, yellowish brown, weakly nut-structured, friable, stone-free loam; pH 6.6.
A_e 1–3 inches, mottled grey and brown stone-free loam; pH 6.8.
B 4–12 inches, medium brown, compact nut-structured, stone-free, clay or clay loam; pH 7.0.
C grey, stratified, stone-free, compact calcareous silt and clay.

A recent study of trends in rural land use has been made of this area by R. G. Putnam (111) while a study of the process of urbanization in Pickering township has been made by Cramm (30); both have been used to supplement our own observations. Previous to 1930, and for the most part until 1940, the Iroquois plain was a general farming area, differing little from the till slopes above it, except for a tendency toward horticulture and the growth of canning crops. The soft stone-free soils of the lacustrine deposits and the diminished frost hazard near the lake were contributing factors. The area near Newcastle was noted as one of the best apple producing areas in Ontario. The sand plain, however, was an area of poor general farms. Since the Second World War the farms have become fewer, larger, and more specialized, while much land has been taken out of agriculture and put to urban uses. The specialties include those which were carried on before, orchards, other horticultural crops and canning crops, and tobacco which has come into the sand plain in a minor way. However, of all the farms studied in the area, over 25 per cent proved to be part-time or residential farms, the operators of which had a full-time job elsewhere.

The area is no longer actually rural. Of its 125,000 acres, about 27,000 are included within the bounds of urban municipalities which now contain about 100,000 people. Largest of these is the city of Oshawa, but there are also Ajax, Pickering, Whitby, Bowmanville, and Newcastle, making an urban zone extending for 30 miles to the east of the limits of Metropolitan Toronto. In addition to those who are actually urban dwellers, there are many suburbanites, dwellers in rural subdivisions, and several thousand who live in scattered rural non-farm residences. The whole area may well be said to lie within the "urban shadow" (111).

Small urban nuclei had been there before, even from the earliest days of settlement, in order to serve the surrounding rural areas. Physiographic influence was strong in early settlement days. Physiography also provided the easy grades for the railways and highways that now link these settlements together and to the metropolitan centre to the west. Physiography alone cannot be credited with the location of General Motors at Oshawa, any more than that of Ford at Oakville; but for both of them the Iroquois lake plain provided reasonably flat building sites. Oshawa and neighbouring areas will undoubtedly go on attracting both industrial and residential land uses until full urbanization takes place. In this way, as in the area west of Toronto, waterside locations will attract certain types of large industries while water supplies and sewage disposal will be comparatively easy to install. It is not difficult to foresee a time when agricultural land uses have been banished forever to be replaced by a city of nearly half a million people.

Newcastle to Trenton. East of Newcastle a group of drumlinized uplands formed islands in Lake Iroquois and now serve to interrupt the continuity of the lake plain. Beyond it there is a small natural basin drained by the Ganaraska River which has already been mentioned. The soils in this area are sandy and rather unproductive under a general farming programme. However, since the Second World War this area has become an outlier of the Ontario tobacco belt, with an annual crop of flue-cured cigarette tobacco from more than 2,000 acres.

To the north of Port Hope and Cobourg, the old shoreline is very much indented. It is a region of many large drumlins, some of which stood out as islands in the old lake while others formed jutting promontories. Against the exposed ends of the drumlins the waves maintained a constant and merciless attack, cutting steep shorecliffs and spreading the debris in the hollows to form a broad offshore terrace. For the most part it is a dry terrace, cut at intervals by deep stream valleys. Wet boulder pavements are rare exceptions. These islands, cliffs, and terraces are well in evidence along Highway 46 from Cobourg to Baltimore and along the county road from Baltimore eastward to Dale. This is a picturesque, general farming section with numerous fieldstone houses. The steep hillsides are mostly used for pasture, while the dry interdrumlin valleys are used for field crops.

From Cobourg to a point a few miles east of Colborne, the Iroquois plain is about three and one-half miles in width and has a peculiar belted pattern. Through the centre, along Highway 2, numerous drumlins may be seen. They are large drumlins, some of them reaching a height of 150 feet or more, while some of the slopes have been over steepened by the waves of post-Iroquois waterplanes. They lie in a northeast-southwest alignment, the hollows between them floored with silt. To the north along the route of Highway 401, the high shoreline of Lake Iroquois may be seen, the old waterplane rising steadily from 565 feet near Baltimore to 600 feet near Biddy Lake. The broad, slightly sloping sand terrace of the old shoreline provides an excellent base for a four-lane highway. Between Highway 2 and Lake Ontario, the land is almost flat, much of it having only a shallow covering of drift over the bedrock. An occasional, long, low, slender drumlin is seen, oriented almost east-west. There is a considerable area of swamps, cedar-forested. The lacustrine deposits here are composed of sand, fine sand, and silt, and where drainage is good the soils are suitable for orchards.

South of Biddy Lake, the Iroquois plain is narrowly constricted, but the old shoreline then swings abruptly northeastward along the base of the Murray Hills to leave a lake plain over six miles wide. The triple belts are still in evidence but they become wider and most semblances of drumlin form are missing from the central belt. Instead, it is a rather hilly, sandy region at the base of which a post-Iroquois shoreline is visible.

This wider area of the old lake plain, extending from Brighton to Trenton is also one of the noted orchard areas along the north shore of Lake Ontario. Tomatoes and other canning crops are also extensively produced. Near Smithfield an important experimental station for orchard practices is maintained by the government of Canada. The Ontario government maintains a fruit inspection service at Brighton. Brighton is also the nearest point of access to Presqu'ile Park, a low rock mass in

Lake Ontario tied to the mainland by a long sand spit enclosing a lovely little bay.

Once the Oshawa urban-industrial complex is left behind the eastward traveller enters one of the oldest settled rural areas in Ontario. Not only was it settled early but it rapidly became and has remained a corridor between more densely settled areas. It also has some of the oldest towns in Ontario, none of which have grown very large. Among them may be mentioned Port Hope (8,000) and Cobourg (11,000) both of which in their time had hoped to be the lake port serving Rice Lake, Peterborough, and the whole upper Trent watershed. They have become small manufacturing centres as well as service centres for their rural umlands. Colborne (1,400), which used to advertise itself as "the oldest apple town in Ontario," and Brighton (2,500) are little more than rural service centres with some interest in the lakeshore recreational areas. Trenton (13,500) serves the lower Trent valley, is a small port at the outlet of the Trent Canal, and has some manufacturing. It is also the site of one of Canada's most important airforce bases.

The Trent embayment. While Lake Iroquois waterplanes continued far to the east of the Trent River, the inundated area has other characteristics which have led us to give names which do not emphasize the relationship of Lake Iroquois. Northward and eastward, however, there is a complex area embodying much of the lower Trent drainage, which bears many marks of the Iroquois inundation and to which we attach the name Trent embayment. It was, indeed, a great bay of Lake Iroquois, containing many islands. Large areas, of course, are made up of wave-washed hillsides or boulder pavements. There are numerous sand plains such as that near Morganstown. Other areas in the Trent lowland have deposits of stratified silt such as those near Codrington, Norham, Campbellford, and elsewhere. Lacustrine deposits of various types are found in the Trent basin all the way upstream to Peterborough. Although in many places the old shorelines are vague, the general "drumlin and clay flat" pattern of the landscape is unmistakably characteristic. The Trent basin also includes the great marshes along Percy Reach as well as Rice Lake which is still an open body of water. Northeast of the Trent, near Stirling, and again near Bailieboro at the western end of Rice Lake, fairly extensive clay plains occur. A wide area of lacustrine deposits occurs along the Ouse. The whole area defies a rigid classification for it is, simultaneously, a part of both the Peterborough drumlin field and the Iroquois lake plain.

In terms of land use it is still largely a general farming area, with perhaps nearly equal emphasis on meat and dairy products.

To some extent the Trent River itself provides a natural geographical focus to this region. Falling 368 feet from Rice Lake to the Bay of Quinte, the Trent and its major branches provided a number of power sites which attracted pioneer urbanization. Hastings (900), Campbellford (3,500), Stirling (1,350), and Frankford (1,650), are the largest urban centres upstream from Trenton. There are eighteen locks on the Trent River in a distance of about 52 miles, with nearly as many dams thrown across the river to provide depth of water for navigation. The locks are 175 feet in length, 33 feet in width, and have a draught of 6.3 feet and are suitable for small craft only. The long stillwaters and widenings behind the dams, however, provide excellent fishing areas and waterside cottage sites, so that this area with its boating and sport fishing resources attracts a great many summer visitors. The building of the dams also enabled the development of considerable hydro-electric power along the Trent River.

One cannot, of course, relate all this development to the fact that the Trent embayment was part of Lake Iroquois. For many things, particularly the course of the Trent River itself, other physiographical factors are much more important. The best agricultural lands, however, are found on lacustrine deposits; without them, settlement would probably have been less successful.

The geographical significance of the Iroquois plain. The Iroquois plain has been described at some length, not only because of its extent and complexity but, also, because its importance within the geographical framework of Ontario warrants full attention. About one-third of all the people in Ontario live within its borders, making it by far the most densely inhabited area. Many others live as close to it as they can get because of the opportunity for employment. By reason of its proximity to Lake Ontario and its facilities for transportation, it became the first settled area while, from the various lake ports, colonization roads were pushed into the interior. But few of the interior areas were as attractive to the new settlers as the old lake plain where the first settlements already had a head start. Here land transportation routes developed because of the easy grades, linking together the lakefront settlements and stimulating the growth of new centres depending entirely upon road and rail facilities. Its desirability is no more evident than in its two largest cities, Toronto and Hamilton, both of which have been forced to spread beyond its borders.

The presence of large cities creates demands for special agricultural products. Even in this technological age, it is desirable to grow them as

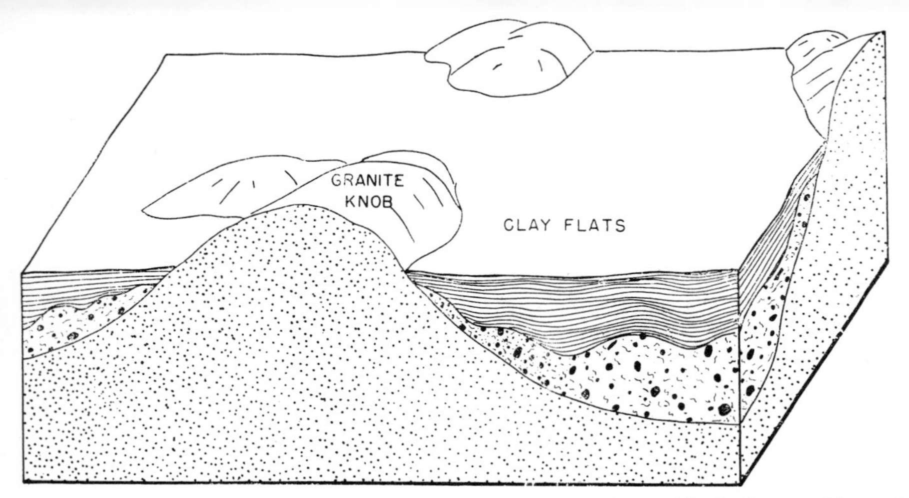

104. *Black diagram illustrating the structure of the rock knobs and clay flats of Leeds county*

close to market as possible. The Iroquois plain thus became, and still remains, an area of specialized farming, where not overwhelmed by suburban growth. Part of its special advantage accrues from the presence of Lake Ontario and its climatic influences but the favourable soil factor is a direct inheritance from glacial Lake Iroquois.

Other points might be mentioned. The old sandbars and offshore aprons of sand are good aquifers, supplying domestic water to farms, villages, and even some towns. The gravel bars have been good sources of road metal and building materials, while the clays of the old lake bed have been used for the manufacture of bricks. The wide-awake observer will be able to note many phenomena which recall the former presence of the old lake.

LEEDS KNOBS AND FLATS

In the St. Lawrence River near Gananoque are the well-known "Thousand Islands," consisting of knobs of granite and other Precambrian rocks between which, by devious channels, the river finds its course. In Leeds and Frontenac counties to the north of the river there are another "thousand islands" of rocks, but the channels between are filled with clay left by the Champlain Sea. This landscape of rock-knobs and clay flats occupies an area of about 385 square miles, a typical section of which is seen along No. 2 Highway between Kingston and Brockville. The rock-knobs of the Frontenac axis are relatively bare because the shallow covering of drift was removed by the waves of the Champlain Sea, and the contrast between them and the deep clay beds is clear and sharp. The clay, itself, is grey to drab in colour and very weakly calcareous. Where drainage is good it gives rise to an acid soil similar to that found in Renfrew county. Half-bog soils are also found.

The largest block of clay land is probably that lying west of Lansdowne village. Another good block is found in the area between Taylor and South Lake and extending well to the west into Pittsburgh township. Clay lands are also found on Wolfe and Howe Islands.

LAND USE AND SETTLEMENT

Dairy farming is the mainstay of this area as it is in most parts of eastern Ontario. The deep clay soils promote excellent yields of hay and oats which are the chief crops. Silage corn is being given increased attention.

Lying between Kingston (65,000) and Brockville (18,000) and having adjoining it, also, Gananoque (5,000), "Gateway to the Thousand Isles," this section is fortunate in having nearby markets and service centres. It is also well provided with railway and highway facilities which give rapid access not only to local centres but to Ottawa, Montreal, and Toronto as well as to the United States by way of the International Bridge at Ivy Lea. The St. Lawrence waterfront is not the only attraction for summer recreation, for the interior of the county also contains a number of excellent lakes along which many cottages have been erected.

105. *Granite knobs and clay flats near Gananoque*

SMITHS FALLS LIMESTONE PLAIN

The largest and most continuous tract of shallow soil over limestone in Southern Ontario, covers nearly 1,400 square miles in Leeds, Grenville, Carleton, and Lanark counties. Between Brockville and Carleton Place, Highway No. 29 passes almost exclusively through this type of country. Smiths Falls lies almost in the centre of it and the Rideau River divides it into two nearly equal portions. The Mississippi is the only other large stream.

The exposed rock strata belong to the Beekmantown formation and include grey limestone, magnesian limestone, blue-grey dolomite, and some calcareous sandstone. The surface has a slight slope towards the northeast, but over large areas is almost level. Locally, a little relief is provided by low ledges and shallow valleys in the rock. North of Carleton Place, there is greater irregularity of the surface due to faulting and clay has been deposited in the depressions, while its most northeasterly extension lies along an upraised fault block which overlooks the Carp valley. To the west, against the Canadian Shield the dolomite does not present a strong scarp similar to that of the Black River formation in central Ontario, but the Rideau Lakes occupy depressions along the Beekmantown border which are comparable to those along the borders of the Trenton and Black River formations in which the Kawartha Lakes are situated. The over-all gradient on the dolomite around Smiths Falls is about five feet per mile, and many depressions are left undrained. The area abounds in bogs which are especially prevalent in the townships of Marlborough, Montague, Beckwith, and Wolford. In the case of Cranberry Lake in the latter township, a small body of open water is left, while around lie hundreds of acres which have been filled with peat to the point where it supports a dense swamp forest.

On the higher parts of the plain are found remnants of old marine beaches which are interesting and valuable features. For the most part they are composed of limestone shingle and sand, often providing the only areas of soil deep enough for cultivation. The gravels have also been used extensively for road construction.

In a region of shallow soil those areas having appreciable depth of soil material are highly important. West of Smiths Falls a few low drumlins appear, while elsewhere there are scattered small patches of till, three to five feet or more in depth. Fairly deep clay deposits occur along the Rideau River, allowing the development of small tracts of first-class soil. Other tracts of clay, not quite so deep, are found near Athens, Perth, and Carleton Place.

For the most part, this plain supported a hardwood forest in which sugar maple was the dominant tree. Evidence seems to indicate that oak and pine were formerly more common than at present. In the poorly drained areas, elm, ash, soft maple, and white cedar are found. In some of the bogs there are larch and black spruce.

The physical characteristics of this region present the agriculturist with great difficulty. Disregarding for the moment, a few small areas of deeper soil, there are hardly any good points to be mentioned. The shallow soils vary greatly in texture from clays to light loams, sands, and even gravels, although they are all classed within the Farmington series. Surface stoniness is common. Drainage is often impeded so that seeding is delayed even though later in the summer they become exceedingly droughty. Water for stock may also be scarce, although there is usually access to a swamp on most farms. In fact, very large areas of this limestone plain are covered with peat and muck deposits which, for the most part, remain under a forest cover. In a few marginal areas, however, the land has been cleared but is used only as rough pasture.

SETTLEMENT AND LAND USE

The "front" of this area was part of the original townships laid out to accommodate the United Empire Loyalists who took up land here in 1781 and following years. Military settlements in the interior included Perth in Lanark county (1815) and Richmond in Carleton county (1819). The real spread of settlement, however, came during and after the building of the Rideau Canal, begun by Colonel John By in 1827.

Early farm life was fraught with extreme hardship and the economy was of the most elementary subsistence type. Because of its harsh climate this area could not, even on its better soils, participate in the wheat enterprise which characterized central and western Ontario during the middle part of the nineteenth century. Some hay, oats, and perhaps potatoes, were produced for the lumbering areas which lay immediately to the north. Agricultural stability of a sort was reached with the introduction of cheesemaking in the latter half of the century. The farm economy was based on summer dairying, the milk being marketed through small local cheese factories which operated during the warmer months only and were closed during the winter. The cows began lactation in the spring to take full advantage of the June flush of pasture on the shallow land. They were allowed to go dry in autumn, and were wintered over on forage produced on patches of deeper soil. Good hay, oats, and sometimes corn were produced on the gravelly loams of the old

106. *Farmsteads consisting of several log buildings are common on the limestone plains of eastern Ontario*

beach ridges which, on some farms were the only soils really deep enough for cultivation. A few hogs, a few sheep, and a small flock of poultry were also found on most farms. On some farms maple syrup provided an important supplementary source of income. This area still contains the greatest concentration of maple groves in the province and the situation might be improved by the reforestation of a great deal of rocky land which is more suited to maple trees than to grass.

The landscape of this area is peculiar. In comparison with the landscape in southwestern Ontario, or with the Winchester clay plains, it presents a rather ragged and unfinished appearance. In fact it retains many elements of the pioneer stage of development. There is the previously mentioned high percentage of uncleared land, including rough pasture, woods, and swamp. Rail, stump, and stone fences are common. Log houses and barns are nearly as plentiful as those of frame construction. In addition there is much evidence of abandonment with many farmsteads falling into ruins, or having disappeared entirely, leaving perhaps a gaunt old-fashioned chimney to mark the site.

Nearly all the stages of the agricultural development of Ontario may be discerned. One encounters the original settler's shanty, now housing chickens to be sure, but still, after 125 years preserving the elements and design of its primitive construction. It is built of logs, rather raggedly notched together at the corners. The eaves are low and there are very few windows. The roof is made of cedar "scoops"–hollow trunks, split and overlapped in the fashion of the classical clay tile. Within is a huge square fire-place built of flat limestone blocks quarried nearby, and beside it still hangs the old iron crane.

The second stage is represented by the house built of squared logs with carefully dovetailed corners, plastered chinks, and shingled roof. On many a farm the shallow soils have not provided the capital necessary for further improvement. The barns of this stage are built of logs also, but unhewn. They are numerous, sometimes from twelve to twenty, on an individual farm. They are small, the size varying according to the length of the logs available to the builder. Sometimes they are placed end to end in a long train, but often haphazardly scattered about a farmyard of two acres or more.

The third stage is reached with the construction of a frame or brick house and a frame barn; but only on the best and deepest of the soils of the limestone plain is this achieved. Combinations are often seen: a neat brick house accompanied by a dozen log barns; or the reverse, an old squared log house and a frame barn; or sometimes, even a modern steel barn. Basement or bank barns are rather rare in eastern Ontario, but in this region one sometimes sees barn foundations built of cedar blocks instead of stones.

The present agricultural situation may be understood by reference to the 1961 census statistics for a few townships in the neighbourhood of Smiths Falls. Slightly over 80 per cent of the land is occupied by farms averaging 235 acres in size, almost twice as large as those of the good clay lands of eastern Ontario. As a rule only about 40 per cent of the land is improved and about 25 per cent is in crop, the latter consisting chiefly of hay, oats, mixed grains, and some corn for silage. This lack of a cash crop, combined with a large area of unimproved pastureland, is in accordance with the extensive type of dairy farming. The average farm has a herd of ten milking cows, ten sheep, half a dozen hogs, and a poultry flock of about seventy. The area is still a cheddar cheese district, but there are fewer and larger factories, and an attempt is being made toward year-round production. The presence of so many submarginal farms, however, brings the average farm income in these townships down to a very low level.

The limestone plains are shared by the umlands of Ottawa and Kingston but, rather surprisingly, they contain several sizable towns including:

Brockville (18,000), Smiths Falls (10,000), Perth (5,400), Carleton Place (4,800), and Almonte (3,300). Merrickville (950), Athens (1,050), and Newboro (300), are incorporated villages. A complete census population analysis of the Smiths Falls plain such as that given for Prince Edward is not possible but the accompanying table deals with the central and most typical part of it. This area of some 640 square miles, immediately surrounding Smiths Falls, is comprised of eight rural township municipalities and five incorporated towns and villages.

Population of the CENTRAL PART OF THE SMITHS FALLS PLAIN 1961			
	Total	37,652	100%
	Urban (4 towns, 1 village)	23,969	63.6
	Rural	13,683	36.4
	farm (1,398 farms)	5,699	15.2
	non-farm	7,984	21.2
	unincorporated villages (3)	950	2.5
area: 640 sq. mi.	small hamlets (9)	884	2.4
density: 59	scattered	6,150	16.3

The limestone plains for years constituted an area of declining rural populations. Many of the townships reached their population peak almost a century ago. Among such may be cited: Kitley with 2,870 people in 1871 and only 1,445 in 1941; Montague with 3,187 in 1871 and only 1,547 in 1941; and Marlborough with 2,260 in 1871 and only 886 in 1941. The unincorporated villages and small hamlets as well as the farms were involved in the decline. The farms have gradually increased in size as smaller and more marginally located farms have ceased to operate. Gradually the farmers are adjusting their operations to the quality of the land at their disposal while adopting new techniques to overcome the tremendous handicap under which they were placed when each settler was given his hundred acres regardless of where it was or of what it consisted. It is a slow process, however, and the adjustment is far from complete.

Something else is also happening. Total population declined until 1931, rural population continued to decline until 1941, while farm population is still declining. On the other hand urban population has slowly and steadily increased and, especially since 1951, rural non-farm population has increased at a more rapid rate than urban population. Physical features—power sites—on the Rideau and Mississippi were responsible for the urban settlements in the first place even though they were also rural service centres. This function decreased as the farm population dropped. Now these places have become small manufacturing centres with job opportunities and people seek cheap building sites nearby.

EDWARDSBURG SAND PLAIN

In Edwardsburg and adjoining townships in Grenville and Dundas counties, the bedrock and most of the boulder clay are covered by beds of sand. The surface of the sand plain is for the most part nearly level or only slightly undulating, although hummocks and ridges appear in places. The sand is undoubtedly glacio-fluvial in origin. Although it has been well spread about by wave action, probably in the late stages of the Champlain Sea, a few undoubted morainic structures remain. As the land emerged from below sea level, a few beach ridges were formed on high ground, notably east of Kemptville. Finally the wind built some of the drier sand into dunes before it was covered with forest. The sand plains thus formed, though not continuous, characterize an area of about 280 square miles.

The Edwardsburg sands have an altitude of 300 to 400 feet and lie within the South Nation watershed. The over-all relief is small and the water table generally stands near the surface, so much so that shallow muck and even peat bogs have frequently developed.

The common forest on this plain is an association of moisture-loving trees including elm, ash, soft maple, and white cedar. Tamarack and black spruce is seen in bogs and on the wetter sand. On the ridges white pine, hard maple, beech, and burr oak are found.

The soils are acid and deficient in all important nutrients. The parent sands contain some lime carbonate, but the bases have been largely leached from the surface horizons of the soils. Podzol and Brown Podzolic profiles appear on higher sites and Ground-Water Podzols on lower sites. Still wetter sites have dark, deep surface horizons underlain by grey and mottled subsoils. Granby sandy loam represents this wet soil on the Grenville county soil map. If drainage could be secured, this would be the best of the sandy types. The Bridgman sand of the former sand dunes is subject to blowing and should be kept in forest.

This area is part of the eastern Ontario dairy belt where small cheese factories used to dot the road corners to take in milk during summer and fall. It still produces milk which is hauled longer distances to larger plants. The region has long had the appearance of having seen better days. There are well-built stone houses along the concession lines, albeit some of them are now abandoned. The quality of buildings is about on a par with that of the adjoining limestone plains, but the farms are gradually becoming larger. The population of Edwardsburg township fell steadily from the end of the nineteenth century until the Second World War, but since then it has risen considerably. However, farm population continued to fall and in 1961 was only about half what it was in 1931. There were also less than half as many farms. Only 65 per cent of the total area is now in farms, and the average size has risen to 145 acres. There are fewer cows, but more of them per farm, and they give more milk. Over 40 per cent of the improved land is used for pasture.

The towns, Prescott (5,400), Kemptville (2,000), and Cardinal (2,000), serve as the principal supply points for this and adjacent areas. Several unincorporated places have been built adjacent to mill sites. Spencerville (400) is the most important of these, being situated on the Ottawa–Prescott highway, where it crosses the South Nation River.

NORTH GOWER DRUMLIN FIELD

The North Gower drumlin field occupies an area of about 150 square miles lying in the townships of North Gower, Marlborough, Osgoode, Nepean, and Gloucester. The drumlins are scattered and some of them incompletely formed, the best group being those clustered about the village of North Gower. For the most part the land between the drumlins is covered with clay or silt deposited by the Champlain Sea. Other features adding to the complexity of the area are various gravel ridges such as those near Kars and Osgoode, a few small areas of partially drumlinized till plain in Osgoode township, and some bare limestone plain in the same vicinity.

Although the area is traversed by the Rideau River and its tributaries, drainage is not well developed and the lowlands are occupied by Half-Bog soils. North Gower clay loam and Osgoode loam are both poorly drained, the latter having more sand in both surface and subsoil than the former. They suffer badly from delayed seeding in wet seasons.

The drumlins of this area have good drainage and exhibit a normal Grey-Brown Podzolic type of profile. The soil is adaptable and its chief limitation is stoniness. In the Carleton county soil report it is called Grenville loam.

While the North Gower drumlin field is not quite so highly developed agriculturally as the Winchester clay plain, nevertheless its achievements are above average for eastern Ontario. North Gower township lies wholly within the field and may be taken as representative of the whole area. It is a small township comprising less than 34,000 acres (53 square miles), of which over 90 per cent is occupied by farms. There are 221 farms, averaging 140 acres in area, which is the same as the average for the whole province. About 81 per cent of the farm land is improved, 52 per cent is under crop, and 25 per cent is improved pasture. Only about 7.5 per cent remains in woods. Hay and oats are the most important crops but silage corn and even some husking corn are grown. Having fairly easy communication with Ottawa, this area is an outlying part of the Ottawa milk shed. Each farm has, on the average, a herd of 18 milking cows plus a few pigs and poultry. The lighter soils near Osgoode constitute an extension of the Gloucester special crop belt and a few fruit and vegetable farms are found. Potatoes are important.

There are no towns or incorporated villages within the area, but there are several unincorporated supply points such as North Gower (400), Manotick (450), and Osgoode Station (700), which serve their own constituencies. These villages also serve as residential areas for many who commute to work in Ottawa. In addition more than 40 per cent of the population lives in scattered non-farm residences. Only 34 per cent lives on farms.

GLENGARRY TILL PLAIN

The Glengarry till plain is a region of low relief forming the drainage divide between the international section of the St. Lawrence and the Ottawa basin, from Prescott to the Quebec boundary. Its area is estimated to be about 935 square miles, including the greater part of Glengarry and Stormont counties and smaller portions of Dundas and Grenville. The surface is undulating to rolling, consisting of long drumlinoidal ridges and a few well-formed drumlins together with intervening clay flats and swamps.

A great many small streams arise within the area. Drainage toward the north are a number of tributaries of the Nation River, including the South Branch, Black Creek, Payne River, and Scotch River. Draining to the St. Lawrence are Hoople Creek, Raisin River, Baudet River, Delisle River, and other shorter streams. To the east the Rigaud River drains to the Ottawa. The drainage pattern is peculiar in that the headwaters of most of these systems flow sluggishly for long distances between the ridges before finding outlets to the master streams.

The till has a loamy texture and contains a high proportion of limestone with admixture of materials derived from the Precambrian rocks to the north and from the Nepean sandstone at the base of Rigaud mountain. The depth to bedrock is seldom over 100 feet, and over much of the area the till is less than 25 feet in depth.

The outstanding characteristic of the region is stoniness. The fields are

107. *Boulder pavement*

either dotted with rock piles or bordered by stone fences. The till itself is very stony, and on the crests of the ridges and drumlins, where it was modified by the waves of the Champlain Sea, there are areas of boulder pavement. The clearing of this sort of land has required a prodigious amount of work, and high wheeled stone-wagons, with steel tongs for clasping the boulders and windlasses to lift them, are still in use. Nevertheless, many of the stoniest areas remain uncleared, either as woodlots or rough pasture. The action of the waves also built numerous bars of sand and gravel which are useful sources of building material. Other deposits of sand and gravel, considered to be of fluvio-glacial origin but modified by wave action, are found near Newington, Finch, Avonmore, and Maxville.

In its natural state the St. Lawrence River bordering this area was marked by numerous rapids, most noted of which was the Long Sault. In order to make the river navigable a series of canals was constructed in the early days and subsequently improved, culminating with the building of the St. Lawrence International Seaway which was opened in 1959. This construction also made it possible to install turbines of 2,200,000 h.p. in the Robert Moses–Robert Saunders Power Plant which is shared by New York and Ontario.

This construction has greatly altered the physical landscape of this area. By raising the water to the level of Lake Ontario much low-lying land was flooded and the shoreline placed squarely against the till ridges where, of course, a certain degree of shoreline erosion will take place. Some of the previous islands have been completely submerged, while new islands and peninsulas have been outlined at higher levels. Much of the new shoreline has been set aside as a provincial park.

The normal soil on this till plain resembles the Grey-Brown Earths of Southwestern Ontario but there is, actually, little mature soil. Even on the ridges, where drainage is good, the highly calcareous material has resisted weathering and the soil profile is shallow with an abundance of free carbonates within the ploughed layer. In this there is a resemblance to the Otonabee soils of the drumlins near Peterborough. The most commonly encountered of these soils in much of the Glengarry till plain, and on the North Gower drumlins as well, is Grenville loam. There are many areas which are gently sloping or almost flat where imperfectly drained, and even poorly drained soils are developed. There are, also, large undrained depressions in which peat and muck are found.

GRENVILLE LOAM

A_h 2–4 inches, dark greyish brown, granular, friable, stony loam; pH –7.0.
A_e 3–5 inches, light brown, friable, slightly stony loam; pH –6.5.
B_1 3–5 inches, light yellowish to medium brown, weakly blocky, friable, loam; pH –6.8.
B_2 4–12 inches, light brown to medium brown or "coffee" brown, compact moderately blocky, slightly stony loam; pH –6.8.
C grey, hard, stony, loamy, calcareous, till.

LAND USE AND SETTLEMENT

These limestone till soils must be rated as good land, in spite of some excessive stoniness and the inclusion of some wet land. Alfalfa, oats, barley, and silage corn are all well adapted. The porous nature of the soil and substrata usually predetermine a supply of well water, a utility that is sometimes lacking on a clay plain.

Dairy farming dominates in this region just as it does in the adjoining clay plains, but there are a number of contrasts to be noted. The farms are somewhat larger, averaging over 150 acres, while in some townships less than 50 per cent of the farm land is improved. There is only a small amount of seeded pasture and most of the grazing is on rough, stony, or wet land. Somewhat more than one-third of the area remains in woods and swamp. Crops occupy between 35 and 40 per cent of the farm land; hay and oats are the most important but some farms have barley or fodder corn. On the average each hundred acres of farm land carries about 20 livestock units, one half of which are milking cows; that is, there are 15 cows per farm. There is much variation within the area, some farms being very well kept, but, in general, agriculture seems to be at a somewhat lower level than on the clay plains. The uncleared land provides a certain supplementary income. Most farmers have a woodlot from which fuel and lumber for their own needs are obtained. There are also a number of maple groves which have given Glengarry a reputation for syrup and sugar.

Population of STORMONT AND GLENGARRY 1961			
	Total	77,086	100%
	Urban (1 city, 1 town, 3 villages)	48,010	62.5
	Rural	29,076	37.5
	farm (2,978 farms)	13,661	17.6
	non-farm	15,415	19.9
	unincorporated villages (13)	5,713	7.4
area: 890 sq. mi.	small hamlets (13)	1,357	1.7
density: 86.5	scattered	8,345	10.8

The population pattern is dominated by the city of Cornwall (45,000) which draws most of its milk supply from the adjacent area and provides a market for small amounts of other farm produce. Cornwall also provides employment for many people in the surrounding unincorporated villages, including the "new towns," the hamlets, and even many of the farm residents. Smaller markets are found in the other towns of the area including Iroquois (1,200), Morrisburg (2,000), and Alexandria (2,600). A score or more of unincorporated villages and small villages act as local supply centres. Most of the farm income, however, comes through the sale of milk to the cheese factory or other industrial dairy plant.

WINCHESTER CLAY PLAIN

The Winchester clay plain lies between the Glengarry till plain and the sand plains of Russell and Prescott counties, and comprises an area of about 360 square miles. It is an area of low relief, lying almost entirely within the drainage basin of the South Nation River. The general flatness of the plain is exemplified in the gradient of the river which descends only 50 feet in 30 miles from South Mountain to Crysler.

Although the clay plains are dominant, the landscape has some complexity. In many places the underlying till protrudes and there are a number of low drumlins. In a few cases there are areas of shallow soil over bedrock and occasional bars, beaches, and boulder pavements are found as in the area to the south. There are also several thousand acres of bog.

The original vegetation of these low-lying plains was of the swamp-forest type, consisting chiefly of red maple, elm, white and black ash, but including many other trees such as burr oak, swamp white oak, yellow birch, white-birch, basswood, black alder, and willows, depending upon degree of drainage.

108. *Stone-wagon seen near Vankleek Hill*

The clays, themselves, fall into two main classes, with various intergrades. Toward the south and west they are grey in colour and fairly high in lime while toward the northeast are found deposits of alternately banded pink and grey clays, similar to those of the lower Ottawa valley. They vary greatly in depth, 60 feet or so of clay underlying the sand in the canyon of the South Nation River below Casselman. The two types of clay result in the formation of two soil catenas.

The soils of the Winchester plain, or South Nation valley, for the most part are poorly or, at least, imperfectly drained. This results in the formation of a glei or mottled plastic clay horizon at the base of the profile.

The most commonly identified soil types of the flat South Nation valley in Dundas, Stormont, and Russell counties are the poorly drained North Gower clay and clay loam. These are the soils which are found throughout the area most often involved in the spring floods of the middle course of the South Nation. On slightly better, but still imperfectly drained locations as, for instance, near Winchester the Carp profile is found, in which a shallow A_e horizon and a brown but mottled B horizon may be seen. Both Carp and North Gower are excellent soils when drainage has been established.

NORTH GOWER CLAY LOAM

A_h 6–8 inches, very dark grey, granular, friable clay loam; pH —7.0.
G_1 4–6 inches, grey and brown mottled, plastic clay loam; pH —7.0.
G_2 12–18 inches, grey and rust-brown mottled, extremely plastic, massive clay; pH – 7.2.
C grey, plastic, massive clay, alkaline but with no free carbonates.

In the northern and eastern parts of the plain where the parent material is composed of massive pink and grey clay the Bearbrook is the dominant series. It is also poorly drained. Bearbrook and similar strongly leached soils are the ones which the French-speaking farmer calls "terre grise"; the surface soil of the cultivated fields is grey and inclined to be very plastic when wet and very hard when dry.

BEARBROOK CLAY

A_h 3–4 inches, very dark brown, granular clay; pH 6.0–6.2.
G 7–12 inches, light brownish grey, mottled with rusty brown, plastic clay; pH 6.6.
C grey and pinkish brown banded, plastic clay.

Municipal ditches have been cut to provide drainage throughout the clay plain and very little uncleared land remains. Fuel is a problem since the forests are gone. At Cannamore the farmers used to cut and dry peat, succeeding in the use of this resource at a domestic level where commercial ventures had failed. A few areas of clay soil have been reclaimed by draining and burning off the overlying peat. It is a slow process in which timothy hay is used to advantage in the development of a friable surface soil.

In spite of the surplus surface water at certain times of the year, the provision of a reliable supply for stock watering and domestic purposes throughout the year poses a considerable problem. Wells often go dry because the tight clay holds little available reserve, even at a depth of 50 feet. Drilled wells are becoming common and it is often necessary to go more than 100 feet into the rock to find a satisfactory supply.

SETTLEMENT AND LAND USE

The Winchester clay plain is one of the outstanding agricultural districts in Ontario. It also remains one of the most rural areas in the province. The county of Dundas, of which this plain provides three-fifths of the area, does not have a single town and has only four incorporated villages. Parts of Prescott and Russell counties have even denser populations, undoubtedly greater than any other non-horticultural area of the province. The accompanying table is an analysis of the population situation in Dundas county.

Population of DUNDAS COUNTY 1961	Total	17,162	100%
	Urban (4 villages)	5,633	32.8
	Rural	11,529	67.2
	farm (1,559 farms)	6,964	40.6
	non-farm	4,565	26.6
	unincorporated villages (2)	669	3.9
area: 380 sq. mi.	small hamlets (15)	1,556	9.0
density: 45	scattered	2,340	13.7

Winchester township lies wholly within the clay plain and its agricultural statistics may be taken as representative of the larger area which includes parts of many other townships. The total area, outside of the incorporated villages is 57,962 acres or 90 square miles. In 1961, 458 farms occupied 57,226 acres or about 99 per cent of its area. The average farm has an area of 125 acres, of which 108 acres (88 per cent) are improved, and 75 acres (60 per cent) in crop. Seeded pasture occupies 24.5 per cent or 30.5 acres. Wasteland amounts to 10 per cent, some of which may be grazed, and only 3 per cent remains in woods. The township contains three large bogs, totalling 7,500 acres, much of which has been cleared. In fact, outside of the bogs there is very little wooded or waste land. The important crops are alfalfa hay, oats, mixed grains, and silage corn. It may be worth noting that this township in eastern Ontario grows more acres of silage corn than any township in the immediate Toronto milk shed in Ontario, York, Peel, and Halton counties; in fact only two townships in Oxford county surpass it in this regard. It does not, however, attempt to grow much husking corn. On the average each hundred acres supports a herd of 16.5 milk cows and enough other livestock to make a total of 34.5 livestock units. Some years ago this was the record for the province but in recent years Winchester has been surpassed by several townships in Waterloo, Perth, and Oxford counties. In terms of milk production, however, Dundas county moves in the front rank, both in terms of yield per cow and in terms of yield per acre. This, in no small measure, is due to the productivity of the Winchester clay plain.

One is struck, also, by the well-kept appearance of the majority of farms. In common with all of eastern Ontario, houses as well as barns are usually built of wood on rather simple lines, but they are neat and commodious. Paint is rather liberally used, white for the houses and red for the barns and outbuildings. Mechanization is fairly complete; tractors and tractor-drawn implements are seen everywhere, and milking machines are common. It is noteworthy, too, that many Winchester farmers find time to mow the lawn and to grow a good garden. The observations noted above apply with equal force in the better parts of the adjoining townships of the clay plain, this whole area being one of the foremost agricultural districts in Ontario.

It is not without its problems, however. One of the greatest of these is drainage. The only outlet is by way of the Nation River, the gradient of which is less than 20 inches per mile. Over a distance of several miles above Cass Bridge—near Winchester—the banks of the river are so low that it annually overflows, flooding an area of about ten square miles. The floods are usually in the early spring when little damage is done to crops. The only summer flood on record was that of July, 1945. Sometimes the spring floods repeat, as in 1946 when some crops in the low sections near the river had to be reseeded three times.

Another problem, not peculiar, of course, to this clay plain, is the difficulty of maintaining passable roads in the absence of nearby gravel deposits. This is of importance when the chief product is fluid milk which must be taken out daily.

Market outlets for the milk are provided by large evaporating plants at Winchester, Chesterville, and Maxville, and by a number of large cheese factories. The Winchester clay plain, like most of eastern Ontario, once had numerous, small cheese factories but they have given way to more efficient plants.

There are no large towns on the plain. The main centres include Winchester (1,430), Chesterville (1,250), both in Winchester township, and Casselman (1,300), Maxville (800), Crysler (500), Finch (400), Moose Creek (450), and St. Isadore (450). None of them detract much from a strictly rural landscape.

LANCASTER FLATS

In the southern part of Glengarry county, for about eight miles back from the St. Lawrence, lies a lowland in which the till plain has been buried under water-laid deposits leaving exposed only the stony crests of a few drumlins and ridges. The water-laid materials range from clay to very fine sand in much the same way as they do on the margins of the Winchester clay plain. This area is but a small portion of a clay plain which extends into the Quebec counties of Vaudreuil and Soulanges. The portion within the province of Ontario has an area of about 160 square miles.

Drainage is toward the St. Lawrence by way of the Raisin River, Baudet River, Sutherland Creek, and a number of other shorter streams. Nevertheless, the land is so flat that the area is poorly drained. The original vegetation was an association of moisture-loving species such as American elm, white ash, and red maple. Writers of the early history of the county present vivid accounts of clearing forests amid the mire of the "Lancaster flats."

The soils belong to the intrazonal Half-Bog group, having rather deep black surface soils underlain by mottled subsoils. In the eastern corner of Lancaster township the texture is mainly silt and fine sand, and the soils resemble those which in Carleton county are classified in the Castor series.

From the standpoint of agricultural development this is just another segment of the eastern Ontario cheese producing region, although some fluid milk is shipped to Montreal. It has, however, a crop specialty worth mentioning. Fibre flax seems to find here the conditions of soil and moisture supply which suit it and at one time it was an important crop, but practically none is being grown at present.

THE OTTAWA VALLEY CLAY PLAINS

The Ottawa valley between Pembroke and Hawkesbury consists of clay plains interrupted by ridges of rock or sand. At Ottawa it is naturally divisible into two parts, upper and lower, each with its own distinctive traits.

In the upper section there is a broad valley with rocky Laurentian uplands rising on either side; on the Quebec side the rocks rise sharply to a height of 600 feet or more, while on the Ontario side the slope is more gradual although it also presents some prominent scarps. Within the valley the bedrocks are further broken so that some of the raised pieces appear above the clay beds. The surface of the beds is level in all but a few areas, but swamps are scarce. The sediments themselves are deep silty clays. Although grey in colour like the limestones that underlie them in part, they are only mildly calcareous and likely derived from the more acidic rocks of the Canadian Shield.

Nine miles south of Pembroke, Highway No. 17 crosses Muskrat River and mounts a limestone outcrop. A companion outcrop lies to the west in Stafford township. At this point the Muskrat River leaves Muskrat Lake, a narrow lake which extends to Cobden nine miles southward. This lake, together with Olmstead, and several smaller lakes in line with them, occupy sections of a preglacial river valley excavated along an old fault line. The western shore of Muskrat Lake is on clay while the opposite shore is formed by an escarpment in Precambrian rocks nearly 200 feet high. The only overburden on the crest of this ridge is a little sand or gravel. Its eastern slope is gradual and the clay beds encroach upon it irregularly.

This "Queen's Line" clay tract slopes towards the Ottawa River, giving good drainage, and it is cut at frequent intervals by gullies above the bluff which closely follows the concession road southeast from Forester Falls. Another gullied area appears east of Beachburg. Toward the north in Westmeath township the clay is shallow and the limestone is uncovered in places. A ridge of deep gravel and sand connects the villages of Beachburg and Westmeath.

109. *Treeless sphagnum bog west of Renfrew*

At Haley Station No. 17 Highway turns to cross another outcrop of Precambrian rocks which extends to Osceola and Dore Lake. The Snake River finds a pass through it at Osceola and drains the clay plains on either side. The Bonnechère River drains the main part of the flats southwest of the ridge.

An area of stratified limestone rises above the clay northwest of Arnprior while to the south are several outcrops of ancient crystalline limestone. Two more stratified limestone mesas appear west of Ottawa. Another mesa of sandstone around Fallowfield and a ridge of Precambrian rocks flank the Carp valley on the north and east. Drainage is slow in the Carp valley and on the clay north of Richmond village.

The Ottawa River is entrenched within a low terrace which is bounded on either side by abrupt bluffs. The bluff southeast of Forester Falls has already been mentioned. Its extension just south of the confluence of the Bonnechère and Ottawa rivers appears as a bold bluff 50 or 60 feet high. Across the river in the vicinity of Shawville, Quebec, the bluff and gullied clay beds above it are especially well represented.

The Connaught Rifle Range, nearer Ottawa, is on the floor of the channel, while the low bluff which borders it may be seen near the Canadian National Railway at the same point. Towards Ottawa, bedrock is frequently exposed. The channel cuts across the northern corner of the Central Experimental Farm grounds and, just to the east, contains the Dow Lake section of the Rideau Canal. At Ottawa the old channel is split, passing on either side of an upraised block of limestone.

East of Ottawa the old river channels are from two to six miles wide and, in places, 60 feet deep. They are floored with clay and silt and bordered by sand plains. An elevated clay plain is also found around Sarsfield and Chartrand. In South Plantagenet, Clarence, and Cumberland townships, the main valley is occupied by the South Nation River and its chief tributary Bear Brook. In Gloucester township it is drained by the headwaters of Green's Creek. These underfit streams have failed to establish complete drainage of the valley and consequently it is occupied by the Mer Bleue bog.

Near Fournier the South Nation River, rather unexpectedly, leaves the broad main channel leading eastward, and enters the Ottawa by way of the narrow valley past the village of Plantagenet. The Azatica Brook took over the drainage of the lower stretch of the main channel but proved so inefficient that another big bog developed in Alfred and Caledonia townships. Both the Mer Bleue and the Alfred are raised sphagnum bogs. Drainage is generally poor in all these channels except along the Ottawa River—indeed the Nation River above Plantagenet periodically overflows its banks, flooding the adjacent flats, and depositing a little alluvium in the process.

The stream-eroded channels through the sand and clay of Clarence township contain some areas of very choppy topography in which small remnants of the upland appear to persist out in the valley. Close examination of these areas indicates that these are the result of land slippage when, or very soon after, these channels were cut and at a time when the clays were over-saturated. These marine clays are sensitive and easily disturbed so that landslips and mudflows readily occur. Another such area in the same region is found along the valley of Bear Brook. Similar phenomena may be seen in clay deposits north of the Ottawa in the province of Quebec. Very similar also were the collapses of the clay which were so troublesome during the building of the St. Lawrence Seaway. While such land has actually been cleared and farmed, its cultivation with modern machinery must be quite an ordeal.

The clay to the east of Ottawa is interbanded, pink and grey. It is finer in texture and lower in lime than the clay in Renfrew county. Free carbonates are lacking in most of the layers. These properties and the generally poorer drainage combine to produce less fertile soils than those to be found above Ottawa.

In the Ottawa valley the clay soils described in the Winchester clay plain are found in Carleton, Russell, and Prescott counties but the proportion of acid soil is greater. Another type of acid clay loam with highly leached surface horizons occurs in Renfrew county. The Renfrew clay loam is more pervious than the acid clay east of Ottawa and is a more productive soil, even though it appears to be more strongly leached and deficient in humus. The poorly drained associates of these grey acid clays have peaty surface horizons underlain by plastic grey clay. The most fertile of these soils is that developed under elm, ash, red maple, and other hardwoods in the Carp valley and the plain north of Richmond. When artificially drained and well handled this is a highly productive and durable soil.

Farming practices vary in the different parts of these clay plains. In the vicinity of Ottawa, the Rideau clay has a high value on account of its location; most farms produce fluid milk for distribution in the city. Practically all of the land has been cleared, and a rather high proportion (30 per cent) is used for pasture. Farms, especially along the highways, are well built upon and appear to be prosperous.

The clay plains of Russell and Prescott are cheese producing areas. The soils are not so productive as those of the Winchester area, nor are farms, on the average, quite so prosperous in appearance. Choice of crops is

110. *Level clay plain, Barr line, Renfrew county*

limited: oats, timothy, and red clover are the mainstays; corn is grown but is seen only about half as frequently as in the other area. On the other hand the area devoted to pasture is greater. Formerly this area produced timothy hay for sale but, more recently, the most important cash crop has been red clover seed. Alfalfa is seldom grown. The clay plain north of Richmond is also a cheese producing area and similarly markets some hogs and poultry.

In the Carp valley, on what appears to be the most fertile soils in the whole Ottawa valley region, a more general type of farming is practised in which small grains and beef cattle appear to be the chief products. The grain elevator at Kinburn gives an indication of the market outlet. There is not a high proportion of pasture on the Carp clay loam, probably because of the availability of rough pasture lands along the ridges which border the valley. A few dairy herds are to be seen and undoubtedly indicate the direction of agricultural development within the next few years. Farm buildings, however, will have to be improved for, though roomy, they are mostly of the older types.

The clay plains of Renfrew county, for the most part, support a general agricultural economy. One sees big square brick or frame houses, rarely stone, mostly accompanied by rather second-rate barns. Often the barns

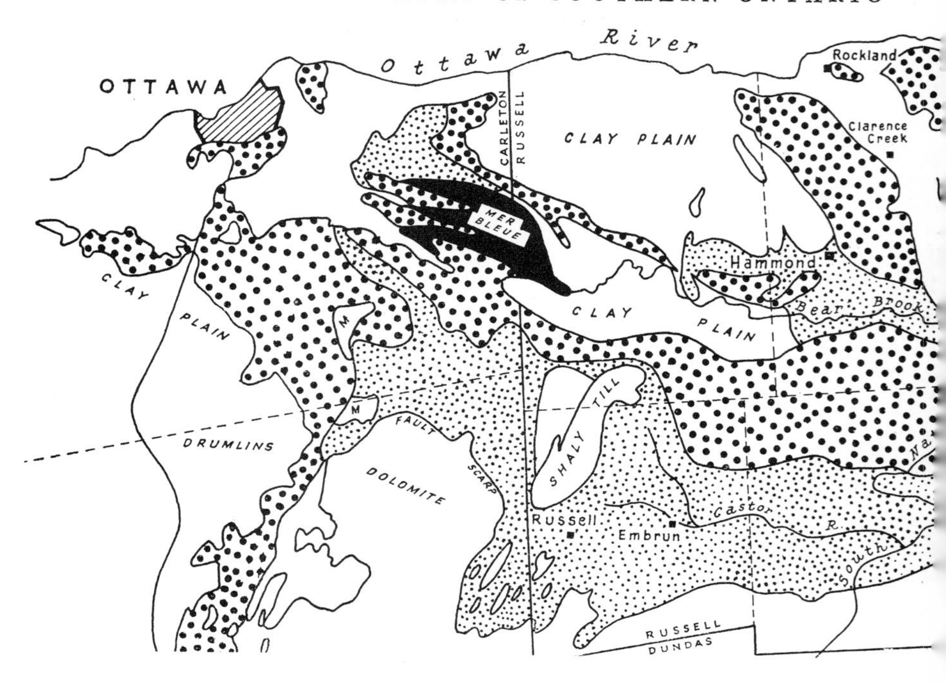

111. *Prescott and Russell sand plains*

are of the "small and numerous" pioneer type. There are many log buildings, pigsties, sheep-folds, etc., even on farms where large frame barns have been built. Fences are usually of wire, but many rail and stump fences are still in service.

Most farms seem to follow an extensive cropping system. There are large areas of spring-sown grain. As a rule, early-sown crops are much better than late-sown because Renfrew is an area of rather light rainfall. Poor drainage, however, often causes late sowing. A striking feature of the agriculture, is the number of good fields of alfalfa for hay or seed which may in part be due to the low rainfall. It would seem that this crop is increasing in the townships of Horton, Ross, and Westmeath. One sees many stacks of alfalfa hay, probably for sale to the alfalfa mill in Renfrew. Another crop which does well is peas; one-third of all the dry peas in Ontario are grown in Renfrew county.

The beef breeds of cattle are commonly seen in the pastures but dairy cattle are increasing. This is especially noticeable near Pembroke, where, also, one sees many silos and good corn crops. Some of the new hybrids seem very well suited to the area. The proportion of cleared pasture on the Renfrew clay is rather small (about 7 or 8 per cent) and crops

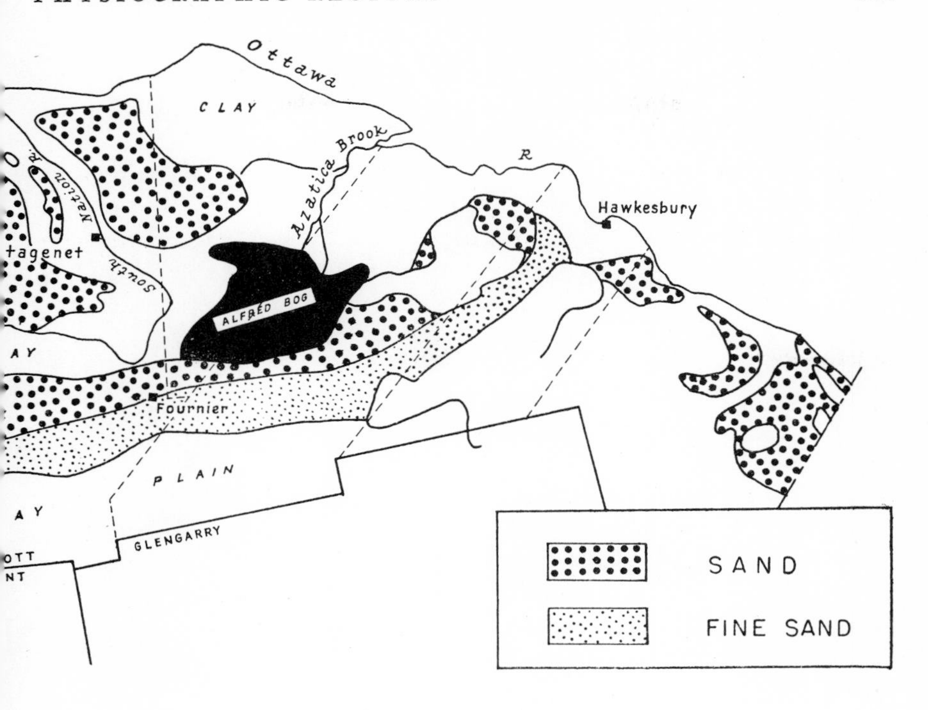

dominate the scene rather than livestock.

Ottawa with its huge population (430,000 in the "greater city") dominates the valley. Of the numerous towns of the lumbering period, only one, Pembroke (17,000) has grown to small-city size and may be counted as a subregional centre. Renfrew (9,000) and Arnprior (5,500) are important in their local umlands as well as having some industries which compete on a wider market. East of Ottawa there is no incorporated centre larger than Rockland (3,100) until Hawkesbury (9,000) is reached. These larger places are manufacturing centres but they serve, also, as farmers' towns in their respective constituencies.

There are many smaller centres in the Ottawa valley clay plains. In the east one might mention L'Orignal (1,200), Alfred (1,200), Plantagenet (900), Chute-à-Blondeau (750), and St. Eugène (650). West of Ottawa are Carp (500), Cobden (950), Beachburg (600), and Westmeath (300). The town of Richmond (1,250) is a peculiar example. Older than Ottawa, it was laid out in 10-acre lots in the hope of establishing people on the basis of part-time handcraftsmanship and part-time husbandry. However, it never became more than a small local supply point until recently. A considerable number of Ottawa commuters now reside in Richmond.

Transportation favours centralization in Ottawa; No. 17 Highway extends from end to end of the clay plains while the main lines of both railways do the same. In the mid-portion where the clay plains are wide, there is a radial road net centred on the city.

Being the capital of the country, Ottawa has grown with great rapidity and shows as yet no signs of slowing down. As a market outlet for the farms of the clay plains, and adjoining sections as well, its importance is bound to increase. There will be further specialization of agriculture within the region and disappearance of the older general crop system of farming.

RUSSELL AND PRESCOTT SAND PLAINS

In the counties of Prescott, Russell, and Carleton, there is a group of large sand plains separated by the clays of the lower Ottawa valley. They consist of one continuous belt, 65 miles in length, from Ottawa to Hawkesbury, together with three fairly large areas lying to the north of it, in Alfred, Plantagenet, and Clarence townships, and a number of smaller sandy remnants dispersed over the clay plains. Taken altogether, these sand plains comprise an area of about 575 square miles.

The sand plains have a level surface whose elevation is approximately 250 feet above sea level, while the bottoms of the intervening clay-floored valleys lie below 200 feet. The boundaries, usually, are abrupt bluffs, although in some places there are intermediate terraces.

Excepting the higher sands south of Ottawa the whole complex was at first a continuous delta built by the Ottawa River and its north-bank tributaries into the Champlain Sea. Formed during one stage, it was cut to pieces later by the Ottawa River when it first rose above sea level. Some sand and gravel of glacio-fluvial origin south of Ottawa is conveniently mapped with the delta.

The depth of the sand varies from 20 to 30 feet, as seen along the edges of the valleys in the north, to ten feet or less in the mid-portions, thinning out along the clay plain to the south. The texture of the sand also varies, being coarse towards the north and grading into fine sand and silt south of the Castor River. Everywhere the sands are underlain by stratified red and grey clay. Both the sand and the clay are low in lime, suggesting an origin in the granitic rocks of the Canadian Shield.

Most of the area lies within the drainage basin of the South Nation River, but smaller parts drain to the Rideau and the Ottawa. Drainage is good for a short distance back from the bordering escarpments but becomes progressively poorer towards the interiors. There are few streams within the sand plains as the water percolates into the sand beds and issues from the bluffs to drain into the clay-floored valleys. The South Nation River cuts a canyon 75 feet deep across the plain from Casselman to Lemieux. Within the valley there are several well-preserved remnants of terraces representing earlier stages of the river. Some of these are wide enough for building sites and narrow fields. It is rather remarkable that the steep escarpments of clay and sand have not been gullied and dissected, but in only a few places does such topography appear, as near Martel Corners. In addition to the steep slopes, springs are often found along the hillsides, thus adding to the difficulties of the farmer.

The original vegetation of these areas included a great many pines on the coarser sands. In wetter locations, American elm, red maple, white ash, black ash, basswood, and yellow birch were found. On wet sands, verging on bog conditions, larch, white spruce, black spruce, alder, and willow are common. In the abandoned fields of the sand plains, white birch, and trembling aspen are the most abundant early invaders. The sand plains, while a unit from the standpoint of physiographic origin, are thus divisible into three land types, which from the practical standpoint influence the trends of land utilization. The types are dry sand plains, wet sand plains, and fine sand and silt plains. With each of these, there is associated certain definite soil types. The sands and fine sands are shown on Figure 111.

The dry sand plains have mature Podzol soils with thin ash-grey horizons. Ground-Water Podzols also occur where the water table is high.

112. *Stream-cut bluff, Hawthorn Station*

In the latter, where the sand is coarse, iron and humus hardpans sometimes develop. All these soils are low in fertility, being deficient in lime, nitrogen, potash, phosphorus, and manganese.

The finer sandy loams, described fully in the Carleton county soil report under the name of Castor, present excellent prospects for improvement. They are mostly poorly drained, even close to streams and bluffs, and require deep drains to effect drainage of the subsoil. With drainage established, a fertility programme involving legumes and fertilizers would be aided by the good physical properties of this soil.

The pattern of land utilization shows some rather striking contrasts. Whereas the fine sands and silts have become fairly prosperous dairy farming areas, the sand plains show a poor development of agriculture. It has long been evident that large areas of sand plain should be withdrawn from general agricultural use. The united counties of Prescott and Russell took an early lead in returning such land to forest. In 1928, upon the recommendation of Mr. Ferdinand LaRose, a block of 100 acres was purchased and planted to young trees. In 1937, by agreement between the county authorities and the Ontario Department of Lands and Forests an area of several square miles was set aside along the borders of Cambridge and Clarence townships to be known as the LaRose Forest. Since then additional parcels have been added to the original plantations and also many separate lots have been reforested under the same management plan. By 1964 the reforested area in the Russell and Prescott sand plains had grown to more than 20,000 acres and some hundreds of acres of the earlier plantings had yielded quantities of poles and fence posts from the first cultural thinning. The forest area is still being enlarged as suitable land is thrown on the market.

The solution of the problem on the areas of fine sand and silt is the amelioration, rather than the abandonment, of land. The provision of good drainage outlets and the raising of fertility levels are important. It is possible, of course, that certain special crops might be introduced to supplement the already well-established dairy economy. Flax is a crop that, at times, has had some success; potatoes do fairly well, and there may be other possibilities such as grain corn.

MUSKRAT LAKE RIDGES

The Ottawa valley is interrupted in several places by prominent rocky ridges which are the protruding crests of fault blocks. One such block lies along the northeast side of Muskrat Lake, presenting a steep scarp towards the southwest, while sloping gently away under a cover of sand to the northeast. A similar ridge borders Olmstead Lake, while a third is found between Haley's Station and Renfrew and extends northwestward beyond Cobden. To this collection of rocky uplands the name Muskrat Lake ridges is applied, but there are other smaller areas which do not appear in Figure 62 although they are outlined on the folded map.

These high fault blocks are all composed of Precambrian rocks, chiefly reddish and reddish-grey gneiss and granite, but with fairly large areas of crystalline limestone. Some of this is dolomitic, such as the area north of Haley's Station where a plant for the extraction of metallic magnesium is located. The tops of these ridges escaped being covered with clay. The scanty overburden being predominantly sand and gravel.

These ridges, in the main, are non-agricultural. Some land has been cleared, but a great deal of farm abandonment has taken place. In a sense, the ridges form a complement to the contiguous clay plains. The latter have been completely deforested and fuel is scarce; consequently, farmers have secured woodlots on the "mountain." Sometimes, in addition to firewood, enough pine logs are cut "to pay the taxes." Another and perhaps somewhat questionable use of this land is as pasture for young cattle and dry cows. It is droughty and the carrying capacity is low. Moreover the grazing of the cattle tends to check the regeneration of the woodlots, which, after all, must be looked upon as the important resource.

PETAWAWA SAND PLAIN

The solid block of settled country in the Ottawa valley ends near Petawawa in a sand plain, on the Ontario side of the river, interrupted by several outcrops of Precambrian rocks. It extends south of Pembroke and covers an estimated 130 square miles. In origin, it is a delta built in the Champlain Sea by the Petawawa, Barrow, Indian, and Ottawa rivers during the Fossmill stage of Lake Algonquin.

The farms are large, averaging over 180 acres each. About 70 per cent of the cleared land (35 per cent of the total) is in crop, mainly hay and oats. Although the farms are generally poor, there are many large square brick and frame houses similar to those in other parts of Renfrew county. It is probable that some of these were built by money made from timber rather than crops and livestock.

Agriculture is no longer of any importance. Of a total rural population of over 10,000, less than 250 actually live on farms. Population is attracted to the area because of the location of the Petawawa Camp. Further north also is the village of Chalk River, the town of Deep River, and the nuclear research station.

SUMMARY

5

Southern Ontario is a glaciated plain underlain, for the most part, by Palaeozoic limestones and shales. The bedrock surface is one of low topographic relief except for the Niagara escarpment and two other, much lower, northeast-facing scarps in the peninsular section and several short, but abrupt, fault-line-scarps in the Ottawa valley. West of the Niagara escarpment, a broad arch of the bedrock forms the Dundalk upland, the crown of which is more than 1,700 feet above sea level. Eastward from the section of the escarpment known as the Caledon Hills, another height of land extends for one hundred and twenty miles through south-central Ontario. It is a great ridge of glacial drift, the highest parts of which stand between 1,000 and 1,300 feet above sea level. Thus the build of Southern Ontario is dominated by two features, the broad half-dome of the Niagara cuesta and the Oak Ridges moraine. A lower arch of the bedrock, known as the Frontenac axis, exposes the granite knobs east of Kingston, while, to the northeast, the St. Lawrence–Ottawa lowland shows little relief with its lowest parts being less than two hundred feet above the sea.

There are broad belts of limestone and dolomite on which very little overburden was left after the melting of the ice. Most extensive of these areas is the rock plain extending northward from Kingston toward Arnprior and Ottawa. Other such areas are found along the low escarpment, westward from Kingston to the Kawartha Lakes and Georgian Bay, and in association with the Niagara cuesta, especially on the Bruce peninsula and Manitoulin Island. The total area of such land types exceeds 4,500 square miles.

Apart from the large tracts of shallow soil on limestones, and a few small areas of shallow drift on shale, the overburden is generally deep and composed of calcareous materials from which fertile soils have developed. The higher inland portions are mostly till plains. The lower lands bordering the Great Lakes were submerged in glacial lakes, and the plain between the St. Lawrence and the Ottawa was inundated by salt water following the disappearance of the last ice sheet. Altogether, these waterlaid sand and clay plains cover nearly half the area of Southern

Ontario. Although this book is concerned with surface beds rather than the underlying drift strata, these lower strata should not be ignored since they were a source of material for the last glacier and from them is derived most of our well water. The uniformly fine-textured tills bordering the Great Lakes derived much of their silt and clay from the older lacustrine beds.

The process of glaciation and deglaciation by the Wisconsin ice sheet created the present configuration of moraines, abandoned spillways, drumlins, eskers, abandoned shorelines, and various types of stillwater sediments. The history of the recession of this ice sheet has been reinterpreted in this edition as a result of new observations and a reassessment of the evidence previously collected and is one of the more important contributions of the study.

The major patterns of physical features in Southern Ontario are shown on the accompanying coloured map. This map is the product of field mapping, supplemented by the examination of aerial photographs. It provides a fairly complete survey of moraines, drumlins, and eskers. Sandy kame-moraines are distinguished from till-moraines and two types of till plains are shown. One is drumlinized, well moulded, and smooth even if no drumlins are present; the other has a subdued morainic topography exhibiting a mosaic of irregular knolls and shallow depressions. The map illustrates the abundance and importance of glacial drainage channels but, because of the scale, the bluffs, terraces, and swamps which are their component parts are not shown. The shorelines of extinct lakes must also be shown without complete detail, but the larger clay plains, sand plains, bogs, and marshes of the old lake basins appear on the map. Although not shown on the map, some other Pleistocene features were discussed in the text: the shallow silty covering, probably wind-deposited, which smooths the surface and reduces the stoniness of the soils in many areas, and areas of sand dunes, sometimes in association with ancient shorelines and sometimes arising in outwash sand plains. Almost all of these dunes are old, were eventually covered with vegetation, and so became stable landforms; but the recent removal of forests has, in some cases, caused them to begin to move again. The ribbed and hummocky relief produced by mudflows in the marine clay beds of the Ottawa valley constitutes an exceptional pattern which has also been dealt with separately.

Most of the surface of Southern Ontario, moulded by the moving ice sheets or veneered by lacustrine deposits, remains relatively unaltered as an assemblage of depositional land forms. The period since the disappearance of the ice sheet and the uncovering of any submerged areas has been too short for stream erosion to assume complete control of the landscape. Nevertheless there are some marks of erosion. The streams, which carry off on the average between 10 and 19 inches of water or between one-third and one-half of the precipitation, have begun to refashion the surface of the land of Southern Ontario. We have not tried to map these erosional forms on the master map as a larger scale would be needed for this purpose. However, the stream valleys have been discussed in some detail with special emphasis on their relation to the Pleistocene deposits and related land forms. In the upland area west of the Niagara escarpment the valleys of the present rivers must be distinguished from the channels made by the meltwaters draining away from the waning glacier. Many of the streams follow the Pleistocene spillways and obviously are misfits. The fact that the spillways often cut through the drift to the bedrock has limited the rate of development of the modern stream valleys. Within the glacial lake plains the stream valleys are all recent and in these areas the amount of dissection due to the streams and their tributaries is related in no small measure to the texture of the lacustrine deposits. The deep beds of silt and fine sand are most readily eroded, and it is on these soils that soil erosion is most serious. Soil erosion and geological erosion are the same process.

In order to point up the variety and contrast of the landscape of Southern Ontario, fifty-two minor regions have been delineated. The Niagara escarpment, with grey dolomite cliffs overlooking long slopes carved in red shale, stands in sharp contrast to the gentler glaciated landscapes of the plains which flank it at both upper and lower levels. But the escarpment itself is somewhat complex; the two deep re-entrant valleys, the Beaver and the Bighead, as well as the broad clay terraces of the Cape Rich foreland, warrant separate descriptions. The Horseshoe and the Dummer moraine systems, with a combined area of about 2,750 square miles, include the roughest and the stoniest of the glaciated surfaces. Less rugged till moraines, separated by smooth till plains, occupy the regions north and south of London. Four sandhill areas, based for the most part on interlobate and terminal kame-moraines, occupy about 860 square miles. While there is some good land in these areas, the dry, hilly sands are subject to both blowing and gully erosion, and constitute a well-known reforestation problem. Much has been done in recent years to take care of this situation, but much still remains to be done. The Peterborough drumlin field and six smaller drumlin fields cover an area of about 3,300 square miles characterized by the presence of oval hills. Undulating till plains, without drumlins, cover at least another 4,000 square miles. The clay plains which were the floors of ancient lakes occur as twelve separate regions with a total area of 7,300 square miles, not counting an estimated 1,000 square miles of clay land scattered throughout the Algonquin and

Iroquois lake plains. The sand plains, amounting to more than 4,500 square miles, border Lake Erie, Lake Ontario, and Lake Simcoe, as well as comprising sizable areas in the Ottawa valley. Six sand plain regions are described, chief among them being the Norfolk tobacco belt.

Regional differentiation is primarily physical, with most attention being given to the topographic form and composition of the surface deposits. These are basic to the classification of the soils which are developed upon them through the agency of climate and vegetation. Our original studies of these areas included observations of a broad reconnaissance nature concerning the soils, not all of which were published. Since that time the Ontario Soil Survey has published detailed reconnaissance reports covering nearly all counties in Southern Ontario, and we have drawn upon this material rather freely in this edition. Major soil types and the pattern of associated soils show a close correspondence with the physical base while, in turn the land use patterns are at least partly correlated.

Southern Ontario is currently experiencing a very rapid increase in population. While most of this increase is in the cities it affects agricultural production in various ways. Comparative census data for 1941 and 1961 have been used to illustrate these changes in the different regions. No longer are the densest livestock populations found in the dairy regions; the beef cattle–hog–poultry areas centred in Waterloo and Perth counties have far outstripped them. The dairy regions are notable for increased milk production per cow rather than an increase in the number of cows. The most important development in field crops, which must be apparent to the casual observer, is the increase in the amount of corn grown. Hybrid corn represents a great break-through in biological technique. Higher yielding hybrids and good short-season hybrids have resulted in much more corn in the old Essex and Kent corn belt and also a northward expansion of the grain corn area. This is one of the few areas in Canada warm enough to accommodate such an expansion. Fortunately, the farmers, seed corn companies, and crop specialists are working together to limit the numerous hybrids to the climatic zones to which each is adapted.

Changes in agricultural patterns constitute only one phase of the land use revolution in southern Ontario; even more spectacular, in many areas, is the disappearance of agriculture in the face of urban expansion. Not by any means all of this is involved with newly built-up city areas; the expanding city casts a long shadow. Scattered out-of-town sites for industry, recreation, water and wildlife conservation constitute demands for land which often can be supplied only from land formerly in agricultural use. Another factor is the increasing need for land on which to base transportation facilities.

Perhaps the most striking change, particularly in sandy and hilly landscapes, is the enormous increase in the use of land for scattered non-farm residential sites. This is by no means a uniform development, being conditioned by distance from large cities or industrial centres of employment. But physiographic factors are important also, particularly as they relate to local water supply and drainage facilities. Clay plains do not invite squatters to the same degree as sand plains. The Niagara escarpment and the Oak Ridges moraine provide the most striking examples of the physiographic attraction for rural residences, but valley sides and shoulders, terraces, raised beaches and other old shoreline features, drumlins, kames, and low morainic ridges have all been occupied. One of the great desires of those who seek a rural residence is to have a view and if possible a stream and some woods nearby. For others cheap land seems to be the only consideration, as, for instance, for those who build in the middle of a swamp.

Thus we have placed considerable emphasis on population density and distribution in these regional discussions, but our main purpose has been to describe the terrestrial surface. This study provides a background for the classification and mapping of the soils of the area. Even in this space age, society is still founded on the soil and proper land use is the key to continued development. The map will serve to indicate the occurrence of gravel, sand, and clay deposits. Those interested in the engineering properties of soils and in water supply may find this book useful. We hope also, that it will stimulate an interest in the land forms of this complex area, which supports more than one quarter of the population of Canada.

GLOSSARY

Barrier beach. A beach, usually of gravel, built across the entrance of a bay.

Baymouth bar. A barrier beach or several of them combined.

Beaches (abandoned beaches). Smooth, horizontal ridges of gravel and sand, often at the boundary between level lake plains below and rolling till plains above. The gravel is usually well rounded, and evenly graded in the different strata.

Boulder clay. Clayey glacial till.

Boulder pavement. A boulder-strewn surface on a wave-cut or stream-cut terrace.

Cap-rock. The hard upper strata of a rock formation.

Catena. A group of soils within one soil zone developed from similar parent material but having differing characteristics due to differences in relief and drainage. From the Latin for chain.

Cuesta. A belt of high ground one side of which is an escarpment and the other a gradual dip-slope.

Dissection. Gullying or the cutting of valleys by streams.

Drift. All unconsolidated mineral material on the bedrock.

Drumlins. Celtic for little hill. Oval hills of glacial till with smooth convex contours. In any area the drumlins all point in the same direction which is considered to be the direction of movement of the glacier which formed them.

Escarpment. A system of steep slopes formed by differential erosion or by a fault (break) in the bedrock.

Esker. A knobby, crooked ridge of coarse gravel and sand considered to be deposited by meltwater in crevasses and tunnels near the front of a glacier.

Fault-scarp. An abrupt slope due to a fault in the bedrock.

Flutings. Parallel grooves and low ridges on a rock surface or on a drumlinized till plain.

Fluvium, flood deposits. Silt, sand, and gravel deposited by running water, often meltwater from a glacier.

Glacial lake. A lake dammed by a glacier.

Glei. A compact, sticky, bluish grey or olive grey soil horizon formed under wet conditions.

Graben. A valley formed by a down-dropped block of bedrock and so bounded by fault-scarps on either side.

Gravel train (valley train). The extended gravel terraces of an abandoned river channel.

Indurated. Hard, often weakly cemented.

Interlobate moraine. A moraine formed between two lobes of the glacier; usually a sandy kame-moraine.

Kames. Knobby hills of irregularly stratified sand and gravel, formed at the edge of a melting glacier.

Kame-moraine. An extended ridge consisting mainly of kames and outwash.
Kettles. Depressions in moraines or other drift having no surface drainage.
Loam. Soil of medium texture, not necessarily surface soil.
Loess. Wind-deposited silt and fine sand.
Meanders. The semi-circular bends in the course of a slow-flowing mature stream.
Mesa. A flat-topped rock formation with abrupt slopes, standing above the surrounding surface.
Moraine. A knobby ridge either of (*a*) boulder clay built by a thrust of a glacier or of (*b*) gravel and sand deposited at the edge of glacier by escaping meltwater.
Outwash. Sand and gravel deposited in stream terraces or broad fans by water draining away from a glacier. The beds are unusually pitted by depressions with no surface drainage.
Ox-bows. The loops of a stream meandering over its flood plain.
Podzol. The type of soil that develops in cool humid climates.
Pondings. Small glacial lakes, sometimes short-lived and so without definite shore features.
Scarp. A bluff or very steep slope in bedrock due to faulting or differential erosion.
Shield (Canadian Shield). The ancient worn-down surface of Precambrian rocks in Canada.
Shorecliff. A wave-cut bluff, usually having a bouldery terrace at its base.
Silt. Mineral particles below sand and above clay sizes.
Spillway. The abandoned channel of a glacial meltwater stream.
Spit. A point of land, often a wave-built sand bar, projecting into a body of water.
Striae. Scratches on the surface of the bedrock produced by the moving glacier.
Terrace. (*a*) River terrace—the beds of gravel and sand left by a stream; (*b*) wave-cut terrace—the strip of level land at the base of a shorecliff.
Till. The heterogeneous mixture of clay, sand, pebbles and boulders deposited directly by a glacier.
Umland. The geographic zone of influence of a town or city.
Varves (varved clay). Regularly stratified clay, dark clayey layers alternating with lighter coloured siltier layers; considered to be annual deposits of winter and summer sediments in glacial lakes.

BIBLIOGRAPHY

1. ALDEN, W. C. The drumlins of Southeastern Wisconsin. *U.S. Geol. Surv. Bull. 273*. 1905.
2. ANTEVS, E. Retreat of the last ice sheet in eastern Canada. *Geol. Surv. Can. Mem. 146*. 1925.
3. —— Climaxes of the last glaciation in North America. *Amer. J. Sci.* Ser. 5. *28*: 304–11. 1934.
4. —— Late quaternary upwarpings of Northeastern North America. *J. Geol. 47*. 1939.
5. BAKER, M. B. Clay and the clay industry of Ontario. *Ann. Rept. Ont. Bur. Mines 15*, Pt. 2: 1–127. 1906.
6. —— The geology of Kingston and vicinity. *Ann. Rept. Ont. Bur. Mines 25*, Pt. 3: 1–71. 1916.
7. BRYDON, J. E. and PATRY, L. M. Mineralogy of Champlain Sea sediments and a Rideau clay profile. *Can. J. Soil Sci. 41*: 169–81. 1961.
8. CALEY, J. F. Palaeozoic Geology of the Toronto–Hamilton area, Ontario. *Geol. Surv. Can. Mem. 224*. 1940.
9. —— Palaeozoic geology of the Brantford area, Ontario. *Geol. Surv. Can. Mem. 226*. 1941.
10. CARMAN, R. S. *Wilmot Creek drainage unit*. Publication of Ont. Forestry Br. 1940.
11. CHAPMAN, D. H. Late-glacial and postglacial history of the Champlain valley. *Amer. J. Sci.* Ser. 5. *34*: 89–124. 1937.
12. CHAPMAN, L. J. An outlet of Lake Algonquin at Fossmill, Ontario. *Geol. Assoc. Can. Proc. 6*, Pt. 2: 61–8. 1954.
13. CHAPMAN, L. J. and DELL, C. I. Revisions in the early history of the retreat of the Wisconsin glacier in Ontario based on the calcite content of sands. *Geol. Assoc. Can. Proc. 15*: 103–8. 1963.
14. CHAPMAN, L. J. and PUTNAM, D. F. The physiography of Eastern Ontario. *Sci. Agric. 20*, 7: 424–41. 1940.
15. —— The soils of South-central Ontario. *Sci. Agric. 18*, 4: 161–97. 1937.
16. —— The moraines of Southern Ontario. *Trans. Roy. Soc. Can. 37*, Sec. 4: 33–41. 1943.
17. —— The physiography of Southwestern Ontario. *Sci. Agric. 24*, 3: 101–25. 1943.
18. —— The recession of the Wisconsin glacier in Southern Ontario. *Trans. Roy. Soc. Can.*, Sec. 4, *43*, 3: 23–52. 1949.
19. COLEMAN, A. P. Marine and fresh water beaches of Ontario. *Bull. Geol. Soc. Amer. 12*: 129–46. 1901.
20. —— Lake Ojibway; last of the great glacial lakes. *Ann. Rept. Ont. Dept. Mines 18*, Pt. 1: 284–93. 1909.
21. —— The extent of Wisconsin glaciation. *Amer. J. Sci.* Ser. 5. *20*: 180–3. 1930.
22. ——The pleistocene of the Toronto region. *Ann. Rept. Ont. Dept. Mines 41*, Pt. 7. 1932.
23. —— Lake Iroquois. *Ann. Rept. Ont. Dept. Mines 45*, Pt. 7. 1936.
24. —— The Iroquois beach in Ontario. *Ann. Rept. Ont. Bur. Mines 13*, Pt. 1: 225–44. 1904.

25. ——— Geology of the north shore of Lake Ontario. *Ann. Rept. Ont. Dept. Mines 45*, Pt. 7: 75–116. 1937.
26. ——— *The last million years.* Toronto: University of Toronto Press. 1941.
27. ——— An interglacial Champlain Sea. *Amer. J. Sci.* Ser. 5, *24*: 311–15. 1932.
28. Comstock, F. M. Ancient lake beaches on the islands of Georgian Bay. *Amer. Geol. 33*: 310–18. 1904.
29. Coventry, A. F. Desiccation in Southern Ontario. *Trans. Roy. Soc. Can.*, Sec. 5, *34*. 1940.
30. Cramm, E. W. R. An analysis of the urban process in Pickering Township. Unpublished M.A. Thesis presented to the School of Graduate Studies, University of Toronto. 1962.
32. Crerar. A. D. The loss of farmland in the metropolitan regions of Canada. Background papers, Resources for To-morrow Conference, Montreal, October, 1961.
32. Deane, R. E. Pleistocene deposits and beaches of Orillia map-area. *Can. Geol. Surv. Paper 46*: 20, 1946.
33. ——— Pleistocene geology of the Lake Simcoe district, Ontario. *Geol. Surv. Can. Mem. 256.* 1950.
34. Dell, Carol I. A study of the mineralogical composition of sand in southern Ontario. *Can. J. Soil Sci. 39*: 185–96. 1959.
35. Dreimanis, A. Wisconsin stratigraphy at Port Talbot on the north shore of Lake Erie, Ontario. *Ohio J. Sci. 58*, 2. 1958.
36. ——— Pleistocene geology of the London-St. Thomas and Port Stanley areas. *Prog. Rept. 1962–3*, Ont. Dept. Mines, Toronto.
37. ——— Lake Warren and the Two Creeks interval. *J. Geol. 72*, 2: 247–50. 1964.
38. Dreimanis, A. *et al.* Heavy mineral studies in tills of Ontario and adjacent areas. *J. Sed. Petrol. 27*, 2: 148–61. 1957.
39. Dreimanis, A., and Terasmae, J. Stratigraphy of Wisconsin glacial deposits of Toronto area. *Proc. Geol. Assoc. Can. 10*: 119–35. 1958.
40. Ells, R. W. Surface deposits, Ottawa and St. Lawrence valleys. *Ann. Rept. Geol. Surv. Can.* 2, 44–51. 1886.
41. ——— *ibid.* 3, 98–101. 1887.
42. ——— Report on the geology of a portion of Eastern Ontario. *Ann. Rept. Geol. Surv. Can. 14*, Pt. J. 1904.
43. ——— Report on the geology and natural resources of the area included in the Northwest Quarter-Sheet. *Geol. Surv. Can. Pub.* 977. 1907.
44. Fairchild, H. L. Pleistocene geology of Western New York. *20th Ann. Rept. of New York State Geol.*: 105–12. 1900.
45. ———Latest and lowest pre-Iroquois channels between Syracuse and Rome. *21st Ann. Rept. New York State Geol.*: 33–47. 1901.
46. ———Glacial waters, Oneida to Little Falls. *22nd Ann. Rept. New York State Geol.*: 17–41. 1902.
47. ——— Drumlin structure and origin. *Bull. Geol. Soc. Amer. 17*: 702–7. 1903.
48. ——— Gilbert Gulf (Marine waters in the Ontario basin). *Bull. Geol. Soc. Amer. 17*: 712–18. 1905.
49. ——— Glacial waters in the Lake Erie basin. *New York State Museum Bull. 106*. 1907.
50. ——— Pleistocene geology of New York State. *Bull. Geol. Soc. Amer. 24*: 133–62. 1913.
51. ——— Post-glacial uplift of Northeastern America. *Bull. Geol. Soc. Amer. 29*: 187–234. 1918.
52. ——— New York drumlins. *Proc. Rochester Acad. Sci.* 7: 1–37. 1929.
53. ——— New York moraines. *Bull. Geol. Soc. Amer. 43*: 627–62. 1932.
54. ——— Closing stages of New York glacial history. *Bull. Geol. Soc. Amer. 43*: 603–26. 1932.
55. Flint, R. F. Eskers and crevasse fillings. *Amer. J. Sci.* Ser. 5. *15*: 410–16. 1928.

56. ——— Growth of the North American ice sheet during the Wisconsin age. *Bull. Geol. Soc. Amer. 54*: 325–62. 1943.
57. ——— *Glacial geology and the pleistocene epoch.* New York: John Wiley & Sons. 1947.
58. ——— Probable Wisconsin substages and Late-Wisconsin events in northeastern United States and southeastern Canada. *Bull. Geol. Soc. Amer. 64*: 897–919. 1953.
59. Fox, W. S. *'Tain't runnin' no more.* London, Ont.: Wendell Holmes.
60. Gadd, N. R. Surficial geology of the Ottawa map-area. Ontario and Quebec. *Geol. Surv. Can. Paper 62–13.* Dept. of Mines & Tech. Surveys, Ottawa. 1963.
61. Gertler, L. O., and Hind-Smith, J. The impact of urban growth on agricultural land. Background papers for the Resources for To-morrow Conference, Montreal, October, 1961.
62. Goldthwait, J. W. An instrumental survey of the shore-lines of the extinct lakes Algonquin and Nipissing in Southwestern Ontario. *Geol. Surv. Can. Mem. 10.* 1910.
63. ——— Physiography of Nova Scotia. *Geol. Surv. Can. Mem. 140*: 1–179. 1924.
64. Grabau, A. W. Geology and paleontology of Niagara Falls and vicinity. *New York State Museum Bull. 45.* 1901.
65. Gravenor, C. P. Surficial geology of the Lindsay-Peterborough area, Ontario. *Geol. Surv. Can. Mem. 288.* Dept. of Mines & Tech. Surv. Ottawa. 1957.
66. Guillet, G. R. *Clay and shale in Ontario. A review.* Ont. Dept. of Mines. 1964.
67. Gwynne, C. S. Swell and swale pattern of the Mankato lobe of the Wisconsin drift plain in Iowa. *J. Geol. 50*: 200–8. 1942.
68. Harcourt, R., Iveson, W. L., and Cline, C. A. Preliminary soil survey of Southwestern Ontario. *Ont. Dept. Agric. Bull. 298*: 1923.
69. Hewitt, D. F. Sand and gravel in Southern Ontario. Ont. Dept. Mines, *Industrial Mineral Rept. 11.* 1963.
70. Hills, A. G., Morwick, F. F., and Richards, N. R. *Soil survey of Carleton county.* Ont. Dept. of Agriculture. 1944.
71. Hobbs, W. H. The glacial anticyclone and the continental glacier of North America. *Proc. Amer. Phil. Soc. 86*, Sec. 3. 1943.
72. Hole, F. D. Correlation of the glacial border drift of North-Central Wisconsin. *Amer. J. Sci. 241*: 498–516. 1943.
73. Holmes, C. Till fabric. *Bull. Geol. Soc. Amer. 52*: 1299–1354. 1941.
74. Hough, J. L. *Geology of the Great Lakes.* Urbana: University of Illinois Press. 1958.
75. ——— The prehistoric Great Lakes of North America. *Am. Scientist 51*: 84–109. 1963.
76. Hunter, A. F. The Algonquin shoreline in Simcoe county, Ontario. *Geol. Surv. Can. Summary Rept. for 1902.*
77. Johnston, W. A. Geology of Lake Simcoe area. *Geol. Surv. Can., Summary Rept. for 1911*: 253–61. 1912.
78. ——— The Trent valley outlet of Lake Algonquin and the deformation of the Algonquin water-plane in Lake Simcoe district, Ontario. *Geol. Surv. Can. Museum Bull. 23.* 1916.
79. ——— Late pleistocene oscillations of sea-level in the Ottawa valley. *Geol. Surv. Can. Museum Bull. 24.* 1916.
80. ——— Pleistocene and recent deposits in the vicinity of Ottawa with a description of the soils. *Geol. Surv. Can. Mem. 101.* 1917.
81. ——— The age of the upper great gorge of Niagara River. *Trans. Roy. Soc. Can.* 3rd ser. 22, Sec. 4: 13–29. 1928.
82. Karrow, P. F. *Pleistocene geology of the Hamilton area.* Ont. Dept. of Mines Geol. Circ. 8. 1959.
83. ——— *Pleistocene geology of the Galt map-area.* Ont. Dept. of Mines, Geol. Circ. 9. 1961.

84. KARROW, P. F., CLARK, J. R., and TERASMAE, J. The age of Lake Iroquois and Lake Ontario. *J. Geol. 69*, 6: 659–67. 1961.
85. KAY, G. F. The relative ages of the Iowan and Illinoian drift sheets. *Amer. J. Sci.* Ser. 5, *16*: 497–518. 1928.
86. KAY, G. M. Ottawa-Bonnechere graben and Lake Ontario homocline. *Bull. Geol. Soc. Amer. 53*, Pt. 4: 585–646. 1942.
87. KINDLE, E. M. Geology of Pelee and adjacent islands. *Ann. Rept. Ont. Dept. of Mines 45*, Pt. 7: 75–116. 1937.
88. KRUEGER, R. R. Land use changes in the Niagara fruit belt. *Geog. Bull. 14*: 5–24. Dept. of Mines and Tech. Surv., Ottawa. 1960.
89. LEE, C. F. The middle Grand River valley: A study in regional geography. University of Toronto Ph.D. Thesis. 1943.
90. LEGGET, R. F. *Soils in Canada.* Roy. Soc. Can. Special Pub. 3. Toronto: University of Toronto Press. 1961.
91. LEIGHTON, M. M. The classification of the Wisconsin glacial stage of North-Central United States. *J. Geol. 68*, 5: 529–52. 1960.
92. LEVERETT, F., and TAYLOR, F. B. The Pleistocene of Indiana and Michigan and the history of the Great Lakes. *U.S. Geol. Surv. Mono. 53.* 1915.
93. ——— Correlations of beaches with moraines in the Huron and Erie Basins. *Amer. J. Sci. 237*: 456–75. 1939.
94. LOGAN, W. Report of its progress from its commencement to 1863. *Ann. Rept. (First) Geol. Surv. Can.* 1863.
95. LOKEN, O. H., and LEAHEY, E. J. Small moraines in southeastern Ontario. *Can. Geog. 8*, 1: 10–21. 1964.
96. MACLACHLAN, D. C. Warren shore line in Ontario and in the thumb of Michigan and its deformation. Unpublished thesis, University of Michigan. 1938.
97. MACCLINTOCK, P. *Pleistocene geology of the St. Lawrence lowland.* N. Y. State Sc. Serv., Report of Investigation No. 10. 1954.
98. MACCLINTOCK, P., and TERASMAE, J. Glacial history of Covey Hill. *J. Geol. 68*, 2: 232–41. 1960.
99. MASON, R. J. Early man and the age of the Champlain Sea. *J. Geol. 68*: 366–76. 1960.
100. MATHER, K. F. The Champlain Sea in the Lake Ontario basin. *J. Geol. 25*: 542–54. 1917.
101. MAYALL, K. M. The natural resources of King township. *Trans. Roy. Can. Inst. 22*, Pt. 2. 1939.
102. MERCIER, R. G., and CHAPMAN, L. J. Peach climate in Ontario. *1955–56 Report,* Horticultural Experiment Station and Products Laboratory, Vineland, Ont.
103. MIRYNECH, E. Pleistocene geology of the Trenton-Campbellford map-area, Ontario. Ph.D. thesis, University of Toronto. 1963.
104. MORWICK, F. F., and HEEG, T. J. The relationship of certain chemical characteristics to the geological origin of some Southern Ontario soils. *Sci. Agric. 19*, 5. 1939.
105. OWEN, E. B. Pleistocene and recent deposits of the Cornwall-Cardinal area, Stormont, Dundas and Grenville counties. *Geol. Surv. of Can. Paper 51–12.* 1951.
106. PUTNAM, D. F. Manitoulin Island. *Geog. Rev. 37*, 4. 1947.
107. PUTNAM, D. F., and CHAPMAN, L. J. The physiography of South-Central Ontario. *Sci. Agric. 16*: 457–77. 1936.
108. ——— The climate of Southern Ontario. *Sci. Agric. 18*: 401–46, 1938.
109. ——— The drumlins of Southern Ontario. *Trans. Roy. Soc. Can. 37*, Sec. 4: 75–88. 1943.
110. PUTNAM, D. F., and REEDS, L. G. *The Don River system: Conservation in South-Central Ontario.* Ont. Dept. Planning & Development. 1946.
111. PUTNAM, R. G. Changes in rural land use patterns on the central Lake Ontario plain. *Can. Geog. 6*: 60–8. 1962.

112. REEDS, L. G. Agricultural geography of the Lindsay-Peterborough region. M.A. Thesis, University of Toronto. 1942.
113. RICHARDS, N. R., and MORWICK, F. F. *Soil survey of Prince Edward County.* Ont. Dept. of Agriculture. 1948.
114. RICHARDS, N. R., CALDWELL, A. G., and MORWICK, F. F. *Soil survey of Essex county.* Ont. Dept. of Agriculture. 1949.
115. RICHARDSON, A. H. *The Ganaraska Watershed.* Ont. Dept. of Planning and Development Pub. 1944.
116. ROBERTS, E. F. *Grand River conservation.* Ont. Dept. of Planning & Development Pub. 1945.
117. SATTERLY, J. Pleistocene glaciation in the Windigo-North Caribou lakes area, Kenora district, Ontario. *Trans. Roy. Can. Inst. 23,* Pt. 1: 75–82. 1940.
118. SHEPARD, F. P. Origin of the Great Lakes basin. *J. Geol. 45*: 76–88. 1937.
119. SMITH, G. The Holland marsh. M.A. Thesis, Dept. of Geog., University of Toronto. 1963.
120. SPENCER, J. W. Discovery of the preglacial outlet of the basin of Lake Erie into that of Lake Ontario; with notes on the origin of the lower Great Lakes. *Proc. Amer. Phil. Soc. 19*: 108. 1881.
121. ——— The Iroquois beach; A chapter in the geological history of Lake Ontario. *Proc. & Trans. Roy. Soc. Can. 7,* Sec. 4: 121–34. 1889.
122. ——— Origins of the basins of the Great Lakes. *Quart. J. Geol. Soc. 46.* 1890.
123. ——— Deformation of the Algonquin beach and the birth of Lake Huron. *Amer. J. Sci. 41*: 12. 1891.
124. ——— High level shores in the region of the Great Lakes and their deformation. *Amer. J. Sci.,* 3rd Ser. *41*: 201–11. 1891.
125. STAUFFER, C. R. The Devonian of southwestern Ontario. *Geol. Surv. Can. Mem. 34.* 1915.
126. STANLEY, G. M. Lower Algonquin beaches of Cape Rich, Georgian Bay. *Bull. Geol. Soc. Amer. 48*: 1665–86. 1937.
127. ——— Impounded early Algonquin beaches at Sucker Creek. *Papers of the Mich. Acad. 23*: 477–95. 1937.
128. ——— Lower Algonquin beaches of Penetanguishene peninsula. *Bull. Geol. Soc. Amer. 47*: 1933–60. 1936.
129. TAYLOR, F. B. Field work on the Pleistocene deposits of southwestern Ontario. *Geol. Surv. Can. Summary Rept. 1909*: 164–7.
130. ——— Field work on the Pleistocene deposits of southwestern Ontario. *Geol. Surv. Can. Summary Rept. 1908*: 103–11.
131. ——— The moraine systems of southwestern Ontario. *Trans. Roy. Can. Inst. 10*: 1–23. 1913.
132. ——— Moraines of the St. Lawrence valley. *J. Geol. 32*: 641–67. 1924.
133. ——— Correlatives of the Port Huron morainic system of Michigan in Ontario and western New York. *Amer. J. Sci. 237*: 375–88. 1939.
134. TERASMAE, J. Notes on the Champlain Sea episode in the St. Lawrence lowlands, Quebec. *Science 130,* 3371: 334–6. 1959.
135. ——— Notes on the Champlain Sea episode in the St. Lawrence lowlands. *Que. Sci. 130*: 334–6. 1959.
136. ——— Contributions to Canadian palynology, No. 2. *Geol. Surv. Can. Bull. 56.* Dept. Mines & Tech. Surv., Canada. 1960.
137. TERASMAE, J., and HUGHES, O. L. Glacial retreat in the North Bay area. *Science 131*: 1444–6. 1960.
138. THWAITES, F. T. *Outline of glacial geology.* Ann Arbor, Mich: Edwards Bros. 1934.
139. ——— Pleistocene of part of northeastern Wisconsin. *Bull. Geol. Soc. Amer. 54*: 87–144. 1943.
140. ——— The origin and significance of pitted outwash. *J. Geol. 34*: 308–19. 1926.
141. TYRELL, J. B. The glaciation of north central Canada. *J. Geol. 6*: 147–60. 1898

142. Upham, W. The structure of drumlins. *Proc. Boston Soc. Nat. Hist. 24*: 228–42. 1890.
143. Veach, J. O. Geology in relation to pedology. *Papers of the Mich. Acad. 23*: 503–5. 1937.
144. Von Engeln, O. D. *Geomorphology*. New York: Macmillan. 1942.
145. Watson, J. W. The geography of the Niagara peninsula. Ph.D. thesis, University of Toronto. 1945.
146. Watt, A. K. Correlation of the Pleistocene geology as seen in the subway with that of the Toronto region. *Geol. Assoc. Can. Proc. 6*, Pt. 2: 69–81. 1954.
147. Watt, A. K. Pleistocene geology and ground water resources of the township of North York, York county. *Ont. Dept. Mines, 64th Ann. Rept.* 1957.
148. Webber, L. R., Morwick, F. F., and Richards, N. R. *Soil survey of Durham county*. Rept. No. 9, Ontario Soil Survey 1946.
149. Williams, M. Y. The Silurian geology and faunas of Ontario peninsula and Manitoulin and adjacent islands. *Geol. Surv. Can. Mem. 111*. 1919.
150. Wilson, A. E., Stewart, J. S., and Caley, J. F. Sedimentary basins of Ontario possible sources of oil and gas. *Trans. Roy. Soc. Can. 3rd Ser.*, Sec. 4, *35*: 1941.
151. Wilson, A. W. G. Physical geology of central Ontario. *Trans. Can. Inst. 7*: 165–83. 1901.
152. ——— A forty mile section of Pleistocene deposits north of Lake Ontario. *Trans. Can. Inst. 8*: 11–21. 1903.
153. ——— Trent River system and St. Lawrence outlet. *Bull. Geol. Soc. Amer. 15*: 211–42. 1904.

INDEX